T0235039

Universitext

Universitext

Universitext is a series of textbooks that presents material from a wide variety of mathematical disciplines at master's level and beyond. The books, often well class-tested by their author, may have an informal, personal even experimental approach to their subject matter. Some of the most successful and established books in the series have evolved through several editions, always following the evolution of teaching curricula, to very polished texts.

Thus as research topics trickle down into graduate-level teaching, first textbooks written for new, cutting-edge courses may make their way into *Universitext*.

More information about this series at http://www.springer.com/series/223

Wei Liu • Michael Röckner

Stochastic Partial Differential Equations: An Introduction

 Springer

Wei Liu
School of Mathematics and Statistics
Jiangsu Normal University
Xuzhou, China

Michael Röckner
Faculty of Mathematics
Bielefeld University
Bielefeld, Germany

ISSN 0172-5939   ISSN 2191-6675   (electronic)
Universitext
ISBN 978-3-319-22353-7   ISBN 978-3-319-22354-4   (eBook)
DOI 10.1007/978-3-319-22354-4

Library of Congress Control Number: 2015953013

Mathematics Subject Classification (2010): 60-XX, 60H15, 60H10, 60H05, 60J60, 60J25, 35-XX,
                              35K58, 35K59, 35Q35, 34-XX, 34F05, 34G20, 47-XX,
                              47J35

Springer Cham Heidelberg New York Dordrecht London
© Springer International Publishing Switzerland 2015

Printed on acid-free paper

Springer International Publishing AG Switzerland is part of Springer Science+Business Media
(www.springer.com)

Contents

Chapter 1
Motivation, Aims and Examples

1.1 Motivation and Aims

In this course we will concentrate on (nonlinear) stochastic partial differential equations (SPDEs) of evolutionary type. All kinds of dynamics with stochastic influence in nature or man-made complex systems can be modeled by such equations. As we shall see from the examples at the end of this section, the state spaces of their solutions are necessarily infinite dimensional, such as spaces of (generalized) functions. In this course the state spaces, denoted by H, will be mostly separable Hilbert spaces, sometimes separable Banach spaces.

There is also enormous research activity on SPDEs where the state spaces are not linear, but rather spaces of measures (particle systems, dynamics in population genetics) or infinite-dimensional manifolds (path or loop spaces over Riemannian manifolds).

There are basically three approaches to analyzing SPDEs: the "martingale (or martingale measure) approach" (cf. [80]), the "semigroup (or mild solution) approach" (cf. [26, 27]) and the "variational approach" (cf. [75]). There is an enormously rich literature on all three approaches which cannot be listed here. We refer instead to the above and the following other monographs on SPDEs: [6, 13, 16, 19, 20, 26–28, 37, 46, 48, 50, 53, 66] and the references therein.

The purpose of this course is to give an introduction to the "variational approach", as self-contained as possible, including the "local case", i.e. where, e.g. the standard (weak) monotonicity conditions only hold locally. In the "global case" this approach was initiated in the pioneering work of E. Pardoux [64, 65] and further developed by N. Krylov and B. Rozovskiĭ in [54] for continuous martingales as integrators in the noise term and later by I. Gyöngy and N. Krylov in [40–42] for not necessarily continuous martingales.

The predecessor [67] of this monograph grew out of a two-semester graduate course given by the second-named author at Purdue University in 2005/2006. This extended edition of [67] is the outcome of a two semester graduate course held at the

© Springer International Publishing Switzerland 2015
W. Liu, M. Röckner, *Stochastic Partial Differential Equations: An Introduction*,
Universitext, DOI 10.1007/978-3-319-22354-4_1

University of Bielefeld in 2012/2013. Prerequisites would be an advanced course in probability theory, covering standard martingale theory, stochastic processes in $\mathbb{R}^d$ and maybe basic stochastic integration, though the latter is not formally required. Since graduate students in probability theory are usually not familiar with the theory of Hilbert spaces or basic linear operator theory, all required material from these areas is included in the text, most of it in the appendices. For the same reason we minimize the general theory of martingales on Hilbert spaces, paying, however, the price that some proofs concerning stochastic integration on Hilbert space are a bit lengthy, since they have to be done "with bare hands".

For simplicity we specialize to the case where the integrator in the noise term is just a cylindrical Wiener process, but everything is spelt out in such a way that it generalizes directly to continuous local martingales. In particular, integrands are always assumed to be predictable rather than just adapted and product measurable. The existence and uniqueness proof (cf. Sect. 4.2) is our personal version of the proof in [54, 75] and is largely taken from [69] presented there in a more general framework. The results on invariant measures (cf. Sect. 4.3) we could not find in the literature for the "variational approach". They are, however, quite straightforward modifications of those in the "semigroup approach" in [27].

To keep this course reasonably self-contained we also include a complete proof of the finite-dimensional case in Chap. 3, which is based on the very focussed and beautiful exposition in [52], which uses the Euler approximation. Among other complementing topics such as Chap. 6, which contains a concise introduction to the "semigroup (or mild solution) approach", the appendices contain a detailed account of the Yamada–Watanabe Theorem on the relation between weak and strong solutions (cf. Appendix E), and a detailed proof of Girsanov's Theorem in infinite dimensions (cf. Appendix I).

The structure of this monograph is, we hope, obvious from the list of contents. Here, we only mention that a substantial part consists of a very detailed introduction to stochastic integration on Hilbert spaces (see Chap. 2), major parts of which (as well as Appendices A–C) are taken from the Diploma thesis of Claudia Prévôt (née Knoche) and Katja Frieler (see [34]), which in turn was based on [26] and supervised by the second named author of this monograph. We would like to thank both of them at this point for their permission to do this. We thank all coauthors of those joint papers which form another component for the basis of this monograph. It was really a pleasure working with them in this exciting area of probability theory. We would also like to thank Nelli Schmelzer, Matthias Stephan, Sven Wiesinger and Lukas Wresch for the excellent type job, as well as the participants of the graduate courses at Purdue and Bielefeld University for checking large parts of the text carefully. Special thanks go to Michael Scheutzow and Byron Schmuland for spotting a number of misprints and small errors in [67]. We also thank Claudia Prévôt for giving permission for this extension of [67]. Furthermore, the first named author acknowledges the financial support from NSFC (No. 11201234, 11571147) and a project funded by PAPD of Jiangsu Higher Education Institutions. The last named author would like to thank the German Science Foundation (DFG) for its financial support through SFB 701 and also Jose Luis da Silva for his hospitality during a very pleasant stay at the University of Madeira where the final proofreading of this monograph was done.

1.2 General Philosophy and Examples

Before starting with the main body of this course, let us briefly recall the general philosophy of describing stochastic dynamics by stochastic differential equations (SDE) in a more heuristic and intuitive way. These usually take values in a space H of (generalized) functions, e.g. on a domain $\Lambda \subset \mathbb{R}^d$, or a differential manifold, a fractal or even merely in an arbitrary measurable space $(\Lambda, \mathcal{B})$. Abstractly, H is a separable Hilbert or Banach space. Then given a map $F : [0, T] \times H \times U \to H$, where $T \in]0, \infty[$ and U is another separable Hilbert space, one considers the differential equations

$$\frac{dX(t)}{dt} = F(t, X(t), \dot{W}(t)) \tag{1.1}$$

on H. Here $\dot{W}(t), t \in [0, T]$, is a U-valued white noise in time, more precisely, the generalised time derivative of a U-valued cylindrical Brownian motion $W(t) = (W_k(t))_{k \in \mathbb{N}}, t \in [0, T]$, on some probability space $(\Omega, \mathcal{F}, P)$. Hence $\dot{W}(t), t \in [0, T]$, are independent centred Gaussian variables with infinite variance, hence in regard to (1.1) the "worst" (Gaussian) random perturbation that can occur in (1.1). Employing a Taylor expansion for F around $0 \in U$ and, neglecting terms of order 2 and higher, turns (1.1) into

$$\frac{dX(t)}{dt} = F(t, X(t), 0) + D_3 F(t, X(t), 0)\dot{W}(t), \tag{1.2}$$

where $D_3 F$ denotes the derivative of F with respect to its third coordinate. Setting

$$A(t, x) := F(t, x, 0), \quad B(t, x) := D_3 F(t, x, 0)$$

and taking into account the non-differentiability of $W(t)$ in t, (1.2) turns into

$$dX(t) = A(t, X(t))dt + B(t, X(t))dW(t), \tag{1.3}$$

to be rigorously understood in integral form. We mention here that the stochastic term in (1.3) is often called or interpreted as a "stochastic force", though the equation is first order. This can, however, be justified by the Kramers–Smoluchowski approximation (see [14, 15, 33, 51, 76]). Furthermore, A is called the *drift* of the equation.

We briefly recall here that the linear pendant of (1.3) is given by the associated Fokker–Planck–Kolmogorov equations obtained as a linearization of (1.3) as follows: Let $\mathcal{F}C_{b,T}^2$ denote the set of all functions $f : [0, T] \times H \to \mathbb{R}$, C^1 in $t \in [0, T]$ and C_b^2 in $x \in H$, depending only on finitely many coordinates of x (with respect to a fixed orthonormal basis of H). Take $f \in \mathcal{F}C_{b,T}^2$ and compose it with the solution of (1.3) for $t \geqslant s$ with initial datum $x \in H$ at times s, denoted by $X(t, s, x)$, i.e. consider $f(t, X(t, s, x))$. Subsequently, take the expectation (with respect to P) to get

$$p_{s,t}(f(t, \cdot))(x) := E\left[f(t, X(t, s, x))\right], \quad s < t.$$

Then apply Itô's formula to find that

$$\frac{\partial}{\partial t} p_{s,t}(f(t, \cdot))(x) = p_{s,t}(Lf(t, \cdot))(x), \quad s < t, x \in H, \tag{1.4}$$

with L being the corresponding Kolmogorov operator given by

$$Lf(t, x) = \frac{\partial}{\partial t} f(t, x) + \frac{1}{2} \mathrm{Tr}(B^*(t, x)B(t, x)D_x^2 f(t, x)) + A(t, x)D_x f(t, x), \quad x \in H, \tag{1.5}$$

where D_x, D_x^2 are the Fréchet derivatives of $f(t, \cdot) : H \to \mathbb{R}$. We note that (1.5) is well-defined (in particular the trace exists) for $f \in \mathcal{F}C_{b,T}^2$. We emphasize that the less $B^*(t, x), B(t, x)$ is degenerate, the less degenerate is the second order part of the operator L. In this sense the noise part in (1.3) makes the equation better ("regularization by noise"). In this course, however, we shall concentrate on infinite dimensional stochastic differential equations as (1.3) which in applications are SPDEs. For a detailed analysis of the corresponding Fokker–Planck–Kolmogorov equations (1.4) we refer to the recent monograph [10].

Now we would like to give a few examples of SPDEs that appear in fundamental applications. We present them in a very brief way and refer to the above-mentioned literature for a more elaborate discussion of these and many more examples and their rôle in the applied sciences. Below we use standard notation, in particular $H_0^{m,2}(\Lambda)$ denotes the Sobolev space of order m in $L^2(\Lambda)$ with Dirichlet boundary condition for an open set $\Lambda \subset \mathbb{R}^d$.

Example 1.2.1 (Stochastic Quantization of the Free Euclidean Quantum Field)

$$dX(t) = (\Delta - m^2)X(t)\,dt + dW(t)$$

on $H \subset \mathcal{S}'(\mathbb{R}^d)$.

- $m \in [0, \infty)$ denotes "mass",
- $(W(t))_{t \geqslant 0}$ is a cylindrical Wiener process on $L^2(\mathbb{R}^d) \subset H$ (with the inclusion being a Hilbert–Schmidt embedding).

Example 1.2.2 (Stochastic Reaction Diffusion Equations)

$$dX(t) = [\Delta X(t) - g(X(t))]\, dt + B(t, X(t))\, dW_t$$

on $H := L^2(\Lambda)$, $\Lambda \subset \mathbb{R}^d$, Λ open, $d \leqslant 3$,

- $B(t, x) : H \to H$ is Hilbert–Schmidt $\forall x \in H, t \geqslant 0$.
- $g : \mathbb{R} \to \mathbb{R}$ is a "locally weakly monotone" function of at most polynomial growth (depending on d),
- $W = (W(t))_{t \geqslant 0}$ is a cylindrical Wiener process on H.

Example 1.2.3 (Stochastic Generalized Burgers Equation)

$$dX(t) = \left[\frac{d^2}{d\xi^2} X(t) - f(X(t)) \frac{d}{d\xi} X(t) + g(X(t)) \right] dt + B(t, X(t))\, dW(t)$$

on $H := L^2(]0, 1[)$.

- $\xi \in]0, 1[$,
- $f : \mathbb{R} \to \mathbb{R}$ is a Lipschitz function,
- $g : \mathbb{R} \to \mathbb{R}$ is as above, of at most cubic growth,
- B and W are as above.

Example 1.2.4 (2D Stochastic Navier–Stokes Equation)

$$dX(t) = P_H\big[\nu \Delta_s X(t) - \langle X(t), \nabla \rangle X(t)\big]\, dt + B(t, X(t))\, dW(t)$$

on $H := \{u \in L^2(\Lambda \to \mathbb{R}^2, d\xi) \mid \operatorname{div} u = 0\}$, $\Lambda \subset \mathbb{R}^2$, Λ open, $\partial\Lambda$ smooth.

- ν denotes viscosity,
- Δ_s denotes the Stokes Laplacian,
- div is taken in the sense of distributions,
- P_H denotes the Helmholtz–Leray projection,
- B and W are as above.

Example 1.2.5 (3D Stochastic Navier–Stokes Equation)

$$dX(t) = P_H\left[\nu \Delta_s X(t) - \langle X(t), \nabla \rangle X(t)\right]\, dt + B(t)\, dW(t)$$

on $H := \{u \in H_0^1(\Lambda \to \mathbb{R}^3; d\xi) \mid \operatorname{div} u = 0\}$, $\Lambda \subset \mathbb{R}^3$, Λ open , $\partial\Lambda$ smooth .

- B and W are as above (but B is independent of $X(t)$).

Example 1.2.6 (Stochastic p-Laplace Equation)

$$dX(t) = \operatorname{div}\left(|\nabla X(t)|^{p-2} \nabla X(t)\right) dt + B(t, X(t))\, dW(t)$$

on $H := L^2(\Lambda)$, $\Lambda \subset \mathbb{R}^d$, Λ open,

- B and W are as above.

Example 1.2.7 (Stochastic Porous Media Equation)

$$dX(t) = \Delta\Psi(X(t))\,dt + B(t,X(t))\,dW(t)$$

on $H :=$ dual of $H_0^1(\Lambda)$, Λ as above.

- $\Psi : \mathbb{R} \to \mathbb{R}$ is "monotone",
- B and W are as above.

Example 1.2.8 (Stochastic Cahn–Hilliard Equation)

$$dX(t) = [-\Delta^2 X(t) + \Delta\varphi(X(t))]\,dt + B(t)\,dW(t)$$

$$\nabla X(t) \cdot n = \nabla(\Delta X(t)) \cdot n = 0 \quad \text{on } \partial\Lambda$$

on $H := L^2(\Lambda)$, Λ as above.

- n denotes the outward unit normal on $\partial\Lambda$,
- $\varphi : \mathbb{R} \to \mathbb{R}$ is C^1, is locally weakly monotone and of at most polynomial growth $p \in [2, \frac{d+4}{d}]$,
- B and W are as above.

Example 1.2.9 (Stochastic Surface Growth Model)

$$dX(t) = [-\frac{\partial^4}{\partial\xi^4}X(t) - \frac{\partial^2}{\partial\xi^2}X(t) + \frac{\partial^2}{\partial\xi^2}(\frac{\partial}{\partial\xi}X(t))^2]\,dt + B(t)\,dW(t)$$

$$X(t) \restriction_{\partial\Lambda} = 0$$

on $H = H_0^{2.2}(\Lambda)$ with $\Lambda :=\,]0, L[$,

- $\xi \in\,]0, L[$,
- B and W are as above.

The general form of these equations with state spaces consisting of functions $\xi \mapsto x(\xi)$, where ξ is a spatial variable, e.g. from a subset of $\mathbb{R}^d$, is as follows:

$$dX(t)(\xi) = A\Big(t, X(t)(\xi), D_\xi X(t)(\xi), D_\xi^2\big(X(t)(\xi)\big)\Big)\,dt$$

$$+ B\Big(t, X(t)(\xi), D_\xi X_t(\xi), D_\xi^2\big(X(t)(\xi)\big)\Big)\,dW(t) .$$

Here D_ξ and D_ξ^2 mean first and second total derivatives, respectively. The stochastic term can be considered as a "perturbation by noise". So, clearly one motivation for studying SPDEs is to get information about the corresponding (unperturbed) deterministic PDE by letting the noise go to zero (e.g. replace B by $\varepsilon \cdot B$ and let $\varepsilon \to 0$) or to understand the different features occurring if one adds the noise term.

If we drop the stochastic term in these equations we get a deterministic PDE of "evolutionary type". Roughly speaking this means we have that the time derivative of the desired solution (on the left) is equal to a non-linear functional of its spatial derivatives (on the right).

For a detailed analysis of Example 1.2.1 and for the non-linear much harder stochastic quantization equations for interacting Euclidean quantum fields we refer to [2] and the recent survey [1] (including the references therein). All other examples above under the respective appropriate conditions are covered by the theory presented in this course and all of them are analyzed in detail here. This is in contrast to [67], where Examples 1.2.2–1.2.5, 1.2.8 and 1.2.9, were not covered, since they only satisfy "local monotonicity conditions" and/or weaker growth conditions and/or "generalized coercivity conditions". All these latter examples are included only in this extended version, more precisely, in the newly added Chap. 5, where global solutions for Examples 1.2.2–1.2.4 are constructed in Sect. 5.1 and local solutions for Examples 1.2.5, 1.2.8 and 1.2.9 in Sect. 5.2 (see also Remark 5.1.11(4) for a number of further examples which, in order to keep it within a reasonable size, have not been included in this course). We would like to stress that Sect. 5.2 in particular contains the presentation of a general technique to construct local solutions to SPDEs on the basis of a classical inequality due to Bihari. Furthermore, we include a study of a "tamed version" of Example 1.2.5 (see Example 5.2.25), for which global solutions exist (at least in the deterministic case), and we show that the solutions in Example 1.2.8 are global, if $B \equiv 0$ or $p \leq 2$.

After having discussed a number of typical examples of nonlinear SPDEs, we would like to address a genuine problem of the theory of SPDEs, namely that in some cases one is interested in perturbing the deterministic PDE by a very rough noise, meaning a noise which is itself no longer function-valued, but only takes values in a space of generalized functions, i.e. in a subspace of the space of Schwarz distributions. One way out is to go to the mild formulation of the SPDE (see Chap. 6) and use the smoothing property of a hopefully "strong enough" linear part of the drift. But if one focuses on the Laplacian, "strong enough" requires that the underlying domain is one dimensional. So, already in two dimensions one can only expect Schwarz distribution-valued solutions to the SPDE, hence the simplest non-linear parts of the drift, as e.g. a power of the solution, have to be defined by renormalization techniques (see [2] and also [1]). Recently, a breakthrough has been achieved in this direction by Martin Hairer in [44] (for which, together with his work on the KPZ-equation [43] and other beautiful results in the field, he was awarded the Fields Medal in 2014). In [44] he develops a whole theory to define and (locally) solve nonlinear SPDEs in a reformulated framework. This theory applies to a number of important (nonlinear) SPDEs with distribution-valued solutions on two- or three-dimensional underlying domains. We refer to [35] for a detailed exposition of this theory.

To conclude this introduction, let us summarize the new parts of this monograph in comparison to [67], adding the references of their respective origin, which are all quite recent papers, except for the new Chap. 6 and the new Appendix F. The new Chap. 5, whose contents has already been summarized in the previous paragraph, is an extended version of [57–59]. Chapter 6 contains a concise introduction to the "semigroup (or mild solution) approach", in particular, addressing the measurability issues occurring in this context and (through the famous "factorization method" in [26]) also the question of when the solutions have continuous sample paths. To complement Chap. 6 we also include Appendix F which is needed there. Both Chap. 6 and Appendix F are elaborations of the corresponding sections in [34], which in turn are based on [26]. Appendix E is a more detailed version of [71] and it contains a complete proof of the Yamada–Watanabe Theorem in infinite dimensions, whereas [67] only contains the finite dimensional case. The new Appendix H contains two elementary proofs for well-known interpolation inequalities which are essential for analyzing the (stochastic) Navier–Stokes equations. The proofs are essentially taken from [61]. Finally, the new Appendix I on Girsanov's Theorem in infinite dimensions is an extended version of Appendix A.1 in [22].

Chapter 2
The Stochastic Integral in General Hilbert Spaces (w.r.t. Brownian Motion)

We fix two separable Hilbert spaces $(U, \langle \ , \ \rangle_U)$ and $(H, \langle \ , \ \rangle)$ with norms $\| \ \|_U$ and $\| \ \|_H$, respectively, where we drop the subscript H in the latter if there is no confusion possible. The first part of this chapter is devoted to the construction of the stochastic Itô integral

$$\int_0^t \Phi(s) \, dW(s) \, , \quad t \in [0, T],$$

where $W(t)$, $t \in [0, T]$, is a Wiener process on U and Φ is a process with values that are linear but not necessarily bounded operators from U to H.

For that we will first have to introduce the notion of the standard Wiener process in infinite dimensions. Then there will be a short section on martingales in general Hilbert spaces. These two concepts are important for the construction of the stochastic integral, which will be explained in the following section.

Following this, we shall collect and prove a number of properties of the stochastic integral, which are necessary for the later chapters.

Finally, we will describe how to transfer the definition of the stochastic integral to the case when $W(t)$, $t \in [0, T]$, is a cylindrical Wiener process. For simplicity we assume that U and H are real Hilbert spaces.

2.1 Infinite-Dimensional Wiener Processes

For a topological space X we denote its Borel σ-algebra by $\mathcal{B}(X)$.

Definition 2.1.1 A probability measure μ on $(U, \mathcal{B}(U))$ is called *Gaussian* if for all $v \in U$ the bounded linear mapping

$$v' : U \to \mathbb{R}$$

© Springer International Publishing Switzerland 2015
W. Liu, M. Röckner, *Stochastic Partial Differential Equations: An Introduction*,
Universitext, DOI 10.1007/978-3-319-22354-4_2

defined by

$$u \mapsto \langle u, v \rangle_U, \quad u \in U,$$

has a Gaussian law, i.e. for all $v \in U$ there exist $m := m(v) \in \mathbb{R}$ and $\sigma := \sigma(v) \in [0, \infty[$ such that, if $\sigma(v) > 0$,

$$\left(\mu \circ (v')^{-1} \right) (A) = \mu(v' \in A) = \frac{1}{\sqrt{2\pi\sigma^2}} \int_A e^{-\frac{(x-m)^2}{2\sigma^2}} \, dx \quad \text{for all } A \in \mathcal{B}(\mathbb{R}),$$

and, if $\sigma(v) = 0$,

$$\mu \circ (v')^{-1} = \delta_{m(v)}.$$

Theorem 2.1.2 *A measure μ on $\left(U, \mathcal{B}(U) \right)$ is Gaussian if and only if*

$$\hat{\mu}(u) := \int_U e^{i \langle u, v \rangle_U} \, \mu(dv) = e^{i \langle m, u \rangle_U - \frac{1}{2} \langle Qu, u \rangle_U}, \quad u \in U,$$

where $m \in U$ and $Q \in L(U)$ is nonnegative, symmetric, with finite trace (see Definition B.0.3; here $L(U)$ denotes the set of all bounded linear operators on U).

In this case μ will be denoted by $N(m, Q)$ where m is called the mean and Q is called the covariance (operator). The measure μ is uniquely determined by m and Q.

Furthermore, for all $h, g \in U$

$$\int \langle x, h \rangle_U \, \mu(dx) = \langle m, h \rangle_U,$$

$$\int \left(\langle x, h \rangle_U - \langle m, h \rangle_U \right) \left(\langle x, g \rangle_U - \langle m, g \rangle_U \right) \mu(dx) = \langle Qh, g \rangle_U,$$

$$\int \| x - m \|_U^2 \, \mu(dx) = \text{tr} \, Q.$$

Proof (Cf. [26]) Obviously, a probability measure with this Fourier transform is Gaussian. Now let us conversely assume that μ is Gaussian. We need the following:

Lemma 2.1.3 *Let v be a probability measure on $(U, \mathcal{B}(U))$. Let $k \in \mathbb{N}$ be such that*

$$\int_U \left| \langle z, x \rangle_U \right|^k v(dx) < \infty \quad \forall z \in U.$$

Then there exists a constant $C = C(k, \nu) > 0$ such that for all $h_1, \ldots, h_k \in U$

$$\int_U \left| \langle h_1, x \rangle_U \cdots \langle h_k, x \rangle_U \right| \, \nu(\mathrm{d}x) \leqslant C \, \|h_1\|_U \cdots \|h_k\|_U.$$

In particular, the symmetric k-linear form

$$U^k \ni (h_1, \ldots, h_k) \mapsto \int \langle h_1, x \rangle_U \cdots \langle h_k, x \rangle_U \, \nu(\mathrm{d}x) \in \mathbb{R}$$

is continuous.

Proof For $n \in \mathbb{N}$ define

$$U_n := \left\{ z \in U \ \middle| \ \int_U |\langle z, x \rangle_U|^k \, \nu(\mathrm{d}x) \leqslant n \right\}.$$

Then each U_n is a closed set in U, since if $z_l \in U_n$, $l \in \mathbb{N}$, and $z \in U$ such that $\lim_{l \to \infty} z_l = z$ in U, then by Fatou's Lemma

$$\int_U |\langle z, x \rangle_U|^k \nu(\mathrm{d}x) \leqslant \liminf_{l \to \infty} \int_U |\langle z_l, x \rangle_U|^k \nu(\mathrm{d}x) \leqslant n.$$

By assumption

$$U = \bigcup_{n=1}^{\infty} U_n.$$

Since U is a complete metric space, by the Baire category theorem, there exists an $n_0 \in \mathbb{N}$ such that U_{n_0} has non-empty interior, so there exists a closed ball (with centre z_0 and radius r_0) $\bar{B}(z_0, r_0) \subset U_{n_0}$. Hence

$$\int_U \left| \langle z_0 + y, x \rangle_U \right|^k \nu(\mathrm{d}x) \leqslant n_0 \quad \forall \, y \in B(0, r_0),$$

therefore for all $y \in \bar{B}(0, r_0)$

$$\int_U |\langle y, x \rangle_U|^k \, \nu(\mathrm{d}x) = \int_U \left| \langle z_0 + y, x \rangle_U - \langle z_0, x \rangle_U \right|^k \nu(\mathrm{d}x)$$

$$\leqslant 2^{k-1} \int_U \left| \langle z_0 + y, x \rangle_U \right|^k \nu(\mathrm{d}x) + 2^{k-1} \int_U |\langle z_0, x \rangle_U|^k \nu(\mathrm{d}x)$$

$$\leqslant 2^k n_0.$$

Applying this for $y := r_0 z$, $z \in U$ with $\|z\|_U = 1$, we obtain

$$\int_U |\langle z, x \rangle_U|^k \, \nu(dx) \leqslant 2^k n_0 r_0^{-k}.$$

Hence, if $h_1, \ldots, h_k \in U \setminus \{0\}$, then by the generalized Hölder inequality

$$\int_U \left| \left\langle \frac{h_1}{\|h_1\|_U}, x \right\rangle_U \cdots \left\langle \frac{h_k}{\|h_k\|_U}, x \right\rangle_U \right| \nu(dx)$$

$$\leqslant \left(\int_U \left| \left\langle \frac{h_1}{\|h_1\|_U}, x \right\rangle_U \right|^k \nu(dx) \right)^{1/k} \cdots \left(\int_U \left| \left\langle \frac{h_k}{\|h_k\|_U}, x \right\rangle_U \right|^k \nu(dx) \right)^{1/k}$$

$$\leqslant 2^k n_0 r_0^{-k},$$

and the assertion follows.   □

Applying Lemma 2.1.3 for $k = 1$ and $\nu := \mu$ we obtain that

$$U \ni h \mapsto \int \langle h, x \rangle_U \, \mu(dx) \in \mathbb{R}$$

is a continuous linear map, hence there exists an $m \in U$ such that

$$\int_U \langle x, h \rangle_U \, \mu(dx) = \langle m, h \rangle \quad \forall \, h \in U.$$

Applying Lemma 2.1.3 for $k = 2$ and $\nu := \mu$ we obtain that

$$U^2 \ni (h_1, h_2) \mapsto \int \langle x, h_1 \rangle_U \langle x, h_2 \rangle_U \, \mu(dx) - \langle m, h_1 \rangle_U \langle m, h_2 \rangle_U$$

is a continuous symmetric bilinear map, hence there exists a symmetric $Q \in L(U)$ such that this map is equal to

$$U^2 \ni (h_1, h_2) \mapsto \langle Q h_1, h_2 \rangle_U.$$

Since for all $h \in U$

$$\langle Q h, h \rangle_U = \int \langle x, h \rangle_U^2 \, \mu(dx) - \left(\int \langle x, h \rangle_U \, \mu(dx) \right)^2 \geqslant 0,$$

Q is positive definite. It remains to prove that Q is trace class (i.e.

$$\operatorname{tr} Q := \sum_{i=1}^{\infty} \langle Qe_i, e_i \rangle_U < \infty$$

for one (hence every) orthonormal basis $\{e_i \mid i \in \mathbb{N}\}$ of U, cf. Appendix B). We may assume without loss of generality that μ has mean zero, i.e. $m = 0$ ($\in U$), since the image measure of μ under the translation $U \ni x \mapsto x - m$ is again Gaussian with mean zero and the same covariance Q. Then we have for all $h \in U$ and all $c \in (0, \infty)$

$$1 - e^{-\frac{1}{2}\langle Qh, h \rangle_U} = \int_U \left(1 - \cos\langle h, x \rangle_U\right) \mu(dx)$$

$$\leqslant \int_{\{\|\cdot\|_U \leqslant c\}} \left(1 - \cos\langle h, x \rangle_U\right) \mu(dx) + 2\mu\left(\{x \in U \mid \|x\|_U > c\}\right)$$

$$\leqslant \frac{1}{2} \int_{\{\|\cdot\|_U \leqslant c\}} \left|\langle h, x \rangle_U\right|^2 \mu(dx) + 2\mu\left(\{x \in U \mid \|x\|_U > c\}\right) \qquad (2.1)$$

(since $1 - \cos x \leqslant \frac{1}{2}x^2$). Defining a positive definite symmetric linear operator Q_c on U by

$$\langle Q_c h_1, h_2 \rangle_U := \int_{\{\|\cdot\|_U \leqslant c\}} \langle h_1, x \rangle_U \cdot \langle h_2, x \rangle_U \, \mu(dx), \quad h_1, h_2 \in U,$$

we even have that Q_c is trace class because for every orthonormal basis $\{e_k \mid k \in \mathbb{N}\}$ of U we have (by monotone convergence)

$$\sum_{k=1}^{\infty} \langle Q_c e_k, e_k \rangle_U = \int_{\{\|\cdot\|_U \leqslant c\}} \sum_{k=1}^{\infty} \langle e_k, x \rangle_U^2 \, \mu(dx) = \int_{\{\|\cdot\|_U \leqslant c\}} \|x\|_U^2 \, \mu(dx)$$

$$\leqslant c^2 < \infty.$$

Claim There exists a $c_0 \in (0, \infty)$ (large enough) so that $Q \leqslant 2\log 4 \, Q_{c_0}$ (meaning that $\langle Qh, h \rangle_U \leqslant 2\log 4 \, \langle Q_{c_0} h, h \rangle_U$ for all $h \in U$).

To prove the claim let c_0 be so big that $\mu\left(\{x \in U \mid \|x\|_U > c_0\}\right) \leqslant \frac{1}{8}$. Let $h \in U$ such that $\langle Q_{c_0} h, h \rangle_U \leqslant 1$. Then (2.1) implies

$$1 - e^{-\frac{1}{2}\langle Qh, h \rangle_U} \leqslant \frac{1}{2} + \frac{1}{4} = \frac{3}{4},$$

hence $4 \geqslant e^{\frac{1}{2}\langle Qh, h \rangle_U}$, so $\langle Qh, h \rangle_U \leqslant 2\log 4$. If $h \in U$ is arbitrary, but $\langle Q_{c_0} h, h \rangle_U \neq 0$, then we apply what we have just proved to $h / \langle Q_{c_0} h, h \rangle_U^{\frac{1}{2}}$ and the claim follows for such h. If, however, $\langle Q_{c_0} h, h \rangle_U = 0$, then for all $n \in \mathbb{N}$, $\langle Q_{c_0} nh, nh \rangle_U = 0 \leqslant 1$, hence by the above $\langle Qh, h \rangle_U \leqslant n^{-2} 2 \log 4$. Therefore, $\langle Qh, h \rangle_U = 0$ and the claim is proved, also for such h.

Since Q_{c_0} has finite trace, so has Q by the claim and the theorem is proved, since the uniqueness part follows from the fact that the Fourier transform is one-to-one.

$\square$

The following result is then obvious.

Proposition 2.1.4 *Let X be a U-valued Gaussian random variable on a probability space $(\Omega, \mathcal{F}, P)$, i.e. there exist $m \in U$ and $Q \in L(U)$ nonnegative, symmetric, with finite trace such that $P \circ X^{-1} = N(m, Q)$.*

Then $\langle X, u \rangle_U$ is normally distributed for all $u \in U$ and the following statements hold:

- $E(\langle X, u \rangle_U) = \langle m, u \rangle_U$ *for all $u \in U$,*
- $E(\langle X - m, u \rangle_U \cdot \langle X - m, v \rangle_U) = \langle Qu, v \rangle_U$ *for all $u, v \in U$,*
- $E(\|X - m\|_U^2) = \operatorname{tr} Q$.

The following proposition will lead to a representation of a U-valued Gaussian random variable in terms of real-valued Gaussian random variables.

Proposition 2.1.5 *If $Q \in L(U)$ is nonnegative, symmetric, with finite trace then there exists an orthonormal basis e_k, $k \in \mathbb{N}$, of U such that*

$$Qe_k = \lambda_k e_k, \quad \lambda_k \geqslant 0, \ k \in \mathbb{N},$$

and 0 is the only accumulation point of the sequence $(\lambda_k)_{k \in \mathbb{N}}$.

Proof See [68, Theorem VI.21; Theorem VI.16 (Hilbert–Schmidt theorem)]. $\square$

Proposition 2.1.6 (Representation of a Gaussian Random Variable) *Let $m \in U$ and $Q \in L(U)$ be nonnegative, symmetric, with $\operatorname{tr} Q < \infty$. In addition, we assume that e_k, $k \in \mathbb{N}$, is an orthonormal basis of U consisting of eigenvectors of Q with corresponding eigenvalues λ_k, $k \in \mathbb{N}$, as in Proposition 2.1.5, numbered in decreasing order.*

Then a U-valued random variable X on a probability space $(\Omega, \mathcal{F}, P)$ is Gaussian with $P \circ X^{-1} = N(m, Q)$ if and only if

$$X = \sum_{k \in \mathbb{N}} \sqrt{\lambda_k} \beta_k e_k + m \quad \text{(as objects in } L^2(\Omega, \mathcal{F}, P; U)),$$

where β_k, $k \in \mathbb{N}$, are independent real-valued random variables with $P \circ \beta_k^{-1} = N(0, 1)$ for all $k \in \mathbb{N}$ with $\lambda_k > 0$. The series converges in $L^2(\Omega, \mathcal{F}, P; U)$.

Proof

1. Let X be a Gaussian random variable with mean m and covariance Q. Below we set $\langle \, , \, \rangle := \langle \, , \, \rangle_U$.

Then $X = \sum_{k \in \mathbb{N}} \langle X, e_k \rangle e_k$ in U where $\langle X, e_k \rangle$ is normally distributed with mean $\langle m, e_k \rangle$ and variance λ_k, $k \in \mathbb{N}$, by Proposition 2.1.4. If we now define

$$\beta_k : \begin{cases} = \frac{\langle X, e_k \rangle - \langle m, e_k \rangle}{\sqrt{\lambda_k}} & \text{if } k \in \mathbb{N} \text{ with } \lambda_k > 0 \\ \equiv 0 \in \mathbb{R} & \text{else,} \end{cases}$$

then we get that $P \circ \beta_k^{-1} = N(0, 1)$ and $X = \sum_{k \in \mathbb{N}} \sqrt{\lambda_k} \beta_k e_k + m$. To prove the independence of β_k, $k \in \mathbb{N}$, we take an arbitrary $n \in \mathbb{N}$ and $a_k \in \mathbb{R}$, $1 \leq k \leq n$, and obtain that for $c := -\sum_{k=1, \lambda_k \neq 0}^{n} \frac{a_k}{\sqrt{\lambda_k}} \langle m, e_k \rangle$

$$\sum_{k=1}^{n} a_k \beta_k = \sum_{\substack{k=1, \\ \lambda_k \neq 0}}^{n} \frac{a_k}{\sqrt{\lambda_k}} \langle X, e_k \rangle + c = \left\langle X, \sum_{\substack{k=1, \\ \lambda_k \neq 0}}^{n} \frac{a_k}{\sqrt{\lambda_k}} e_k \right\rangle + c$$

which is normally distributed since X is a Gaussian random variable. Therefore we have that β_k, $k \in \mathbb{N}$, form a Gaussian family. Hence, to get the independence, we only have to check that the covariance of β_i and β_j, $i, j \in \mathbb{N}$, $i \neq j$, with $\lambda_i \neq 0 \neq \lambda_j$, is equal to zero. But this is clear since

$$E(\beta_i \beta_j) = \frac{1}{\sqrt{\lambda_i \lambda_j}} E\big(\langle X - m, e_i \rangle \langle X - m, e_j \rangle\big) = \frac{1}{\sqrt{\lambda_i \lambda_j}} \langle Q e_i, e_j \rangle$$

$$= \frac{\lambda_i}{\sqrt{\lambda_i \lambda_j}} \langle e_i, e_j \rangle = 0$$

for $i \neq j$.

Furthermore, the series $\sum_{k=1}^{n} \sqrt{\lambda_k} \beta_k e_k$, $n \in \mathbb{N}$, converges in $L^2(\Omega, \mathcal{F}, P; U)$ since the space is complete and

$$E\left(\left\| \sum_{k=m}^{n} \sqrt{\lambda_k} \beta_k e_k \right\|_U^2 \right) = \sum_{k=m}^{n} \lambda_k E\big(|\beta_k|^2\big) = \sum_{k=m}^{n} \lambda_k.$$

Since $\sum_{k \in \mathbb{N}} \lambda_k = \operatorname{tr} Q < \infty$ this expression becomes arbitrarily small for m and n large enough.

2. Let e_k, $k \in \mathbb{N}$, be an orthonormal basis of U such that $Q e_k = \lambda_k e_k$, $k \in \mathbb{N}$, and let β_k, $k \in \mathbb{N}$, be a family of independent real-valued Gaussian random variables with mean 0 and variance 1. Then it is clear that the series $\sum_{k=1}^{n} \sqrt{\lambda_k} \beta_k e_k + m$, $n \in \mathbb{N}$, converges to $X := \sum_{k \in \mathbb{N}} \sqrt{\lambda_k} \beta_k e_k + m$ in $L^2(\Omega, \mathcal{F}, P; U)$ (see part 1). Now we fix $u \in U$ and get that

$$\left\langle \sum_{k=1}^{n} \sqrt{\lambda_k} \beta_k e_k + m, u \right\rangle = \sum_{k=1}^{n} \sqrt{\lambda_k} \beta_k \langle e_k, u \rangle + \langle m, u \rangle$$

is normally distributed for all $n \in \mathbb{N}$ and the sequence converges in $L^2(\Omega, \mathcal{F}, P)$. This implies that the limit $\langle X, u \rangle$ is also normally distributed where

$$E(\langle X, u \rangle) = E\left(\sum_{k \in \mathbb{N}} \sqrt{\lambda_k} \beta_k \langle e_k, u \rangle + \langle m, u \rangle\right)$$

$$= \lim_{n \to \infty} E\left(\sum_{k=1}^{n} \sqrt{\lambda_k} \beta_k \langle e_k, u \rangle\right) + \langle m, u \rangle = \langle m, u \rangle$$

and concerning the covariance we obtain that

$$E\Big(\big(\langle X, u \rangle - \langle m, u \rangle\big)\big(\langle X, v \rangle - \langle m, v \rangle\big)\Big)$$

$$= \lim_{n \to \infty} E\left(\sum_{k=1}^{n} \sqrt{\lambda_k} \beta_k \langle e_k, u \rangle \sum_{k=1}^{n} \sqrt{\lambda_k} \beta_k \langle e_k, v \rangle\right)$$

$$= \sum_{k \in \mathbb{N}} \lambda_k \langle e_k, u \rangle \langle e_k, v \rangle = \sum_{k \in \mathbb{N}} \langle Q e_k, u \rangle \langle e_k, v \rangle$$

$$= \sum_{k \in \mathbb{N}} \langle e_k, Q u \rangle \langle e_k, v \rangle = \langle Q u, v \rangle$$

for all $u, v \in U$.   □

By part 2 of this proof we finally get the following existence result.

Corollary 2.1.7 *Let Q be a nonnegative and symmetric operator in $L(U)$ with finite trace and let $m \in U$. Then there exists a Gaussian measure $\mu = N(m, Q)$ on $(U, \mathcal{B}(U))$.*

Let us give an alternative, more direct proof of Corollary 2.1.7 without using Proposition 2.1.6. For the proof we need the following exercise.

Exercise 2.1.8 Consider $\mathbb{R}^\infty$ with the product topology. Let $\mathcal{B}(\mathbb{R}^\infty)$ denote its Borel σ-algebra. Prove:

(i) $\mathcal{B}(\mathbb{R}^\infty) = \sigma(\pi_k \mid k \in \mathbb{N})$, where $\pi_k : \mathbb{R}^\infty \to \mathbb{R}$ denotes the projection on the k-th coordinate.

(ii) $l^2(\mathbb{R}) \left(:= \left\{(x_k)_{k \in \mathbb{N}} \in \mathbb{R}^\infty \mid \sum_{k=1}^{\infty} x_k^2 < \infty\right\}\right) \in \mathcal{B}(\mathbb{R}^\infty)$.

(iii) $\mathcal{B}(\mathbb{R}^\infty) \cap l^2(\mathbb{R}) = \sigma(\pi_k|_{l^2} \mid k \in \mathbb{N})$.

(iv) Let $l^2(\mathbb{R})$ be equipped with its natural norm

$$\|x\|_{l^2} := \left(\sum_{k=1}^{\infty} x_k^2\right)^{\frac{1}{2}}, \quad x = (x_k)_{k \in \mathbb{N}} \in l^2(\mathbb{R}),$$

and let $\mathcal{B}(l^2(\mathbb{R}))$ be the corresponding Borel σ-algebra. Then:

$$\mathcal{B}(l^2(\mathbb{R})) = \mathcal{B}(\mathbb{R}^\infty) \cap l^2(\mathbb{R}).$$

Alternative Proof of Corollary 2.1.7 It suffices to construct $N(0, Q)$, since $N(m, Q)$ is the image measure of $N(0, Q)$ under translation with m. For $k \in \mathbb{N}$ consider the normal distribution $N(0, \lambda_k)$ on $\mathbb{R}$ and let ν be their product measure on $(\mathbb{R}^\infty, \mathcal{B}(\mathbb{R}^\infty))$, i.e.

$$\nu = \prod_{k \in \mathbb{N}} N(0, \lambda_k) \quad \text{on } (\mathbb{R}^\infty, \mathcal{B}(\mathbb{R}^\infty)).$$

Here $\lambda_k, k \in \mathbb{N}$, are as in Proposition 2.1.5. Since the map $g : \mathbb{R}^\infty \to [0, \infty]$ defined by

$$g(x) := \sum_{k=1}^\infty x_k^2, \quad x = (x_k)_{k \in \mathbb{N}} \in \mathbb{R}^\infty,$$

is $\mathcal{B}(\mathbb{R}^\infty)$-measurable, we may calculate

$$\int_{\mathbb{R}^\infty} g(x) \, \nu(dx) = \sum_{k=1}^\infty \int x_k^2 \, N(0, \lambda_k)(dx_k) = \sum_{k=1}^\infty \lambda_k < \infty.$$

Therefore, using Exercise 2.1.8(ii), we obtain $\nu(l^2(\mathbb{R})) = 1$. Restricting ν to $\mathcal{B}(\mathbb{R}^\infty) \cap l^2(\mathbb{R})$, by Exercise 2.1.8(iv) we get a probability measure, let us call it $\tilde{\mu}$, on $(l^2(\mathbb{R}), \mathcal{B}(l^2(\mathbb{R})))$. Now take the orthonormal basis $\{e_k \mid k \in \mathbb{N}\}$ from Proposition 2.1.5 and consider the corresponding canonical isomorphism $I : l^2(\mathbb{R}) \to U$ defined by

$$I(x) = \sum_{k=1}^\infty x_k e_k, \quad x = (x_k)_{k \in \mathbb{N}} \in l^2(\mathbb{R}).$$

It is then easy to check that the image measure

$$\mu := \tilde{\mu} \circ I^{-1} \quad \text{on } (U, \mathcal{B}(U))$$

is the desired measure, i.e. $\mu = N(0, Q)$. □

After these preparations we will give the definition of a standard Q-Wiener process. To this end we fix an element $Q \in L(U)$, nonnegative, symmetric and with finite trace and a positive real number T.

Definition 2.1.9 A U-valued stochastic process $W(t)$, $t \in [0, T]$, on a probability space $(\Omega, \mathcal{F}, P)$ is called a (standard) Q-Wiener process if:

- $W(0) = 0$,
- W has P-a.s. continuous trajectories,
- the increments of W are independent, i.e. the random variables

$$W(t_1), \ W(t_2) - W(t_1), \ldots, \ W(t_n) - W(t_{n-1})$$

  are independent for all $0 \leqslant t_1 < \cdots < t_n \leqslant T$, $n \in \mathbb{N}$,
- the increments have the following Gaussian laws:

$$P \circ \big(W(t) - W(s)\big)^{-1} = N\big(0, (t - s)Q\big) \quad \text{for all } 0 \leqslant s \leqslant t \leqslant T.$$

Similarly to the existence of Gaussian measures the existence of a Q-Wiener process in U can be reduced to the real-valued case. This is the content of the following proposition.

Proposition 2.1.10 (Representation of the Q-Wiener Process) *Let e_k, $k \in \mathbb{N}$, be an orthonormal basis of U consisting of eigenvectors of Q with corresponding eigenvalues λ_k, $k \in \mathbb{N}$. Then a U-valued stochastic process $W(t)$, $t \in [0, T]$, is a Q-Wiener process if and only if*

$$W(t) = \sum_{k \in \mathbb{N}} \sqrt{\lambda_k} \beta_k(t) e_k, \quad t \in [0, T], \tag{2.2}$$

where β_k, $k \in \{n \in \mathbb{N} \mid \lambda_n > 0\}$, are independent real-valued Brownian motions on a probability space $(\Omega, \mathcal{F}, P)$. The series even converges in $L^2\big(\Omega, \mathcal{F}, P; C([0, T], U)\big)$, and thus always has a P-a.s. continuous version. (Here the space $C\big([0, T], U\big)$ is equipped with the sup norm.) In particular, for any Q as above there exists a Q-Wiener process on U.

Proof

1. Let $W(t)$, $t \in [0, T]$, be a Q-Wiener process in U.
   Since $P \circ W(t)^{-1} = N(0, tQ)$, we see as in the proof of Proposition 2.1.6 that

$$W(t) = \sum_{k \in \mathbb{N}} \sqrt{\lambda_k} \beta_k(t) e_k, \quad t \in [0, T],$$

  with

$$\beta_k(t) : \begin{cases} = \dfrac{\langle W(t), e_k \rangle}{\sqrt{\lambda_k}} & \text{if } k \in \mathbb{N} \text{ with } \lambda_k > 0 \\ \equiv 0 & \text{else,} \end{cases}$$

for all $t \in [0, T]$. Furthermore, $P \circ \beta_k^{-1}(t) = N(0, t)$, $k \in \mathbb{N}$, and $\beta_k(t)$, $k \in \mathbb{N}$, are independent for each fixed $t \in [0, T]$.

Now we fix $k \in \mathbb{N}$. First we show that $\beta_k(t)$, $t \in [0, T]$, is a Brownian motion:

If we take an arbitrary partition $0 = t_0 \leqslant t_1 < \cdots < t_n \leqslant T$, $n \in \mathbb{N}$, of $[0, T]$ we get that

$$\beta_k(t_1), \ \beta_k(t_2) - \beta_k(t_1), \ldots, \ \beta_k(t_n) - \beta_k(t_{n-1})$$

are independent for each $k \in \mathbb{N}$ since for $1 \leqslant j \leqslant n$

$$\beta_k(t_j) - \beta_k(t_{j-1}) = \begin{cases} \frac{1}{\sqrt{\lambda_k}} \langle W(t_j) - W(t_{j-1}), e_k \rangle & \text{if } \lambda_k > 0, \\ 0 & \text{else.} \end{cases}$$

Moreover, we obtain that for the same reason $P \circ \big(\beta_k(t) - \beta_k(s)\big)^{-1} = N(0, t-s)$ for $0 \leqslant s \leqslant t \leqslant T$.

In addition,

$$t \mapsto \frac{1}{\sqrt{\lambda_k}} \langle W(t), e_k \rangle = \beta_k(t)$$

is P-a.s. continuous for all $k \in \mathbb{N}$.

Secondly, it remains to prove that β_k, $k \in \mathbb{N}$, are independent.

We take $k_1, \ldots, k_n \in \mathbb{N}$, $n \in \mathbb{N}$, $k_i \neq k_j$ if $i \neq j$ and an arbitrary partition $0 = t_0 \leqslant t_1 \leqslant \ldots \leqslant t_m \leqslant T$, $m \in \mathbb{N}$.

Then we have to show that

$$\sigma\big(\beta_{k_1}(t_1), \ldots, \beta_{k_1}(t_m)\big), \ldots, \sigma\big(\beta_{k_n}(t_1), \ldots, \beta_{k_n}(t_m)\big)$$

are independent.

We will prove this by induction with respect to m:

If $m = 1$ it is clear that $\beta_{k_1}(t_1), \ldots, \beta_{k_n}(t_1)$ are independent as observed above. Thus, we now take a partition $0 = t_0 \leqslant t_1 \leqslant \ldots \leqslant t_{m+1} \leqslant T$ and assume that

$$\sigma\big(\beta_{k_1}(t_1), \ldots, \beta_{k_1}(t_m)\big), \ldots, \sigma\big(\beta_{k_n}(t_1), \ldots, \beta_{k_n}(t_m)\big)$$

are independent. We note that

$$\sigma\big(\beta_{k_i}(t_1), \ldots, \beta_{k_i}(t_m), \ \beta_{k_i}(t_{m+1})\big)$$
$$= \sigma\big(\beta_{k_i}(t_1), \ldots, \beta_{k_i}(t_m), \ \beta_{k_i}(t_{m+1}) - \beta_{k_i}(t_m)\big), \quad 1 \leqslant i \leqslant n,$$

and that

$$\beta_{k_i}(t_{m+1}) - \beta_{k_i}(t_m) = \begin{cases} \frac{1}{\sqrt{\lambda_{k_i}}} \langle W(t_{m+1}) - W(t_m), e_{k_i} \rangle_U & \text{if } \lambda_k > 0, \\ 0 & \text{else,} \end{cases}$$

$1 \leq i \leq n$, are independent since they are pairwise orthogonal in $L^2(\Omega, \mathcal{F}, P; \mathbb{R})$ and since $W(t_{m+1}) - W(t_m)$ is a Gaussian random variable. If we take $A_{i,j} \in \mathcal{B}(\mathbb{R})$, $1 \leq i \leq n$, $1 \leq j \leq m+1$, then because of the independence of $\sigma(W(s) \mid s \leq t_m)$ and $\sigma(W(t_{m+1}) - W(t_m))$ we get that

$$P\Big(\bigcap_{i=1}^{n} \{\beta_{k_i}(t_1) \in A_{i,1}, \ldots, \beta_{k_i}(t_m) \in A_{i,m},$$

$$\beta_{k_i}(t_{m+1}) - \beta_{k_i}(t_m) \in A_{i,m+1}\}\Big)$$

$$= P\Big(\underbrace{\bigcap_{i=1}^{n} \bigcap_{j=1}^{m} \{\beta_{k_i}(t_j) \in A_{i,j}\}}_{\in \sigma(W(s) \mid s \leq t_m)} \cap \underbrace{\bigcap_{i=1}^{n} \{\beta_{k_i}(t_{m+1}) - \beta_{k_i}(t_m) \in A_{i,m+1}\}}_{\in \sigma(W(t_{m+1}) - W(t_m))}\Big)$$

$$= P\Big(\bigcap_{i=1}^{n} \bigcap_{j=1}^{m} \{\beta_{k_i}(t_j) \in A_{i,j}\}\Big) \cdot P\Big(\bigcap_{i=1}^{n} \{\beta_{k_i}(t_{m+1}) - \beta_{k_i}(t_m) \in A_{i,m+1}\}\Big)$$

$$= \Big(\prod_{i=1}^{n} P\Big(\bigcap_{j=1}^{m} \{\beta_{k_i}(t_j) \in A_{i,j}\}\Big)\Big)$$

$$\cdot \Big(\prod_{i=1}^{n} P\{\beta_{k_i}(t_{m+1}) - \beta_{k_i}(t_m) \in A_{i,m+1}\}\Big)$$

$$= \prod_{i=1}^{n} P\Big(\bigcap_{j=1}^{m} \{\beta_{k_i}(t_j) \in A_{i,j}\} \cap \{\beta_{k_i}(t_{m+1}) - \beta_{k_i}(t_m) \in A_{i,m+1}\}\Big)$$

and therefore the assertion follows.

2. If we define

$$W(t) := \sum_{k \in \mathbb{N}} \sqrt{\lambda_k} \beta_k(t) e_k, \quad t \in [0, T],$$

where β_k, $k \in \mathbb{N}$, are independent real-valued continuous Brownian motions then it is clear that $W(t)$, $t \in [0, T]$, is well-defined in $L^2(\Omega, \mathcal{F}, P; U)$. Besides, it is obvious that the process $W(t)$, $t \in [0, T]$, starts at zero and that

$$P \circ \big(W(t) - W(s)\big)^{-1} = N\big(0, (t - s)Q\big), \quad 0 \leqslant s < t \leqslant T,$$

by Proposition 2.1.6. It is also clear that the increments are independent.

Thus it remains to show that the above series converges in $L^2\big(\Omega, \mathcal{F}, P; C([0, T], U)\big)$. To this end we set

$$W^N(t, \omega) := \sum_{k=1}^{N} \sqrt{\lambda_k} \beta_k(t, \omega) e_k$$

for all $(t, \omega) \in \Omega_T := [0, T] \times \Omega$ and $N \in \mathbb{N}$. Then W^N, $N \in \mathbb{N}$, is P-a.s. continuous and we have that for $M < N$

$$E\Big(\sup_{t \in [0,T]} \big\| W^N(t) - W^M(t) \big\|_U^2 \Big) = E\Big(\sup_{t \in [0,T]} \sum_{k=M+1}^{N} \lambda_k \beta_k^2(t) \Big)$$

$$\leqslant \sum_{k=M+1}^{N} \lambda_k E\Big(\sup_{t \in [0,T]} \beta_k^2(t) \Big) \leqslant 4 \sum_{k=M+1}^{N} \lambda_k \sup_{t \in [0,T]} E\big(\beta_k(t)^2\big) \leqslant 4T \sum_{k=M+1}^{N} \lambda_k$$

because of Doob's maximal inequality for nonnegative real-valued submartingales. As $\sum_{k \in \mathbb{N}} \lambda_k = \mathrm{tr}\, Q < \infty$, the assertion follows. ∎

Definition 2.1.11 (Normal Filtration) A filtration $\mathcal{F}_t$, $t \in [0, T]$, on a probability space $(\Omega, \mathcal{F}, P)$ is called normal if:

- $\mathcal{F}_0$ contains all elements $A \in \mathcal{F}$ with $P(A) = 0$ and
- $\mathcal{F}_t = \mathcal{F}_{t+} = \bigcap_{s > t} \mathcal{F}_s$ for all $t \in [0, T[$.

Definition 2.1.12 (Q-Wiener Process with Respect to a Filtration) A Q-Wiener process $W(t)$, $t \in [0, T]$, is called a Q-Wiener process with respect to a filtration $\mathcal{F}_t$, $t \in [0, T]$, if:

- $W(t)$, $t \in [0, T]$, is adapted to $\mathcal{F}_t$, $t \in [0, T]$, and
- $W(t) - W(s)$ is independent of $\mathcal{F}_s$ for all $0 \leqslant s \leqslant t \leqslant T$.

In fact it is possible to show that any U-valued Q-Wiener process $W(t)$, $t \in [0, T]$, is a Q-Wiener process with respect to a normal filtration:

We define

$$\mathcal{N} := \big\{ A \in \mathcal{F} \mid P(A) = 0 \big\}, \quad \mathcal{F}_t^0 := \sigma\big(W(s) \mid s \leqslant t\big)$$

and $\quad \tilde{\mathcal{F}}_t^0 := \sigma(\mathcal{F}_t^0 \cup \mathcal{N}).$

Then it is clear that

$$\mathcal{F}_t := \bigcap_{s>t} \tilde{\mathcal{F}}_s^0, \quad t \in [0, T[, \; \mathcal{F}_T := \tilde{\mathcal{F}}_T^0, \tag{2.3}$$

is a normal filtration, called the *normal filtration associated to* $W(t)$, $t \in [0, T]$, and we get:

Proposition 2.1.13 *Let* $W(t)$, $t \in [0, T]$, *be an arbitrary* U-*valued* Q-*Wiener process on a probability space* $(\Omega, \mathcal{F}, P)$. *Then it is a* Q-*Wiener process with respect to the normal filtration* $\mathcal{F}_t$, $t \in [0, T]$, *given by* (2.3).

Proof It is clear that $W(t)$, $t \in [0, T]$, is adapted to $\mathcal{F}_t$, $t \in [0, T]$. Hence we only have to verify that $W(t) - W(s)$ is independent of $\mathcal{F}_s$, $0 \leqslant s < t \leqslant T$. But if we fix $0 \leqslant s < t \leqslant T$ it is clear that $W(t) - W(s)$ is independent of $\tilde{\mathcal{F}}_s$ since

$$\sigma\big(W(t_1), W(t_2), \ldots, W(t_n)\big)$$
$$= \sigma\big(W(t_1), W(t_2) - W(t_1), \ldots, W(t_n) - W(t_{n-1})\big)$$

for all $0 \leqslant t_1 < t_2 < \cdots < t_n \leqslant s$. Of course, $W(t) - W(s)$ is then also independent of $\tilde{\mathcal{F}}_s^0$. To prove now that $W(t) - W(s)$ is independent of $\mathcal{F}_s$ it is enough to show that

$$P\big(\{W(t) - W(s) \in A\} \cap B\big) = P\big(W(t) - W(s) \in A\big) \cdot P(B)$$

for any $B \in \mathcal{F}_s$ and any closed subset $A \subset U$ as $\mathcal{E} := \{A \subset U \mid A \text{ closed}\}$ generates $\mathcal{B}(U)$ and is stable under finite intersections. But we have

$$P\big(\{W(t) - W(s) \in A\} \cap B\big)$$
$$= E\Big(1_A \circ \big(W(t) - W(s)\big) \cdot 1_B\Big)$$
$$- \lim_{n \to \infty} E\Big(\big[\big(1 - n \operatorname{dist}\big(W(t) - W(s), A\big)\big) \vee 0\big] 1_B\Big)$$
$$= \lim_{n \to \infty} \lim_{m \to \infty} E\Big(\big[\big(1 - n \operatorname{dist}\big(W(t) - W(s + \tfrac{1}{m}), A\big)\big) \vee 0\big] 1_B\Big)$$
$$= \lim_{n \to \infty} \lim_{m \to \infty} E\Big(\big(1 - n \operatorname{dist}\big(W(t) - W(s + \tfrac{1}{m}), A\big)\big) \vee 0\Big) \cdot P(B)$$
$$= P\big(W(t) - W(s) \in A\big) \cdot P(B),$$

since $W(t) - W(s + \tfrac{1}{m})$ is independent of $\tilde{\mathcal{F}}_{s+\frac{1}{m}}^0 \supset \mathcal{F}_s$ for all $m \in \mathbb{N}$. $\qquad\square$

2.2 Martingales in General Banach Spaces

Analogously to the real-valued case it is possible to define the conditional expectation of any Bochner integrable random variable with values in an arbitrary separable Banach space $(E, \| \ \|)$. This result is formulated in the following proposition.

Proposition 2.2.1 (Existence of the Conditional Expectation) *Assume that E is a separable real Banach space. Let X be a Bochner integrable E-valued random variable defined on a probability space $(\Omega, \mathcal{F}, P)$ and let $\mathcal{G}$ be a σ-field contained in $\mathcal{F}$.*

Then there exists a unique, up to a set of P-probability zero, Bochner integrable E-valued random variable Z, measurable with respect to $\mathcal{G}$, such that

$$\int_A X \, dP = \int_A Z \, dP \quad \text{for all } A \in \mathcal{G}. \tag{2.4}$$

The random variable Z is denoted by $E(X \mid \mathcal{G})$ and is called the conditional expectation *of X given $\mathcal{G}$. Furthermore,*

$$\big\| E(X \mid \mathcal{G}) \big\| \leqslant E\big(\|X\| \mid \mathcal{G} \big).$$

Proof (Cf. [26, Proposition 1.10, p. 27]) Let us first show the uniqueness.

Since E is a separable Banach space, there exist $l_n \in E^*$, $n \in \mathbb{N}$, separating the points of E. Suppose that Z_1, Z_2 are Bochner integrable, $\mathcal{G}$-measurable mappings from Ω to E such that

$$\int_A X \, dP = \int_A Z_1 \, dP = \int_A Z_2 \, dP \quad \text{for all } A \in \mathcal{G}.$$

Then for $n \in \mathbb{N}$ by Proposition A.2.2

$$\int_A \big(l_n(Z_1) - l_n(Z_2) \big) \, dP = 0 \quad \text{for all } A \in \mathcal{G}.$$

Applying this with $A := \{ l_n(Z_1) > l_n(Z_2) \}$ and $A := \{ l_n(Z_1) < l_n(Z_2) \}$ it follows that $l_n(Z_1) = l_n(Z_2)$ P-a.s., so

$$\Omega_0 := \bigcap_{n \in \mathbb{N}} \{ l_n(Z_1) = l_n(Z_2) \}$$

has P-measure one. Since l_n, $n \in \mathbb{N}$, separate the points of E; it follows that $Z_1 = Z_2$ on Ω_0.

To show the existence we first assume that X is a simple function. So, there exist $x_1, \ldots, x_N \in E$ and pairwise disjoint sets $A_1, \ldots, A_N \in \mathcal{F}$ such that

$$X = \sum_{k=1}^{N} x_k 1_{A_k}.$$

Define

$$Z := \sum_{k=1}^{N} x_k E(1_{A_k} \mid \mathcal{G}).$$

Then obviously Z is $\mathcal{G}$-measurable and satisfies (2.4). Furthermore,

$$\|Z\| \leq \sum_{k=1}^{N} \|x_k\| E(1_{A_k} \mid \mathcal{G}) = E\left(\sum_{k=1}^{N} \|x_k\| 1_{A_k} \;\middle|\; \mathcal{G}\right) = E(\|X\| \mid \mathcal{G}). \tag{2.5}$$

Taking the expectation we get

$$E(\|Z\|) \leq E(\|X\|). \tag{2.6}$$

For general X take simple functions X_n, $n \in \mathbb{N}$, as in Lemma A.1.4 and define Z_n as above with X_n replacing X. Then by (2.6) for all $n, m \in \mathbb{N}$

$$E(\|Z_n - Z_m\|) \leq E(\|X_n - X_m\|),$$

so $Z := \lim_{n \to \infty} Z_n$ exists in $L^1(\Omega, \mathcal{F}, P; E)$. Therefore, for all $A \in \mathcal{G}$

$$\int_A X \, dP = \lim_{n \to \infty} \int_A X_n \, dP = \lim_{n \to \infty} \int_A Z_n \, dP = \int_A Z \, dP.$$

Clearly, Z can be chosen $\mathcal{G}$-measurable, since so are the Z_n. Furthermore, by (2.5)

$$\left\| E(X \mid \mathcal{G}) \right\| = \|Z\| = \lim_{n \to \infty} \|Z_n\| \leq \lim_{n \to \infty} E(\|X_n\| \mid \mathcal{G}) = E(\|X\| \mid \mathcal{G}),$$

where the limits are taken in $L^1(P)$.   □

Later we will need the following result:

Proposition 2.2.2 *Let $(E_1, \mathcal{E}_1)$ and $(E_2, \mathcal{E}_2)$ be two measurable spaces and $\Psi : E_1 \times E_2 \to \mathbb{R}$ be a bounded measurable function. Let X_1 and X_2 be two random variables on $(\Omega, \mathcal{F}, P)$ with values in $(E_1, \mathcal{E}_1)$ and $(E_2, \mathcal{E}_2)$ respectively, and let $\mathcal{G} \subset \mathcal{F}$ be a fixed σ-field.*

Assume that X_1 is $\mathcal{G}$-measurable and X_2 is independent of $\mathcal{G}$, then for P-a.e. $\omega \in \Omega$

$$E\big(\Psi(X_1, X_2) \mid \mathcal{G}\big)(\omega) = E\left[\Psi(X_1(\omega), X_2)\right].$$

Proof A simple exercise or see [26, Proposition 1.12, p. 29]. □

Remark 2.2.3 The previous proposition can be easily extended to the case where the function Ψ is not necessarily bounded but nonnegative.

Definition 2.2.4 Let $M(t)$, $t \geq 0$, be a stochastic process on $(\Omega, \mathcal{F}, P)$ with values in a separable Banach space E, and let $\mathcal{F}_t$, $t \geq 0$, be a filtration on $(\Omega, \mathcal{F}, P)$.
The process M is called an $\mathcal{F}_t$-*martingale*, if:

- $E\big(\|M(t)\|\big) < \infty$ for all $t \geq 0$,
- $M(t)$ is $\mathcal{F}_t$-measurable for all $t \geq 0$,
- $E\big(M(t) \mid \mathcal{F}_s\big) = M(s)$ P-a.s. for all $0 \leq s \leq t < \infty$.

Remark 2.2.5 Let M be as above such that $E(\|M(t)\|) < \infty$ for all $t \in [0, T]$. Then M is an $\mathcal{F}_t$-martingale if and only if $l(M)$ is an $\mathcal{F}_t$-martingale for all $l \in E^*$. In particular, results like optional stopping etc. extend to E-valued martingales.

There is the following connection to real-valued submartingales.

Proposition 2.2.6 *If $M(t)$, $t \geq 0$, is an E-valued $\mathcal{F}_t$-martingale and $p \in [1, \infty)$, then $\big\|M(t)\big\|^p$, $t \geq 0$, is a real-valued $\mathcal{F}_t$-submartingale.*

Proof Since E is separable there exist $l_k \in E^*$, $k \in \mathbb{N}$, such that $\|z\| = \sup_{k \in \mathbb{N}} l_k(z)$ for all $z \in E$. Then by Proposition A.2.2 for $s < t$

$$E\big(\|M_t\| \mid \mathcal{F}_s\big) \geq \sup_k E\big(l_k(M_t) \mid \mathcal{F}_s\big)$$
$$= \sup_k l_k\big(E(M_t \mid \mathcal{F}_s)\big)$$
$$= \sup_k l_k(M_s) = \|M_s\|.$$

This proves the assertion for $p = 1$. Then Jensen's inequality implies the assertion for all $p \in [1, \infty)$. □

Theorem 2.2.7 (Maximal Inequality) *Let $p > 1$ and let E be a separable Banach space.*
If $M(t)$, $t \in [0, T]$, is a right-continuous E-valued $\mathcal{F}_t$-martingale, then

$$\left(E\Big(\sup_{t \in [0,T]} \|M(t)\|^p\Big)\right)^{\frac{1}{p}} \leq \frac{p}{p-1} \sup_{t \in [0,T]} \Big(E\big(\|M(t)\|^p\big)\Big)^{\frac{1}{p}}$$
$$= \frac{p}{p-1}\Big(E\big(\|M(T)\|^p\big)\Big)^{\frac{1}{p}}.$$

Proof The inequality is a consequence of the previous proposition and Doob's maximal inequality for nonnegative real-valued submartingales.   □

Remark 2.2.8 We note that in the inequality in Theorem 2.2.7 the first norm is the standard norm on $L^p(\Omega, \mathcal{F}, P; L^\infty([0, T]; E))$, whereas the second is the standard norm on $L^\infty([0, T]; L^p(\Omega, \mathcal{F}, P; E))$. So, for right-continuous E-valued $\mathcal{F}_t$-martingales these two norms are equivalent.

Now we fix $0 < T < \infty$ and denote by $\mathcal{M}_T^2(E)$ the space of all E-valued continuous, square integrable martingales $M(t)$, $t \in [0, T]$. This space will play an important role with regard to the definition of the stochastic integral. We will especially use the following fact.

Proposition 2.2.9 *The space $\mathcal{M}_T^2(E)$ equipped with the norm*

$$\|M\|_{\mathcal{M}_T^2} := \sup_{t \in [0, T]} \left(E\left(\|M(t)\|^2 \right) \right)^{\frac{1}{2}} = \left(E\left(\|M(T)\|^2 \right) \right)^{\frac{1}{2}}$$

$$\leqslant \left(E\left(\sup_{t \in [0, T]} \|M(t)\|^2 \right) \right)^{\frac{1}{2}} \leqslant 2 \cdot E\left(\|M(T)\|^2 \right)^{\frac{1}{2}}$$

is a Banach space.

Proof By the Riesz–Fischer theorem the space $L^2(\Omega, \mathcal{F}, P; L^\infty([0, T], E))$ is complete. So, we only have to show that $\mathcal{M}_T^2$ is closed. But this is obvious since even $L^1(\Omega, \mathcal{F}, P; E)$-limits of martingales are martingales.   □

Proposition 2.2.10 *Let $T > 0$ and $W(t)$, $t \in [0, T]$, be a U-valued Q-Wiener process with respect to a normal filtration $\mathcal{F}_t$, $t \in [0, T]$, on a probability space $(\Omega, \mathcal{F}, P)$. Then $W(t)$, $t \in [0, T]$, is a continuous square integrable $\mathcal{F}_t$-martingale, i.e. $W \in \mathcal{M}_T^2(U)$.*

Proof The continuity is clear by definition and for each $t \in [0, T]$ we have that $E(\|W(t)\|_U^2) = t \operatorname{tr} Q < \infty$ (see Proposition 2.1.4). Hence let $0 \leqslant s \leqslant t \leqslant T$ and $A \in \mathcal{F}_s$. Then we get by Proposition A.2.2 that

$$\left\langle \int_A W(t) - W(s) \, dP, u \right\rangle_U = \int_A \langle W(t) - W(s), u \rangle_U \, dP$$

$$= P(A) \int \langle W(t) - W(s), u \rangle_U \, dP = 0$$

for all $u \in U$ as $\mathcal{F}_s$ is independent of $W(t) - W(s)$ and $E(\langle W(t) - W(s), u \rangle_U) = 0$ for all $u \in U$. Therefore,

$$\int_A W(t) \, dP = \int_A W(s) \, dP, \qquad \text{for all } A \in \mathcal{F}_s.$$

                                                                                      □

2.3 The Definition of the Stochastic Integral

For the whole section we fix a positive real number T and a probability space $(\Omega, \mathcal{F}, P)$ and we define $\Omega_T := [0, T] \times \Omega$ and $P_T := dt \otimes P$ where dx is the Lebesgue measure.

Moreover, we let $Q \in L(U)$ be symmetric, nonnegative and with finite trace and consider a Q-Wiener process $W(t)$, $t \in [0, T]$, with respect to a normal filtration $\mathcal{F}_t$, $t \in [0, T]$.

2.3.1 Scheme of the Construction of the Stochastic Integral

Step 1: First we consider a certain class $\mathcal{E}$ of elementary $L(U, H)$-valued processes and define the mapping

$$\text{Int}: \mathcal{E} \to \mathcal{M}_T^2(H) =: \mathcal{M}_T^2$$
$$\Phi \mapsto \int_0^t \Phi(s) \, dW(s), \quad t \in [0, T].$$

Step 2: We prove that there is a certain norm on $\mathcal{E}$ such that

$$\text{Int}: \mathcal{E} \to \mathcal{M}_T^2$$

is an isometry. Since $\mathcal{M}_T^2$ is a Banach space this implies that Int can be extended to the abstract completion $\bar{\mathcal{E}}$ of $\mathcal{E}$. This extension remains isometric and it is unique.

Step 3: We give an explicit representation of $\bar{\mathcal{E}}$.

Step 4: We show how the definition of the stochastic integral can be extended by localization.

2.3.2 The Construction of the Stochastic Integral in Detail

Step 1: First we define the class $\mathcal{E}$ of all elementary processes as follows.

Definition 2.3.1 (Elementary Process) An $L = L(U, H)$-valued process $\Phi(t)$, $t \in [0, T]$, on $(\Omega, \mathcal{F}, P)$ with normal filtration $\mathcal{F}_t$, $t \in [0, T]$, is said to be *elementary* if there exist $0 = t_0 < \cdots < t_k = T$, $k \in \mathbb{N}$, such that

$$\Phi(t) = \sum_{m=0}^{k-1} \Phi_m 1_{]t_m, t_{m+1}]}(t), \quad t \in [0, T],$$

where:

- $\Phi_m : \Omega \rightarrow L(U, H)$ is $\mathcal{F}_{t_m}$-measurable, w.r.t. the strong Borel σ-algebra on $L(U, H), 0 \leqslant m \leqslant k - 1$,
- Φ_m takes only a finite number of values in $L(U, H), 0 \leqslant m \leqslant k - 1$.

We now define

$$\text{Int}(\Phi)(t) := \int_0^t \Phi(s) \, dW(s) := \sum_{m=0}^{k-1} \Phi_m \big(W(t_{m+1} \wedge t) - W(t_m \wedge t) \big), \quad t \in [0, T],$$

(this is obviously independent of the representation) for all $\Phi \in \mathcal{E}$.

Proposition 2.3.2 *Let $\Phi \in \mathcal{E}$. Then the stochastic integral $\int_0^t \Phi(s) \, dW(s)$, $t \in [0, T]$, defined above, is a continuous square integrable martingale with respect to $\mathcal{F}_t, t \in [0, T]$, i.e.*

$$\text{Int} : \mathcal{E} \rightarrow \mathcal{M}_T^2.$$

Proof Let $\Phi \in \mathcal{E}$ be given by

$$\Phi(t) = \sum_{m=0}^{k-1} \Phi_m 1_{]t_m, t_{m+1}]}(t), \quad t \in [0, T],$$

as in Definition 2.3.1. Then it is clear that

$$t \mapsto \int_0^t \Phi(s) \, dW(s) = \sum_{m=0}^{k-1} \Phi_m \big(W(t_{m+1} \wedge t) - W(t_m \wedge t) \big)$$

is P-a.s. continuous because of the continuity of the Wiener process and the continuity of $\Phi_m(\omega) : U \rightarrow H, 0 \leqslant m \leqslant k - 1, \omega \in \Omega$. In addition, we get for each summand that

$$\left\| \Phi_m \big(W(t_{m+1} \wedge t) - W(t_m \wedge t) \big) \right\|$$

$$\leqslant \| \Phi_m \|_{L(U, H)} \left\| W(t_{m+1} \wedge t) - W(t_m \wedge t) \right\|_U.$$

Since $W(t), t \in [0, T]$, is square integrable and $\omega \rightarrow \| \Phi_m(\omega) \|_{L(U, H)}$ is bounded (because $\Phi_m(\Omega)$ is finite), this implies that $\int_0^t \Phi(s) \, dW(s)$ is square integrable for each $t \in [0, T]$.

To prove the martingale property we take $0 \leq s \leq t \leq T$ and a set A from $\mathcal{F}_s$. If $\{\Phi_m(\omega) \mid \omega \in \Omega\} := \{L_1^m, \ldots, L_{k_m}^m\}$, we obtain by Proposition A.2.2 and the martingale property of the Wiener process that

$$\int_A \sum_{m=0}^{k-1} \Phi_m \big(W(t_{m+1} \wedge t) - W(t_m \wedge t)\big) \, dP$$

$$= \sum_{\substack{0 \leq m \leq k-1, \\ t_{m+1} < s}} \int_A \Phi_m \big(W(t_{m+1} \wedge s) - W(t_m \wedge s)\big) \, dP$$

$$+ \sum_{\substack{0 \leq m \leq k-1, \\ s \leq t_{m+1}}} \sum_{j=1}^{k_m} \int_{A \cap \{\Phi_m = L_j^m\}} L_j^m \big(W(t_{m+1} \wedge t) - W(t_m \wedge t)\big) \, dP$$

$$= \sum_{\substack{0 \leq m \leq k-1, \\ t_{m+1} < s}} \int_A \Phi_m \big(W(t_{m+1} \wedge s) - W(t_m \wedge s)\big) \, dP$$

$$+ \sum_{\substack{0 \leq m \leq k-1, \\ s \leq t_{m+1}}} \sum_{j=1}^{k_m} L_j^m \underbrace{\int_{A \cap \{\Phi_m = L_j^m\}} \big(W(t_{m+1} \wedge t) - W(t_m \wedge t)\big) \, dP}_{\in \mathcal{F}_{s \vee t_m}}$$

$$= \sum_{\substack{0 \leq m \leq k-1, \\ t_{m+1} < s}} \int_A \Phi_m \big(W(t_{m+1} \wedge s) - W(t_m \wedge s)\big) \, dP$$

$$+ \sum_{\substack{0 \leq m \leq k-1, \\ t_m < s \leq t_{m+1}}} \sum_{j=1}^{k_m} L_j^m \int_{A \cap \{\Phi_m = L_j^m\}} \big(W(t_{m+1} \wedge s) - W(t_m \wedge s)\big) \, dP$$

$$= \int_A \sum_{m=0}^{k-1} \Phi_m \big(W(t_{m+1} \wedge s) - W(t_m \wedge s)\big) \, dP.$$

$\square$

Step 2: To verify the assertion that there is a norm on $\mathcal{E}$ such that $\mathrm{Int} : \mathcal{E} \to \mathcal{M}_T^2$ is an isometry, we have to introduce the following notion.

Definition 2.3.3 (Hilbert–Schmidt Operator) Let e_k, $k \in \mathbb{N}$, be an orthonormal basis of U. An operator $A \in L(U, H)$ is called Hilbert–Schmidt if

$$\sum_{k \in \mathbb{N}} \langle Ae_k, Ae_k \rangle < \infty.$$

In Appendix B we take a close look at this notion. So here we only summarize the results which are important for the construction of the stochastic integral.

The definition of a Hilbert–Schmidt operator and the number

$$\|A\|_{L_2} := \left(\sum_{k \in \mathbb{N}} \|Ae_k\|^2 \right)^{\frac{1}{2}}$$

are independent of the choice of the basis (see Remark B.0.6(i)). Moreover, the space $L_2(U, H)$ of all Hilbert–Schmidt operators from U to H equipped with the inner product

$$\langle A, B \rangle_{L_2} := \sum_{k \in \mathbb{N}} \langle Ae_k, Be_k \rangle$$

is a separable Hilbert space (see Proposition B.0.7). Later, we will use the fact that $\|A\|_{L_2(U,H)} = \|A^*\|_{L_2(H,U)}$, where A^* is the adjoint operator of A (see Remark B.0.6(i)). Furthermore, the composition of a Hilbert–Schmidt operator with a bounded linear operator is again Hilbert–Schmidt.

Besides we recall the following fact.

Proposition 2.3.4 *If $Q \in L(U)$ is nonnegative and symmetric then there exists exactly one element $Q^{\frac{1}{2}} \in L(U)$ nonnegative and symmetric such that $Q^{\frac{1}{2}} \circ Q^{\frac{1}{2}} = Q$.*

If, in addition, $\operatorname{tr} Q < \infty$ we have that $Q^{\frac{1}{2}} \in L_2(U)$, $\|Q^{\frac{1}{2}}\|_{L_2}^2 = \operatorname{tr} Q$ and that $L \circ Q^{\frac{1}{2}} \in L_2(U, H)$ for all $L \in L(U, H)$.

Proof [68, Theorem VI.9, p. 196]. □

After these preparations we simply calculate the $\mathcal{M}_T^2$-norm of

$$\int_0^t \Phi(s) \, dW(s), \ t \in [0, T],$$

and get the following result.

Proposition 2.3.5 *If $\Phi = \sum_{m=0}^{k-1} \Phi_m 1_{]t_m, t_{m+1}]}$ is an elementary $L(U, H)$-valued process then*

$$\left\| \int_0^{\cdot} \Phi(s) \, dW(s) \right\|_{\mathcal{M}_T^2}^2 = E\left(\int_0^T \|\Phi(s) \circ Q^{\frac{1}{2}}\|_{L_2}^2 \, ds \right) =: \|\Phi\|_T^2 \ (\text{``Itô-isometry''}).$$

Proof If we set $\Delta_m := W(t_{m+1}) - W(t_m)$ then we get that

$$\left\| \int_0^{\cdot} \Phi(s) \, dW(s) \right\|_{\mathcal{M}_T^2}^2 = E\left(\left\| \int_0^T \Phi(s) \, dW(s) \right\|_H^2 \right) = E\left(\left\| \sum_{m=0}^{k-1} \Phi_m \Delta_m \right\|_H^2 \right)$$

$$= E\left(\sum_{m=0}^{k-1} \|\Phi_m \Delta_m\|_H^2 \right) + 2E\left(\sum_{0 \leqslant m < n \leqslant k-1} \langle \Phi_m \Delta_m, \Phi_n \Delta_n \rangle_H \right).$$

Claim 1:

$$E\Big(\sum_{m=0}^{k-1}\|\Phi_m\Delta_m\|_H^2\Big) = \sum_{m=0}^{k-1}(t_{m+1}-t_m)E\big(\|\Phi_m\circ Q^{\frac{1}{2}}\|_{L_2}^2\big)$$

$$= \int_0^T E\big(\|\Phi(s)\circ Q^{\frac{1}{2}}\|_{L_2}^2\big)\,ds.$$

To prove this we take an orthonormal basis $f_k,\ k\in\mathbb{N}$, of H and get by the Parseval identity and Levi's monotone convergence theorem that

$$E\big(\|\Phi_m\Delta_m\|_H^2\big) = \sum_{l\in\mathbb{N}}E\big(\langle\Phi_m\Delta_m,f_l\rangle_H^2\big) = \sum_{l\in\mathbb{N}}E\Big(E\big(\langle\Delta_m,\Phi_m^*f_l\rangle_U^2\mid\mathcal{F}_{t_m}\big)\Big).$$

Taking an orthonormal basis $e_k,\ k\in\mathbb{N}$, of U we obtain that

$$\Phi_m^*f_l = \sum_{k\in\mathbb{N}}\langle f_l,\Phi_m e_k\rangle_H e_k.$$

Since $\langle f_l,\Phi_m e_k\rangle_H$ is $\mathcal{F}_{t_m}$-measurable, this implies that $\Phi_m^*f_l$ is $\mathcal{F}_{t_m}$-measurable by Proposition A.1.3. Using the fact that $\sigma(\Delta_m)$ is independent of $\mathcal{F}_{t_m}$ we obtain by Proposition 2.2.2 that for P-a.e. $\omega\in\Omega$

$$E\big(\langle\Delta_m,\Phi_m^*f_l\rangle_U^2\mid\mathcal{F}_{t_m}\big)(\omega) = E\Big(\langle\Delta_m,\Phi_m^*(\omega)f_l\rangle_U^2\Big)$$

$$= (t_{m+1}-t_m)\Big\langle Q\big(\Phi_m^*(\omega)f_l\big),\Phi_m^*(\omega)f_l\Big\rangle_U,$$

since $E\big(\langle\Delta_m,u\rangle_U^2\big) = (t_{m+1}-t_m)\langle Qu,u\rangle_U$ for all $u\in U$. Thus, the symmetry of $Q^{\frac{1}{2}}$ finally implies that

$$E\big(\|\Phi_m\Delta_m\|_H^2\big) = \sum_{l\in\mathbb{N}}E\Big(E\big(\langle\Delta_m,\Phi_m^*f_l\rangle_U^2\mid\mathcal{F}_{t_m}\big)\Big)$$

$$= (t_{m+1}-t_m)\sum_{l\in\mathbb{N}}E\big(\langle Q\Phi_m^*f_l,\Phi_m^*f_l\rangle_U\big)$$

$$= (t_{m+1}-t_m)\sum_{l\in\mathbb{N}}E\big(\|Q^{\frac{1}{2}}\Phi_m^*f_l\|_U^2\big)$$

$$= (t_{m+1}-t_m)E\Big(\big\|\big(\Phi_m\circ Q^{\frac{1}{2}}\big)^*\big\|_{L_2(H,U)}^2\Big)$$

$$= (t_{m+1}-t_m)E\Big(\|\Phi_m\circ Q^{\frac{1}{2}}\|_{L_2(U,H)}^2\Big).$$

Hence the first assertion is proved and it only remains to verify the following claim.

Claim 2:

$$E\big(\langle \Phi_m \Delta_m, \Phi_n \Delta_n \rangle_H\big) = 0 , \quad 0 \leqslant m < n \leqslant k - 1.$$

But this can be proved in a similar way to Claim 1:

$$E\big(\langle \Phi_m \Delta_m, \Phi_n \Delta_n \rangle_H\big) = E\Big(E\big(\langle \Phi_n^* \Phi_m \Delta_m, \Delta_n \rangle_U \mid \mathcal{F}_{t_n}\big)\Big)$$

$$= \int E\Big(\langle \Phi_n^*(\omega) \Phi_m(\omega) \Delta_m(\omega), \Delta_n \rangle_U\Big) P(d\omega) = 0,$$

since $E\big(\langle u, \Delta_n \rangle_U\big) = 0$ for all $u \in U$ (see Proposition 2.2.2). Hence the assertion follows.   □

Hence the right norm on $\mathcal{E}$ has been identified. But strictly speaking $\| \ \|_T$ is only a seminorm on $\mathcal{E}$. Therefore, we have to consider equivalence classes of elementary processes with respect to $\| \ \|_T$ to get a norm on $\mathcal{E}$. For simplicity we will not change the notation but stress the following fact.

Remark 2.3.6 If two elementary processes Φ and $\tilde{\Phi}$ belong to one equivalence class with respect to $\| \ \|_T$ it does not follow that they are equal P_T-a.e. because their values only have to correspond on $Q^{\frac{1}{2}}(U)$ P_T-a.e.

Thus we have finally shown that

$$\text{Int} : \big(\mathcal{E}, \| \ \|_T\big) \to \big(\mathcal{M}_T^2, \| \ \|_{\mathcal{M}_T^2}\big)$$

is an isometric transformation. Since $\mathcal{E}$ is dense in the abstract completion $\bar{\mathcal{E}}$ of $\mathcal{E}$ with respect to $\| \ \|_T$ it is clear that there is a unique isometric extension of Int to $\bar{\mathcal{E}}$.

Step 3: To give an explicit representation of $\bar{\mathcal{E}}$ it is useful, at this moment, to introduce the subspace $U_0 := Q^{\frac{1}{2}}(U)$ with the inner product given by

$$\langle u_0, v_0 \rangle_0 := \big\langle Q^{-\frac{1}{2}} u_0, Q^{-\frac{1}{2}} v_0 \big\rangle_U,$$

$u_0, v_0 \in U_0$, where $Q^{-\frac{1}{2}}$ is the pseudo inverse of $Q^{\frac{1}{2}}$ in the case that Q is not one-to-one. Then we get by Proposition C.0.3(i) that $(U_0, \langle \ , \ \rangle_0)$ is again a separable Hilbert space.

The separable Hilbert space $L_2(U_0, H)$ is called L_2^0. By Proposition C.0.3(ii) we know that $Q^{\frac{1}{2}} g_k, k \in \mathbb{N}$, is an orthonormal basis of $(U_0, \langle \ , \ \rangle_0)$ if $g_k, k \in \mathbb{N}$, is an orthonormal basis of $\big(\text{Ker } Q^{\frac{1}{2}}\big)^{\perp}$. This basis can be supplemented to a basis of U

by elements of $\operatorname{Ker} Q^{\frac{1}{2}}$. Thus we obtain that

$$\|L\|_{L_2^0} = \left\|L \circ Q^{\frac{1}{2}}\right\|_{L_2} \quad \text{for each } L \in L_2^0.$$

Define $L(U, H)_0 := \left\{T|_{U_0} \mid T \in L(U, H)\right\}$. Since $Q^{\frac{1}{2}} \in L_2(U)$ it is clear that $L(U, H)_0 \subset L_2^0$ and that the $\| \ \|_T$-norm of $\Phi \in \mathcal{E}$ can be written in the following way:

$$\|\Phi\|_T = \left(E\left(\int_0^T \|\Phi(s)\|_{L_2^0}^2 \, ds\right)\right)^{\frac{1}{2}}.$$

We also need the following σ-field:

$$\mathcal{P}_T := \sigma\left(\left\{]s, t] \times F_s \mid 0 \leqslant s < t \leqslant T, \ F_s \in \mathcal{F}_s\right\} \cup \left\{\{0\} \times F_0 \mid F_0 \in \mathcal{F}_0\right\}\right)$$

$$= \sigma\left(Y : \Omega_T \to \mathbb{R} \mid Y \text{ is left-continuous and adapted to}\right.$$

$$\left.\mathcal{F}_t, \ t \in [0, T]\right).$$

Let $\tilde{H}$ be an arbitrary separable Hilbert space. If $Y : \Omega_T \to \tilde{H}$ is $\mathcal{P}_T/\mathcal{B}(\tilde{H})$-measurable it is called $(\tilde{H}\text{-})$predictable.

If, for example, the process Y itself is continuous and adapted to $\mathcal{F}_t, \ t \in [0, T]$, then it is predictable.

So, we are now able to characterize $\bar{\mathcal{E}}$.

Claim There is an explicit representation of $\bar{\mathcal{E}}$ and it is given by

$$\mathcal{N}_W^2(0, T; H) := \left\{\Phi : [0, T] \times \Omega \to L_2^0 \mid \Phi \text{ is predictable and } \|\Phi\|_T < \infty\right\}$$

$$= L^2\left([0, T] \times \Omega, \mathcal{P}_T, \ dt \otimes P; L_2^0\right).$$

For simplicity we also write $\mathcal{N}_W^2(0, T)$ or $\mathcal{N}_W^2$ instead of $\mathcal{N}_W^2(0, T; H)$.

To prove this claim we first notice the following facts:

1. Since $L(U, H)_0 \subset L_2^0$ and since any $\Phi \in \mathcal{E}$ is L_2^0-predictable by construction we have that $\mathcal{E} \subset \mathcal{N}_W^2$.
2. Because of the completeness of L_2^0 we get by Appendix A that

$$\mathcal{N}_W^2 = L^2(\Omega_T, \mathcal{P}_T, P_T; L_2^0)$$

is also complete.

Therefore $\mathcal{N}_W^2$ is at least a candidate for a representation of $\bar{\mathcal{E}}$. Thus it only remains to show that $\mathcal{E}$ is a dense subset of $\mathcal{N}_W^2$. But this is formulated in Proposition 2.3.8 below, which can be proved with the help of the following lemma.

Lemma 2.3.7 *There is an orthonormal basis of L_2^0 consisting of elements of $L(U, H)_0$. This especially implies that $L(U, H)_0$ is a dense subset of L_2^0.*

Proof Since Q is symmetric, nonnegative and $\operatorname{tr} Q < \infty$ we know by Lemma 2.1.5 that there exists an orthonormal basis e_k, $k \in \mathbb{N}$, of U such that $Q e_k = \lambda_k e_k$, $\lambda_k \geq 0$, $k \in \mathbb{N}$. In this case $Q^{\frac{1}{2}} e_k = \sqrt{\lambda_k} e_k$, $k \in \mathbb{N}$ with $\lambda_k > 0$, is an orthonormal basis of U_0 (see Proposition C.0.3(ii)).

If f_k, $k \in \mathbb{N}$, is an orthonormal basis of H then by Proposition B.0.7 we know that

$$f_j \otimes \sqrt{\lambda_k} e_k = f_j \langle \sqrt{\lambda_k} e_k, \cdot \rangle_{U_0} = \frac{1}{\sqrt{\lambda_k}} f_j \langle e_k, \cdot \rangle_U, \quad j, k \in \mathbb{N}, \ \lambda_k > 0,$$

form an orthonormal basis of L_0^2 consisting of operators in $L(U, H)_0$. $\qquad\square$

Proposition 2.3.8 *If $\Phi \in \mathcal{N}_W^2$ then there exists a sequence Φ_n, $n \in \mathbb{N}$, of $L(U, H)_0$-valued elementary processes such that*

$$\|\Phi - \Phi_n\|_T \longrightarrow 0 \quad as \ n \to \infty.$$

Proof

Step 1: If $\Phi \in \mathcal{N}_W^2$ there exists a sequence of simple random variables $\Phi_n = \sum_{k=1}^{M_n} L_k^n 1_{A_k^n}$, $A_k^n \in \mathcal{P}_T$ and $L_k^n \in L_2^0$, $n \in \mathbb{N}$, such that

$$\|\Phi - \Phi_n\|_T \longrightarrow 0 \quad as \ n \to \infty.$$

As L_2^0 is a separable Hilbert space, this is a simple consequence of Lemma A.1.4 and Lebesgue's dominated convergence theorem.

Thus the assertion is reduced to the case that $\Phi = L 1_A$ where $L \in L_2^0$ and $A \in \mathcal{P}_T$.

Step 2: Let $A \in \mathcal{P}_T$ and $L \in L_2^0$. Then there exists a sequence L_n, $n \in \mathbb{N}$, in $L(U, H)_0$ such that

$$\|L 1_A - L_n 1_A\|_T \longrightarrow 0 \quad as \ n \to \infty.$$

This result is obvious by Lemma 2.3.7 and thus now we only have to consider the case when $\Phi = L 1_A$, $L \in L(U, H)_0$ and $A \in \mathcal{P}_T$.

Step 3: If $\Phi = L 1_A$, $L \in L(U, H)_0$, $A \in \mathcal{P}_T$, then there is a sequence Φ_n, $n \in \mathbb{N}$, of elementary $L(U, H)_0$-valued processes in the sense of Definition 2.3.1 such that

$$\|L 1_A - \Phi_n\|_T \longrightarrow 0 \quad as \ n \longrightarrow \infty.$$

To show this it is sufficient to prove that for any $\varepsilon > 0$ there is a finite union $\Lambda := \bigcup_{n=1}^{N} A_n$ of pairwise disjoint predictable rectangles

$$A_n \in \left\{]s,t] \times F_s \mid 0 \leqslant s < t \leqslant T, F_s \in \mathcal{F}_s \right\} \cup \left\{ \{0\} \times F_0 \mid F_0 \in \mathcal{F}_0 \right\} =: \mathcal{A}$$

such that

$$P_T \big((A \setminus \Lambda) \cup (\Lambda \setminus A) \big) < \varepsilon.$$

For then we get that $\sum_{n=1}^{N} L1_{A_n}$ differs from an elementary process by a function of type $L1_{\{0\} \times F_0}$ with $F_0 \in \mathcal{F}_0$, which has $\| \ \|_T$-norm zero, and that

$$\left\| L1_A - \sum_{n=1}^{N} L1_{A_n} \right\|_T^2 = E\left(\int_0^T \left\| L\left(1_A - \sum_{n=1}^{N} 1_{A_n}\right) \right\|_{L_2^0}^2 ds \right) \leqslant \varepsilon \|L\|_{L_2^0}^2.$$

Hence we define

$$\mathcal{K} := \left\{ \bigcup_{i \in I} A_i \mid I \text{ is finite and } A_i \in \mathcal{A}, i \in I \right\}.$$

Then $\mathcal{K}$ is an algebra and any element in $\mathcal{K}$ can be written as a finite disjoint union of elements in $\mathcal{A}$. Now let $\mathcal{G}$ be the family of all $A \in \mathcal{P}_T$ which can be approximated by elements of $\mathcal{K}$ in the above sense. Then $\mathcal{G}$ is a Dynkin system, because obviously $\Omega_T \in \mathcal{K} \subset \mathcal{G}$, and $A^c \in \mathcal{G}$ if $A \in \mathcal{G}$. Furthermore, if $A_i \in \mathcal{G}$, $i \in \mathbb{N}$, pairwise disjoint, and $\epsilon > 0$, then there exists an $N \in \mathbb{N}$ such that

$$P_T \left(\bigcup_{i=N+1}^{\infty} A_i \right) = \sum_{i=N+1}^{\infty} P_T(A_i) < \frac{\epsilon}{2}$$

and $\Lambda_i \in \mathcal{K}$ such that $P_T \left((A_i \backslash \Lambda_i) \cup (\Lambda_i \backslash A_i) \right) < \frac{\epsilon}{2^{i+2}}$. Hence $\Lambda := \bigcup_{i=1}^{N} \Lambda_i \in \mathcal{K}$ and

$$P_T \left(\bigcup_{i=1}^{\infty} A_i \backslash \bigcup_{i=1}^{N} \Lambda_i \right) \cup \left(\bigcup_{i=1}^{N} \Lambda_i \backslash \bigcup_{i=1}^{\infty} A_i \right) \leqslant P_T \left(\bigcup_{i=N+1}^{\infty} A_i \right) + P_T \left(\bigcup_{i=1}^{N} A_i \backslash \Lambda_i \right)$$

$$+ P_T \left(\bigcup_{i=1}^{N} \Lambda_i \backslash A_i \right) < \epsilon.$$

Therefore $\mathcal{P}_T = \sigma(\mathcal{K}) = \mathcal{D}(\mathcal{K}) \subset \mathcal{G}$ as $\mathcal{K} \subset \mathcal{G}$. $\qquad \square$

Step 4: Finally, by the so-called localization procedure we shall extend the definition of the stochastic integral to the linear space

$$\mathcal{N}_W(0, T; H) := \left\{ \Phi : \Omega_T \to L_2^0 \,\middle|\, \Phi \text{ is predictable with} \right.$$

$$\left. P\left(\int_0^T \|\Phi(s)\|_{L_2^0}^2 \, ds < \infty\right) = 1 \right\}.$$

For simplicity we also write $\mathcal{N}_W(0, T)$ or $\mathcal{N}_W$ instead of $\mathcal{N}_W(0, T; H)$ and $\mathcal{N}_W$ is called the class of *stochastically integrable* processes on $[0, T]$.

The extension is done in the following way:

For $\Phi \in \mathcal{N}_W$ we define

$$\tau_n := \inf\left\{ t \in [0, T] \,\middle|\, \int_0^t \|\Phi(s)\|_{L_2^0}^2 \, ds > n \right\} \wedge T. \tag{2.7}$$

Then by the right-continuity of the filtration $\mathcal{F}_t$, $t \in [0, T]$, we get that

$$\{\tau_n \leqslant t\} = \bigcap_{m \in \mathbb{N}} \left\{ \tau_n < t + \frac{1}{m} \right\}$$

$$= \bigcap_{m \in \mathbb{N}} \underbrace{\bigcup_{q \in [0, t + \frac{1}{m}[\cap \mathbb{Q}} \underbrace{\left\{ \int_0^q \|\Phi(s)\|_{L_2^0}^2 \, ds > n \right\}}_{\in \mathcal{F}_q \text{ by the real Fubini theorem}}}_{\in \mathcal{F}_{t + \frac{1}{m}} \text{ and decreasing in } m} \in \mathcal{F}_t.$$

Therefore τ_n, $n \in \mathbb{N}$, is an increasing sequence of stopping times with respect to $\mathcal{F}_t$, $t \in [0, T]$, such that

$$E\left(\int_0^T \|1_{]0, \tau_n]}(s)\Phi(s)\|_{L_2^0}^2 \, ds\right) \leqslant n < \infty.$$

In addition, the processes $1_{]0, \tau_n]}\Phi$, $n \in \mathbb{N}$, are still L_2^0-predictable since $1_{]0, \tau_n]}$ is left-continuous and $(\mathcal{F}_t)$-adapted or since

$$]0, \tau_n] := \{(s, \omega) \in \Omega_T \mid 0 < s \leqslant \tau_n(\omega)\}$$

$$= \left(\{(s, \omega) \in \Omega_T \mid \tau_n(\omega) < s \leqslant T\} \cup \{0\} \times \Omega\right)^c$$

$$= \left(\bigcup_{q \in \mathbb{Q}} (]q, T] \times \underbrace{\{\tau_n \leqslant q\}}_{\in \mathcal{F}_q}) \cup \{0\} \times \Omega\right)^c \in \mathcal{P}_T.$$

Thus we get that the stochastic integrals

$$\int_0^t 1_{]0,\tau_n]}(s)\Phi(s)\,dW(s), \quad t \in [0,T],$$

are well-defined for all $n \in \mathbb{N}$. For arbitrary $t \in [0,T]$ we set

$$\int_0^t \Phi(s)\,dW(s) := \int_0^t 1_{]0,\tau_n]}(s)\Phi(s)\,dW(s) \text{ on } \{\tau_n \geq t\}. \tag{2.8}$$

(Note that the sequence τ_n, $n \in \mathbb{N}$, even reaches T P-a.s., in the sense that for P-a.e. $\omega \in \Omega$ there exists an $n(\omega) \in \mathbb{N}$ such that $\tau_n(\omega) = T$ for all $n \geq n(\omega)$.)
To show that this definition is consistent we have to prove that for arbitrary natural numbers $m < n$ and $t \in [0,T]$

$$\int_0^t 1_{]0,\tau_m]}(s)\Phi(s)\,dW(s) = \int_0^t 1_{]0,\tau_n]}(s)\Phi(s)\,dW(s) \quad P\text{-a.s.}$$

on $\{\tau_m \geq t\} \subset \{\tau_n \geq t\}$. This result follows from the following lemma, which also implies that the process in (2.8) is a continuous H-valued local martingale.

Lemma 2.3.9 *Assume that $\Phi \in \mathcal{N}_W^2$ and that τ is an $\mathcal{F}_t$-stopping time such that $P(\tau \leq T) = 1$. Then there exists a P-null set $N \in \mathcal{F}$ independent of $t \in [0,T]$ such that*

$$\int_0^t 1_{]0,\tau]}(s)\Phi(s)\,dW(s) = \mathrm{Int}\big(1_{]0,\tau]}\Phi\big)(t) = \mathrm{Int}(\Phi)(\tau \wedge t)$$

$$= \int_0^{\tau \wedge t} \Phi(s)\,dW(s) \quad \text{on } N^c \text{ for all } t \in [0,T].$$

Proof Since both integrals which appear in the equation are P-a.s. continuous we only have to prove that they are equal P-a.s. at any fixed time $t \in [0,T]$.

Step 1: We first consider the case that $\Phi \in \mathcal{E}$ and that τ is a simple stopping time which means that it takes only a finite number of values.
Let $0 = t_0 < t_1 < \cdots < t_k \leq T$, $k \in \mathbb{N}$, and

$$\Phi = \sum_{m=0}^{k-1} \Phi_m 1_{]t_m,t_{m+1}]}$$

where $\Phi_m : \Omega \to L(U,H)$ is $\mathcal{F}_{t_m}$-measurable and only takes a finite number of values for all $0 \leq m \leq k-1$.

If τ is a simple stopping time, there exists an $n \in \mathbb{N}$ such that $\tau(\Omega) = \{a_0, \dots, a_n\}$ and

$$\tau = \sum_{j=0}^{n} a_j 1_{A_j},$$

where $0 \leq a_j < a_{j+1} \leq T$ and $A_j = \{\tau = a_j\} \in \mathcal{F}_{a_j}$. Then we get that $1_{]\tau,T]}\Phi$ is an elementary process since

$$1_{]\tau,T]}(s)\Phi(s) = \sum_{m=0}^{k-1} \Phi_m 1_{]t_m,t_{m+1}] \cap]\tau,T]}(s)$$

$$= \sum_{m=0}^{k-1} \sum_{j=0}^{n} 1_{A_j} \Phi_m 1_{]t_m,t_{m+1}] \cap]a_j,T]}(s)$$

$$= \sum_{m=0}^{k-1} \sum_{j=0}^{n} \underbrace{1_{A_j} \Phi_m}_{\mathcal{F}_{t_m \vee a_j}\text{-measurable}} 1_{]t_m \vee a_j, t_{m+1} \vee a_j]}(s)$$

and concerning the integral we are interested in, we obtain that

$$\int_0^t 1_{]0,\tau]}(s)\Phi(s)\,dW(s) = \int_0^t \Phi(s)\,dW(s) - \int_0^t 1_{]\tau,T]}(s)\Phi(s)\,dW(s)$$

$$= \sum_{m=0}^{k-1} \Phi_m\big(W(t_{m+1} \wedge t) - W(t_m \wedge t)\big)$$

$$- \sum_{m=0}^{k-1} \sum_{j=0}^{n} 1_{A_j} \Phi_m\Big(W\big((t_{m+1} \vee a_j) \wedge t\big) - W\big((t_m \vee a_j) \wedge t\big)\Big)$$

$$= \sum_{m=0}^{k-1} \Phi_m\big(W(t_{m+1} \wedge t) - W(t_m \wedge t)\big)$$

$$- \sum_{m=0}^{k-1} \sum_{j=0}^{n} 1_{A_j} \Phi_m\Big(W\big((t_{m+1} \vee \tau) \wedge t\big) - W\big((t_m \vee \tau) \wedge t\big)\Big)$$

$$= \sum_{m=0}^{k-1} \Phi_m\big(W(t_{m+1} \wedge t) - W(t_m \wedge t)\big)$$

$$- \sum_{m=0}^{k-1} \Phi_m\Big(W\big((t_{m+1} \vee \tau) \wedge t\big) - W\big((t_m \vee \tau) \wedge t\big)\Big)$$

$$= \sum_{m=0}^{k-1} \Phi_m \Big(W(t_{m+1} \wedge t) - W(t_m \wedge t)$$

$$- W\big((t_{m+1} \vee \tau) \wedge t\big) + W\big((t_m \vee \tau) \wedge t\big)\Big)$$

$$= \sum_{m=0}^{k-1} \Phi_m \Big(W(t_{m+1} \wedge \tau \wedge t) - W(t_m \wedge \tau \wedge t)\Big) = \int_0^{t \wedge \tau} \Phi(s)\, dW(s).$$

Step 2:   Now we consider the case when Φ is still an elementary process while τ is an arbitrary stopping time with $P(\tau \leq T) = 1$.
Then there exists a sequence

$$\tau_n = \sum_{k=0}^{2^n-1} T(k+1)2^{-n} 1_{]Tk2^{-n}, T(k+1)2^{-n}]} \circ \tau, \quad n \in \mathbb{N},$$

of simple stopping times such that $\tau_n \downarrow \tau$ as $n \to \infty$ and because of the continuity of the stochastic integral we get that

$$\int_0^{\tau_n \wedge t} \Phi(s)\, dW(s) \xrightarrow{n \to \infty} \int_0^{\tau \wedge t} \Phi(s)\, dW(s) \quad P\text{-a.s.}$$

Besides, we obtain (even for non-elementary processes Φ) that

$$\big\| 1_{]0,\tau_n]}\Phi - 1_{]0,\tau]}\Phi \big\|_T^2 = E\left(\int_0^T 1_{]\tau,\tau_n]}(s)\|\Phi(s)\|_{L_2^0}^2\, ds \right) \xrightarrow{n \to \infty} 0,$$

which by the definition of the integral implies that

$$E\left(\left\| \int_0^t 1_{]0,\tau_n]}(s)\Phi(s)\, dW(s) - \int_0^t 1_{]0,\tau]}(s)\Phi(s)\, dW(s) \right\|^2 \right) \xrightarrow{n \to \infty} 0$$

for all $t \in [0, T]$. As by Step 1

$$\int_0^t 1_{]0,\tau_n]}(s)\Phi(s)\, dW(s) = \int_0^{\tau_n \wedge t} \Phi(s)\, dW(s), \quad n \in \mathbb{N},\ t \in [0, T],$$

the assertion follows.

Step 3:   Finally we generalize the statement to arbitrary $\Phi \in \mathcal{N}_W^2(0, T)$:
If $\Phi \in \mathcal{N}_W^2(0, T)$, then there exists a sequence of elementary processes $\Phi_n, n \in \mathbb{N}$, such that

$$\| \Phi_n - \Phi \|_T \xrightarrow{n \to \infty} 0 .$$

By the definition of the stochastic integral this means that

$$\int_0^{\cdot} \Phi_n(s) \, dW(s) \xrightarrow{n \to \infty} \int_0^{\cdot} \Phi(s) \, dW(s) \quad \text{in } \mathcal{M}_T^2 .$$

Hence it follows that there is a subsequence n_k, $k \in \mathbb{N}$, and a P-null set $N \in \mathcal{F}$ independent of $t \in [0, T]$ such that

$$\int_0^t \Phi_{n_k}(s) \, dW(s) \xrightarrow{k \to \infty} \int_0^t \Phi(s) \, dW(s) \quad \text{on } N^c$$

for all $t \in [0, T]$ and therefore we get for all $t \in [0, T]$ that

$$\int_0^{\tau \wedge t} \Phi_{n_k}(s) \, dW(s) \xrightarrow{k \to \infty} \int_0^{\tau \wedge t} \Phi(s) \, dW(s) \quad P\text{-a.s.}$$

In addition, it is clear that

$$\| 1_{]0,\tau]} \Phi_n - 1_{]0,\tau]} \Phi \|_T \xrightarrow[n \to \infty]{} 0$$

which implies that for all $t \in [0, T]$

$$E \left(\left\| \int_0^t 1_{]0,\tau]}(s) \Phi_n(s) \, dW(s) - \int_0^t 1_{]0,\tau]}(s) \Phi(s) \, dW(s) \right\|^2 \right) \xrightarrow{n \to \infty} 0.$$

As by Step 2

$$\int_0^t 1_{]0,\tau]}(s) \Phi_{n_k}(s) \, dW(s) = \int_0^{\tau \wedge t} \Phi_{n_k}(s) \, dW(s) \quad P\text{-a.s.}$$

for all $k \in \mathbb{N}$ the assertion follows.   □

Therefore, for $m < n$ on $\{ \tau_m \geq t \} \subset \{ \tau_n \geq t \}$

$$\int_0^t 1_{]0,\tau_n]}(s) \Phi(s) \, dW(s) = \int_0^{\tau_m \wedge t} 1_{]0,\tau_n]}(s) \Phi(s) \, dW(s)$$

$$= \int_0^t 1_{]0,\tau_m]}(s) 1_{]0,\tau_n]}(s) \Phi(s) \, dW(s) = \int_0^t 1_{]0,\tau_m]}(s) \Phi(s) \, dW(s) \quad P\text{-a.s.,}$$

where we used Lemma 2.3.9 for the second equality. Hence the definition is consistent.

Remark 2.3.10 Let $\Phi \in \mathcal{N}_W$ and τ_n, $n \in \mathbb{N}$, as in 2.3.1. In fact it is easy to see that the definition of the stochastic integral for $\Phi \in \mathcal{N}_W$ does not depend on the choice of τ_n, $n \in \mathbb{N}$. We shall show this in several steps. So, let if σ_n, $n \in \mathbb{N}$, be another sequence of stopping times such that $\sigma_n \uparrow T$ as $n \to \infty$ and $1_{]0,\sigma_n]}\Phi \in \mathcal{N}_W^2$ for all $n \in \mathbb{N}$. Then:

(i)

$$\int_0^t \Phi(s)\, dW(s) = \lim_{n\to\infty} \int_0^t 1_{]0,\sigma_n]}(s)\Phi(s)\, dW(s) \quad \text{for all } t \in [0, T] \, P\text{-a.s.}$$

(ii) Lemma 2.3.9 holds for all $\Phi \in \mathcal{N}_W$.

(iii)

$$\int_0^t \Phi(s)\, dW(s) = \int_0^t 1_{]0,\sigma_n]}(s)\Phi(s)\, dW(s) \text{ on } \{\sigma_n \geq t\} P\text{-a.s.}$$

Proof

(i) Let $t \in [0, T]$. Then we get that on the set $\{\tau_m \geq t\}$

$$\int_0^t \Phi(s)\, dW(s) = \int_0^t 1_{]0,\tau_m]}(s)\Phi(s)\, dW(s)$$

$$= \lim_{n\to\infty} \int_0^{t\wedge\sigma_n} 1_{]0,\tau_m]}(s)\Phi(s)\, dW(s)$$

$$= \lim_{n\to\infty} \int_0^{t\wedge\tau_m} 1_{]0,\sigma_n]}(s)\Phi(s)\, dW(s)$$

$$= \lim_{n\to\infty} \int_0^t 1_{]0,\sigma_n]}(s)\Phi(s)\, dW(s) \quad P\text{-a.s.},$$

$\square$

where we used Lemma 2.3.9 twice for the third equality. Letting $m \to \infty$ assertion (i) follows.

(ii) Let τ be as in Lemma 2.3.9. Then P-a.s. for all $t \in]0, T]$

$$\int_0^{\tau\wedge t} \Phi(s)\, dW(s) = \lim_{n\to\infty} \int_0^{\tau\wedge t} 1_{]0,\sigma_n]}(s)\Phi(s)\, dW(s)$$

$$= \lim_{n\to\infty} \int_0^t 1_{]0,\tau]}(s) 1_{]0,\sigma_n]}(s)\Phi(s)\, dW(s)$$

$$= \int_0^t 1_{]0,\tau]}(s)\Phi(s)\, dW(s),$$

where we used (i) for the first and last equality and Lemma 2.3.9 for the second.

(iii) By (ii) we have P-a.s. on $\{\sigma_n \geq t\}$

$$\int_0^t \Phi(s) \, dW(s) = \int_0^{\sigma_n \wedge t} \Phi(s) \, dW(s)$$

$$= \int_0^t 1_{]0,\sigma_n]}(s)\Phi(s) \, dW(s).$$

2.4   Properties of the Stochastic Integral

Let T be a positive real number and $W(t)$, $t \in [0, T]$, a Q-Wiener process as described at the beginning of the previous section.

Lemma 2.4.1 *Let Φ be a L_2^0-valued stochastically integrable process, $(\tilde{H}, \| \ \|_{\tilde{H}})$ a further separable Hilbert space and $L \in L(H, \tilde{H})$.*
Then the process $L(\Phi(t))$, $t \in [0, T]$, is an element of $\mathcal{N}_W(0, T; \tilde{H})$ and

$$L\left(\int_0^T \Phi(t) \, dW(t)\right) = \int_0^T L(\Phi(t)) \, dW(t) \quad P\text{-a.s.}$$

Proof Since Φ is a stochastically integrable process and

$$\left\|L(\Phi(t))\right\|_{L_2(U_0,\tilde{H})} \leq \|L\|_{L(H,\tilde{H})} \|\Phi(t)\|_{L_2^0},$$

it is obvious that $L(\Phi(t))$, $t \in [0, T]$, is $L_2(U_0, \tilde{H})$-predictable and

$$P\left(\int_0^T \left\|L(\Phi(t))\right\|^2_{L_2(U_0,\tilde{H})} \, dt < \infty\right) = 1.$$

Step 1:   As the first step we consider the case that Φ is an elementary process, i.e.

$$\Phi(t) = \sum_{m=0}^{k-1} \Phi_m 1_{]t_m,t_{m+1}]}(t), \quad t \in [0, T],$$

where $0 = t_0 < t_1 < \cdots < t_k = T$, $\Phi_m : \Omega \to L(U, H)$ $\mathcal{F}_{t_m}$-measurable with $|\Phi_m(\Omega)| < \infty$ for $0 \leq m \leq k$. Then

$$L\left(\int_0^T \Phi(t) \, dW(t)\right) = L\left(\sum_{m=0}^{k-1} \Phi_m\big(W(t_{m+1}) - W(t_m)\big)\right)$$

$$= \sum_{m=0}^{k-1} L\big(\Phi_m\big(W(t_{m+1}) - W(t_m)\big)\big) = \int_0^T L(\Phi(t)) \, dW(t).$$

Step 2: Now let $\Phi \in \mathcal{N}_W^2(0,T)$. Then there exists a sequence Φ_n, $n \in \mathbb{N}$, of elementary processes with values in $L(U,H)_0$ such that

$$\|\Phi_n - \Phi\|_T = \left(E\left(\int_0^T \|\Phi_n(t) - \Phi(t)\|_{L_2^0}^2 \, dt \right) \right)^{\frac{1}{2}} \xrightarrow{n \to \infty} 0.$$

Then $L(\Phi_n)$, $n \in \mathbb{N}$, is a sequence of elementary processes with values in $L(U,\tilde{H})_0$ and

$$\|L(\Phi_n) - L(\Phi)\|_T \leqslant \|L\|_{L(H,\tilde{H})} \|\Phi_n - \Phi\|_T \xrightarrow{n \to \infty} 0.$$

By the definition of the stochastic integral, Step 1 and the continuity of L we get that there is a subsequence n_k, $k \in \mathbb{N}$, such that

$$\int_0^T L\big(\Phi(t)\big) \, dW(t) = \lim_{k \to \infty} \int_0^T L\big(\Phi_{n_k}(t)\big) \, dW(t)$$

$$= \lim_{k \to \infty} L\left(\int_0^T \Phi_{n_k}(t) \, dW(t) \right) = L\left(\lim_{k \to \infty} \int_0^T \Phi_{n_k}(t) \, dW(t) \right)$$

$$= L\left(\int_0^T \Phi(t) \, dW(t) \right) \quad P\text{-a.s.}$$

Step 3: Finally let $\Phi \in \mathcal{N}_W(0,T)$.
Let τ_n, $n \in \mathbb{N}$, be a sequence of stopping times as in (2.7). Then $1_{]0,\tau_n]} L(\Phi) \in \mathcal{N}_W^2(0,T,\tilde{H})$ for all $n \in \mathbb{N}$ and we obtain by Remark 2.3.10 and Step 2 (selecting a subsequence if necessary)

$$\int_0^T L\big(\Phi(t)\big) \, dW(t) = \lim_{n \to \infty} \int_0^T 1_{]0,\tau_n]}(t) L\big(\Phi(t)\big) \, dW(t)$$

$$= \lim_{n \to \infty} L\left(\int_0^T 1_{]0,\tau_n]}(t) \Phi(t) \, dW(t) \right) = L\left(\lim_{n \to \infty} \int_0^T 1_{]0,\tau_n]}(t) \Phi(t) \, dW(t) \right)$$

$$= L\left(\int_0^T \Phi(t) \, dW(t) \right) \quad P\text{-a.s.}$$

$\square$

Below $\mathcal{B}_s(L(U_0,H))$ denotes the Borel σ-algebra generated by the strong topology on $L(U_0,H)$.

Lemma 2.4.2 *Let $\Phi : \Omega_T \to L(U_0,H)$ be $\mathcal{P}_T \backslash \mathcal{B}_s(L(U_0,H))$-measurable and $f : \Omega_T \to H$ be $\mathcal{P}_T \backslash \mathcal{B}(H)$-measurable such that*

$$\int_0^T \|\Phi^*(t,\omega) f(t,\omega)\|_{U_0}^2 \, dt < \infty \quad P\text{-a.s.}$$

Set

$$\int_0^T \langle f(t), \Phi(t) \, dW(t) \rangle := \int_0^T \tilde{\Phi}_f(t) \, dW(t) \tag{2.9}$$

with

$$\tilde{\Phi}_f(t)(u) := \langle f(t), \Phi(t)u \rangle, \; u \in U_0.$$

Then the stochastic integral in (2.9) is well-defined as a continuous $\mathbb{R}$-valued stochastic process. More precisely, $\tilde{\Phi}_f$ is a $\mathcal{P}_T / \mathcal{B}(L_2(U_0, \mathbb{R}))$-measurable map from $[0, T] \times \Omega$ to $L_2(U_0, \mathbb{R})$, and

$$\|\tilde{\Phi}_f(t, \omega)\|_{L_2(U_0, \mathbb{R})} = \|\Phi^*(t, \omega)f(t, \omega)\|_{U_0}.$$

Proof Let e_k, $k \in \mathbb{N}$, be an orthonormal basis of U_0. Then for all $(t, \omega) \in [0, T] \times \Omega$

$$\begin{aligned}
\|\tilde{\Phi}_f(t, \omega)\|_{L_2(U_0, \mathbb{R})}^2 &= \sum_{k=1}^\infty \langle f(t, \omega), \Phi(t, \omega)e_k \rangle^2 \\
&= \sum_{k=1}^\infty \langle \Phi^*(t, \omega)f(t, \omega), e_k \rangle_{U_0}^2 \\
&= \|\Phi^*(t, \omega)f(t, \omega)\|_{U_0}^2.
\end{aligned}$$

Now all assertions follow.   □

Lemma 2.4.3 *Let $\Phi \in \mathcal{N}_W(0, T)$ and ζ_n, $n \in \mathbb{N}$, a sequence in $C([0, T], H)$ which converges uniformly to ζ. Then there exists a subsequence ζ_{n_k}, $k \in \mathbb{N}$, such that*

$$\int_0^T \langle \zeta_{n_k}(t), \Phi(t) \, dW(t) \rangle \xrightarrow{k \to \infty} \int_0^T \langle \zeta(t), \Phi(t) \, dW(t) \rangle \quad P\text{-a.s.}$$

Proof

Step 1:   Let $\Phi \in \mathcal{N}_W^2(0, T)$.
   Then we get that

$$\|\tilde{\Phi}_{\zeta_n} - \tilde{\Phi}_\zeta\|_T \leq \sup_{t \in [0, T]} \|\zeta_n(t) - \zeta(t)\| \, \|\Phi\|_T$$

and therefore we get by the isometry that

$$\int_0^T \tilde{\Phi}_{\zeta_n}(t) \, dW(t) \xrightarrow{n \to \infty} \int_0^T \tilde{\Phi}_\zeta(t) \, dW(t)$$

in $L^2(\Omega, \mathcal{F}, P; \mathbb{R})$ which implies that there is a subsequence n_k, $k \in \mathbb{N}$, such that

$$\int_0^T \langle \zeta_{n_k}(t), \Phi(t) \, dW(t) \rangle \xrightarrow{k \to \infty} \int_0^T \langle \zeta(t), \Phi(t) \, dW(t) \rangle \quad P\text{-a.s.}$$

Step 2: Let $\Phi \in \mathcal{N}_W(0, T)$.
As in Step 4 of the definition of the stochastic integral we define the stopping
times

$$\tau_m := \inf \left\{ t \in [0, T] \,\middle|\, \int_0^t \|\Phi(s)\|_{L_2}^2 \, ds > m \right\} \wedge T.$$

Then the process $1_{]0,\tau_m]}\Phi$ is in $\mathcal{N}_W^2(0, T; H)$ for all $m \in \mathbb{N}$. By Step 1 and a
diagonal argument we get the existence of a subsequence n_k, $k \in \mathbb{N}$, such that

$$\int_0^T \langle \zeta_{n_k}(t), 1_{]0,\tau_m]}(t)\Phi(t) \, dW(t) \rangle \xrightarrow{k \to \infty} \int_0^T \langle \zeta(t), 1_{]0,\tau_m]}(t)\Phi(t) \, dW(t) \rangle \quad P\text{-a.s.}$$

for all $m \in \mathbb{N}$. Hence, by the definition of the stochastic integral, we obtain by
Lemma 2.3.9 that

$$\int_0^T \langle \zeta(t), \Phi(t) \, dW(t) \rangle = 1_{\bigcup\limits_{m=1}^\infty \{\tau_{m-1} < T \leq \tau_m\}} \int_0^T \langle \zeta(t), \Phi(t) \, dW(t) \rangle$$

$$= \sum_{m=1}^\infty 1_{\{\tau_{m-1} < T \leq \tau_m\}} \int_0^{T \wedge \tau_m} \langle \zeta(t), \Phi(t) \, dW(t) \rangle$$

$$= \sum_{m=1}^\infty 1_{\{\tau_{m-1} < T \leq \tau_m\}} \int_0^T \langle \zeta(t), 1_{]0,\tau_m]}(t)\Phi(t) \, dW(t) \rangle$$

$$= \sum_{m=1}^\infty 1_{\{\tau_{m-1} < T \leq \tau_m\}} \lim_{k \to \infty} \int_0^T \langle \zeta_{n_k}(t), 1_{]0,\tau_m]}(t)\Phi(t) \, dW(t) \rangle$$

$$= \lim_{k \to \infty} \sum_{m=1}^\infty 1_{\{\tau_{m-1} < T \leq \tau_m\}} \int_0^T \langle \zeta_{n_k}(t), 1_{]0,\tau_m]}(t)\Phi(t) \, dW(t) \rangle$$

$$= \lim_{k \to \infty} \int_0^T \langle \zeta_{n_k}(t), \Phi(t) \, dW(t) \rangle \quad P\text{-a.s.}$$

$\square$

Lemma 2.4.4 *Let $\Phi \in \mathcal{N}_W(0, T; H)$ and $M(t) := \int_0^t \Phi(s) \, dW(s)$, $t \in [0, T]$. Define*

$$\langle M \rangle_t := \int_0^t \|\Phi(s)\|_{L_2^0}^2 \, ds, \ t \in [0, T].$$

Then $\langle M \rangle$ is the unique continuous increasing $(\mathcal{F}_t)$-adapted process starting at zero such that $\|M(t)\|^2 - \langle M \rangle_t$, $t \in [0, T]$, is a local martingale. If $\Phi \in \mathcal{N}_W^2(0, T)$, then for any sequence

$$I_l := \{0 = t_0^l < t_1^l < \ldots < t_{k_l}^l = T\}, \, l \in \mathbb{N},$$

of partitions with

$$\max_i(t_i^l - t_{i-1}^l) \to 0 \text{ as } l \to \infty$$

$$\lim_{l \to \infty} E\left(\left| \sum_{t_{j+1}^l \le t} \|M(t_{j+1}^l) - M(t_j^l)\|^2 - \langle M \rangle_t \right|\right) = 0 \text{ for all } t \in [0, T].$$

Proof For $n \in \mathbb{N}$ let τ_n be as in (2.7) and τ an $\mathcal{F}_t$-stopping time with $P[\tau \le T] = 1$. Then by Remark 2.3.10 for $\sigma := \tau \wedge \tau_n$,

$$E\left(\left\| \int_0^\sigma \Phi(s) \, dW(s) \right\|^2\right) = E\left(\left\| \int_0^T 1_{]0,\sigma]}\Phi(s) \, dW(s) \right\|^2\right)$$

$$= E\left(\int_0^T \|1_{]0,\sigma]}\Phi(s)\|_{L_2^0}^2 \, ds\right)$$

$$= E\left(\int_0^\sigma \|\Phi(s)\|_{L_2^0}^2 \, ds\right),$$

and the first assertion follows, because the uniqueness is obvious, since any real-valued local martingale of bounded variation is constant.

To prove the second assertion we fix an orthonormal basis $\{e_i | i \in \mathbb{N}\}$ of H and note that by the theory of real-valued martingales we have for each $i \in \mathbb{N}$

$$\lim_{l \to \infty} E\left(\left| \sum_{t_{j+1}^l \le t} \langle e_i, M(t_{j+1}^l) - M(t_j^l) \rangle_H^2 - \int_0^t \|\Phi(s)^* e_i\|_{U_0}^2 \, ds \right|\right) = 0, \quad (2.10)$$

since by the first part of the assertion and Lemmas 2.4.1 and 2.4.2

$$\left\langle \int_0^\cdot \langle e_i, \Phi(s) \, dW(s) \rangle_H \right\rangle_t = \int_0^t \|\Phi(s)^* e_i\|_{U_0}^2 \, ds, \, t \in [0, T].$$

Furthermore, for all $i, l \in \mathbb{N}$

$$
E\left(\left|\sum_{t_{j+1}^l \leqslant t} \langle e_i, M(t_{j+1}^l) - M(t_j^l)\rangle_H^2 - \int_0^t \|\Phi(s)^* e_i\|_{U_0}^2 \, ds\right|\right)
$$

$$
\leqslant \sum_{t_{j+1}^l \leqslant t} E\left[\left(\int_{t_j^l}^{t_{j+1}^l} \langle e_i, \Phi(s) \, dW(s)\rangle_H\right)^2\right] + E\left(\int_0^t \|\Phi(s)^* e_i\|_{U_0}^2 \, ds\right)
$$

$$
= \sum_{t_{j+1}^l \leqslant t} E\left(\int_{t_j^l}^{t_{j+1}^l} \|\Phi(s)^* e_i\|_{U_0}^2 \, ds\right) + E\left(\int_0^t \|\Phi(s)^* e_i\|_{U_0}^2 \, ds\right)
$$

$$
\leqslant 2E\left(\int_0^t \|\Phi(s)^* e_i\|_{U_0}^2 \, ds\right) \tag{2.11}
$$

which is summable over $i \in \mathbb{N}$. Here we used the isometry property of Int in the second to last step. But

$$
E\left(\left|\sum_{t_{j+1}^l \leqslant t} \|M(t_{j+1}^l) - M(t_j^l)\|^2 - \int_0^t \|\Phi(s)\|_{L_2^0}^2 \, ds\right|\right)
$$

$$
= E\left(\left|\sum_{i=1}^\infty \left(\sum_{t_{j+1}^l \leqslant t} \langle e_i, M(t_{j+1}^l) - M(t_j^l)\rangle_H^2 - \int_0^t \|\Phi(s)^* e_i\|_{U_0}^2 \, ds\right)\right|\right)
$$

$$
\leqslant \sum_{i=1}^\infty E\left(\left|\sum_{t_{j+1}^l \leqslant t} \langle e_i, M(t_{j+1}^l) - M(t_j^l)\rangle_H^2 - \int_0^t \|\Phi(s)^* e_i\|_{U_0}^2 \, ds\right|\right)
$$

where we used Remark B.0.6(i) in the first step. Hence the second assertion follows by the Lebesgue dominated convergence theorem from (2.10) and (2.11).   □

We conclude this section with the following useful result:

Proposition 2.4.5 *Let* $\Phi \in \mathcal{N}_W^2(0, T; H)$. *Then P-a.s.*

$$
\int_0^t \Phi(s) \, dW(s) = \sum_{k=1}^\infty \sqrt{\lambda_k} \int_0^t \Phi(s)(e_k) \, d\beta_k(s), \quad t \in [0, T],
$$

where $\lambda_k, e_k, \beta_k, k \in \mathbb{N}$, *are as in the representation* (2.2) *of Proposition 2.1.10 for our Q-Wiener process* $W(t)$, $t \in [0, T]$, *on U and the sum on the right-hand side converges in* $L^2(\Omega, \mathcal{F}, P; C([0, T]; H))$.

Proof For $n \in \mathbb{N}$ define

$$Q_{\leq n} := \sum_{k=1}^{n} \lambda_k \langle e_k, \cdot \rangle_U e_k,$$

$$Q_{>n} := \sum_{k=n+1}^{\infty} \lambda_k \langle e_k, \cdot \rangle_U e_k,$$

$$W_{\leq n} := \sum_{k=1}^{n} \sqrt{\lambda_k} \beta_k(t) e_k, \quad t \in [0, T],$$

$$W_{>n}(t) := \sum_{k=n+1}^{\infty} \sqrt{\lambda_k} \beta_k(t) e_k, \quad t \in [0, T].$$

Then $Q_{\leq n}, Q_{>n}$ are symmetric, nonnegative definite operators on U with finite trace and $W_{\leq n}(t), W_{>n}(t), t \in [0, T]$, are $Q_{\leq n}$- and $Q_{>n}$-Wiener processes on U respectively. Furthermore, since by construction our H-valued stochastic integrals are linear in the integrators, it then follows by Proposition 2.2.9 that

$$\mathbb{E}[\sup_{t \in [0,T]} \left\| \int_0^t \Phi(s) \, dW(s) - \int_0^t \Phi(s) \, dW_{\leq n}(s) \right\|^2]$$

$$\leq 2 \, \mathbb{E} \, [\left\| \int_0^T \Phi(s) \, dW_{>n}(s) \right\|^2]$$

$$= 2 \, \mathbb{E} \, [\int_0^T \| \Phi(s) \|^2_{L_2(Q_{>n}^{1/2}(U), H)} \, ds]$$

$$= 2 \, \mathbb{E} \, [\int_0^T \sum_{k=n+1}^{\infty} \lambda_k \| \Phi(s)(e_k) \|^2_H \, ds],$$

which converges to zero as $n \longrightarrow \infty$ by Lebesgue's dominated convergence theorem, since $\Phi \in \mathcal{N}_W^2(0, T; H)$. But clearly for $t \in [0, T]$

$$\int_0^t \Phi(s) \, dW_{\leq n}(s) = \sum_{k=1}^{n} \sqrt{\lambda_k} \int_0^t \Phi(s)(e_k) \, d\beta_k(s)$$

by construction of our H-valued stochastic integrals. Hence the assertion is proved.

$\square$

2.5 The Stochastic Integral for Cylindrical Wiener Processes

Until now we have considered the case that $W(t)$, $t \in [0, T]$, was a standard Q-Wiener process where $Q \in L(U)$ was nonnegative, symmetric and with finite trace. We could integrate processes in

$$\mathcal{N}_W := \Big\{ \Phi : \Omega_T \to L_2(Q^{\frac{1}{2}}(U), H) \mid \Phi \text{ is predictable and}$$

$$P\left(\int_0^T \|\Phi(s)\|_{L_2^0}^2 \ ds < \infty \right) = 1 \Big\}.$$

In fact it is possible to extend the definition of the stochastic integral to the case when Q is not necessarily of finite trace. To this end we first have to introduce the concept of cylindrical Wiener processes.

2.5.1 Cylindrical Wiener Processes

Let $Q \in L(U)$ be nonnegative definite and symmetric. We recall that in the case when Q is of finite trace, a Q-Wiener process has the following representation:

$$W(t) = \sum_{k \in \mathbb{N}} \beta_k(t) e_k, \quad t \in [0, T],$$

where e_k, $k \in \mathbb{N}$, is an orthonormal basis of $Q^{\frac{1}{2}}(U) = U_0$ and β_k, $k \in \mathbb{N}$, is a family of independent real-valued Brownian motions. The series converges in $L^2(\Omega, \mathcal{F}, P; U)$, because the inclusion $U_0 \subset U$ defines a Hilbert–Schmidt embedding from $(U_0, \langle \ , \ \rangle_0)$ to $(U, \langle \ , \ \rangle)$. In the case when Q is no longer of finite trace one loses this convergence. Nevertheless, it is possible to define the Wiener process.

To this end we need a further Hilbert space $(U_1, \langle \ , \ \rangle_1)$ and a Hilbert–Schmidt embedding

$$J : (U_0, \langle \ , \ \rangle_0) \to (U_1, \langle \ , \ \rangle_1).$$

Remark 2.5.1 $(U_1, \langle \ , \ \rangle_1))$ and J as above always exist; e.g. choose $U_1 := U$ and $\alpha_k \in]0, \infty[$, $k \in \mathbb{N}$, such that $\sum_{k=1}^{\infty} \alpha_k^2 < \infty$. Define $J : U_0 \to U$ by

$$J(u) := \sum_{k=1}^{\infty} \alpha_k \langle u, e_k \rangle_0 \, e_k, \quad u \in U_0.$$

Then J is one-to-one and Hilbert–Schmidt.

Then the process given by the following proposition is called a *cylindrical Q-Wiener process* in U.

Proposition 2.5.2 *Let e_k, $k \in \mathbb{N}$ be an orthonormal basis of $U_0 = Q^{\frac{1}{2}}(U)$ and β_k, $k \in \mathbb{N}$, a family of independent real-valued Brownian motions. Define $Q_1 := JJ^*$. Then $Q_1 \in L(U_1)$, Q_1 is nonnegative definite and symmetric with finite trace and the series*

$$W(t) = \sum_{k=1}^{\infty} \beta_k(t) J e_k, \quad t \in [0, T], \tag{2.12}$$

converges in $\mathcal{M}_T^2(U_1)$ and defines a Q_1-Wiener process on U_1. Moreover, we have that $Q_1^{\frac{1}{2}}(U_1) = J(U_0)$ and for all $u_0 \in U_0$

$$\|u_0\|_0 = \|Q_1^{-\frac{1}{2}} J u_0\|_1 = \|J u_0\|_{Q_1^{\frac{1}{2}}(U_1)},$$

i.e. $J : U_0 \to Q_1^{\frac{1}{2}}(U_1)$ is an isometry.

Proof

Step 1: We prove that $W(t)$, $t \in [0, T]$, defined in (2.12) is a Q_1-Wiener process in U_1.
If we set $\xi_j(t) := \beta_j(t) J(e_j), j \in \mathbb{N}$, we obtain that $\xi_j(t)$, $t \in [0, T]$, is a continuous U_1-valued martingale with respect to

$$\mathcal{G}_t := \sigma\left(\bigcup_{j \in \mathbb{N}} \sigma(\beta_j(s)|s \leqslant t)\right),$$

$t \in [0, T]$, since

$$E(\beta_j(t) \mid \mathcal{G}_s) = E(\beta_j(t) \mid \sigma(\beta_j(u)|u \leqslant s)) = \beta_j(s) \quad \text{for all } 0 \leqslant s < t \leqslant T$$

as $\sigma\big(\sigma(\beta_j(u)|u \leqslant s) \cup \sigma(\beta_j(t))\big)$ is independent of

$$\sigma\left(\bigcup_{\substack{k \in \mathbb{N} \\ k \neq j}} \sigma(\beta_k(u)|u \leqslant s)\right).$$

Then it is clear that

$$W_n(t) := \sum_{j=1}^{n} \beta_j(t) J(e_j), \quad t \in [0, T],$$

is also a continuous U_1-valued martingale with respect to $\mathcal{G}_t$, $t \in [0, T]$. In addition, we obtain that

$$E\left(\sup_{t\in[0,T]} \left\| \sum_{j=n}^{m} \beta_j(t)J(e_j) \right\|_1^2\right) \leqslant 4 \sup_{t\in[0,T]} E\left(\left\| \sum_{j=n}^{m} \beta_j(t)J(e_j) \right\|_1^2\right)$$

$$= 4T \sum_{j=n}^{m} \|J(e_j)\|_1^2, \quad m \geqslant n \geqslant 1.$$

Note that $\|J\|_{L_2(U_0,U_1)}^2 = \sum_{j\in\mathbb{N}} \|J(e_j)\|_1^2 < \infty$. Therefore, we get the convergence of $W_n(t)$, $t \in [0, T]$, in $\mathcal{M}_T^2(U_1)$, hence the limit $W(t)$, $t \in [0, T]$, is P-a.s. continuous.

Now we want to show that $P \circ (W(t) - W(s))^{-1} = N(0, (t-s)JJ^*)$. Analogously to the second part of the proof of Proposition 2.1.6 we get that $\langle W(t) - W(s), u_1 \rangle_1$ is normally distributed for all $0 \leqslant s < t \leqslant T$ and $u_1 \in U_1$. It is easy to see that the mean is equal to zero and concerning the covariance of $\langle W(t) - W(s), u_1 \rangle_1$ and $\langle W(t) - W(s), v_1 \rangle_1$, $u_1, v_1 \in U_1$, we obtain that

$$E(\langle W(t) - W(s), u_1 \rangle_1 \langle W(t) - W(s), v_1 \rangle_1)$$

$$= \sum_{k\in\mathbb{N}} (t - s)\langle Je_k, u_1 \rangle_1 \langle Je_k, v_1 \rangle_1$$

$$= (t - s) \sum_{k\in\mathbb{N}} \langle e_k, J^*u_1 \rangle_0 \langle e_k, J^*v_1 \rangle_0$$

$$= (t - s)\langle J^*u_1, J^*v_1 \rangle_0 = (t - s)\langle JJ^*u_1, v_1 \rangle_1.$$

Thus, it only remains to show that the increments of $W(t)$, $t \in [0, T]$, are independent but this can be done in the same way as in the proof of Proposition 2.1.10.

Step 2:   We prove that $\mathrm{Im}\, Q_1^{\frac{1}{2}} = J(U_0)$ and that $\|u_0\|_0 = \|Q_1^{-\frac{1}{2}} Ju_0\|_1$ for all $u_0 \in U_0$.

Since $Q_1 = JJ^*$, by Corollary C.0.6 we obtain that $Q_1^{\frac{1}{2}}(U_1) = J(U_0)$ and that $\|Q_1^{-\frac{1}{2}} u_1\|_1 = \|J^{-1}u_1\|_0$ for all $u_1 \in J(U_0)$. We now replace u_1 by $J(u_0)$, $u_0 \in U_0$, to get the last assertion, because $J : U_0 \to U_1$ is one-to-one. $\square$

2.5.2   The Definition of the Stochastic Integral for Cylindrical Wiener Processes

We fix $Q \in L(U)$ nonnegative, symmetric but not necessarily of finite trace. After the preparations of the previous section we are now able to define the stochastic integral with respect to a cylindrical Q-Wiener process $W(t)$, $t \in [0, T]$.

Basically we integrate with respect to the standard U_1-valued Q_1-Wiener process given by Proposition 2.5.2. In this sense we first get that a process $\Phi(t)$, $t \in [0, T]$, is integrable with respect to $W(t)$, $t \in [0, T]$, if it takes values in $L_2(Q_1^{\frac{1}{2}}(U_1), H)$, is predictable and if

$$
P\left(\int_0^T \|\Phi(s)\|^2_{L_2(Q_1^{\frac{1}{2}}(U_1), H)} \, ds < \infty \right) = 1.
$$

But in addition, we have by Proposition 2.5.2 that $Q_1^{\frac{1}{2}}(U_1) = J(U_0)$ and that

$$
\langle Ju_0, Jv_0 \rangle_{Q_1^{\frac{1}{2}}(U_1)} = \langle Q_1^{-\frac{1}{2}} Ju_0, Q_1^{-\frac{1}{2}} Jv_0 \rangle_1 = \langle u_0, v_0 \rangle_0
$$

for all $u_0, v_0 \in U_0$ (by polarization). In particular, it follows that Je_k, $k \in \mathbb{N}$, is an orthonormal basis of $Q_1^{\frac{1}{2}}(U_1)$. Hence we get that

$$
\Phi \in L_2^0 = L_2(Q^{\frac{1}{2}}(U), H) \iff \Phi \circ J^{-1} \in L_2(Q_1^{\frac{1}{2}}(U_1), H)
$$

since

$$
\begin{aligned}
\|\Phi\|^2_{L_2^0} &= \sum_{k \in \mathbb{N}} \langle \Phi e_k, \Phi e_k \rangle \\
&= \sum_{k \in \mathbb{N}} \langle \Phi \circ J^{-1}(Je_k), \Phi \circ J^{-1}(Je_k) \rangle = \|\Phi \circ J^{-1}\|^2_{L_2(Q_1^{\frac{1}{2}}(U_1), H)}.
\end{aligned}
$$

Now we define

$$
\int_0^t \Phi(s) \, dW(s) := \int_0^t \Phi(s) \circ J^{-1} \, dW(s), \quad t \in [0, T]. \tag{2.13}
$$

Then the class of all integrable processes is given by

$$
\mathcal{N}_W = \left\{ \Phi : \Omega_T \to L_2^0 \mid \Phi \text{ predictable and } P\left(\int_0^T \|\Phi(s)\|^2_{L_2^0} \, ds < \infty \right) = 1 \right\}
$$

as in the case where $W(t)$, $t \in [0, T]$, is a standard Q-Wiener process in U.

Remark 2.5.3

1. We note that the stochastic integral defined in (2.13) is independent of the choice of $(U_1, \langle \ , \ \rangle_1)$ and J. This follows by construction, since by (2.12) for elementary processes (2.13) does not depend on J.

2. If $Q \in L(U)$ is nonnegative, symmetric and with finite trace the standard Q-Wiener process can also be considered as a cylindrical Q-Wiener process by setting $J = I : U_0 \to U$ where I is the identity map. In this case both definitions of the stochastic integral coincide.

Exercise 2.5.4   Prove the analogue of Proposition 2.4.5, when $W(t), t \in [0, T]$, is a cylindrical Wiener process.

Finally, we note that since the stochastic integrals in this chapter all have a standard Wiener process as integrator, we can drop the predictability assumption on $\Phi \in \mathcal{N}_W$ and (as we shall do in subsequent chapters) just assume progressive measurability, i.e. $\Phi|_{[0,t] \times \Omega}$ is $\mathcal{B}([0,t]) \otimes \mathcal{F}_t / \mathcal{B}(L_2^0)$-measurable for all $t \in [0, T]$, at least if $(\Omega, \mathcal{F}, P)$ is complete (otherwise we consider its completion) (cf. [81, Theorem 6.3.1]). We used the above framework so that it easily extends to more general Hilbert-space-valued martingales as integrators replacing the standard Wiener process. The details are left to the reader.

Chapter 3
SDEs in Finite Dimensions

This chapter is an extended version of [52, Sect. 1].

3.1 Main Result and A Localization Lemma

Let $(\Omega, \mathcal{F}, P)$ be a complete probability space and $\mathcal{F}_t$, $t \in [0, \infty[$, a normal filtration. Let $(W_t)_{t \geq 0}$ be a standard Wiener process on $\mathbb{R}^{d_1}$, $d_1 \in \mathbb{N}$, with respect to $\mathcal{F}_t$, $t \in [0, \infty[$. So, in the terminology of the previous section $U := \mathbb{R}^{d_1}$, $Q := I$. The role of the Hilbert space H there will be taken by $\mathbb{R}^d$, $d \in \mathbb{N}$.

Let $M(d \times d_1, \mathbb{R})$ denote the set of all real $d \times d_1$-matrices. Let the following maps $\sigma = \sigma(t, x, \omega)$, $b = b(t, x, \omega)$ be given:

$$\sigma : [0, \infty[\times \mathbb{R}^d \times \Omega \to M(d \times d_1, \mathbb{R}),$$

$$b : [0, \infty[\times \mathbb{R}^d \times \Omega \to \mathbb{R}^d$$

such that both are continuous in $x \in \mathbb{R}^d$ for each fixed $t \in [0, \infty[$, $w \in \Omega$, and progressively measurable, i.e. for each t their restriction to $[0, t] \times \Omega$ is $\mathcal{B}([0, t]) \otimes \mathcal{F}_t$-measurable, for each fixed $x \in \mathbb{R}^d$. We note that then both σ and b restricted to $[0, t] \times \mathbb{R}^d \times \Omega$ are $\mathcal{B}([0, t]) \otimes \mathcal{B}(\mathbb{R}^d) \otimes \mathcal{F}_t$-measurable for every $t \in [0, \infty[$. In particular, for every $x \in \mathbb{R}^d$, $t \in [0, \infty[$ both are $\mathcal{F}_t$-measurable. We also assume that the following integrability conditions hold:

$$\int_0^T \sup_{|x| \leq R} \{ \| \sigma(t, x) \|^2 + |b(t, x)| \} \, dt < \infty \text{ on } \Omega, \tag{3.1}$$

for all $T, R \in [0, \infty[$. Here $| \; |$ denotes the Euclidean distance on $\mathbb{R}^d$ and

$$\|\sigma\|^2 := \sum_{i=1}^{d} \sum_{j=1}^{d_1} |\sigma_{ij}|^2. \tag{3.2}$$

$\langle \, , \, \rangle$ below denotes the Euclidean inner product on $\mathbb{R}^d$.

Theorem 3.1.1 *Let b, σ be as above satisfying (3.1). Assume that on Ω for all $t, R \in [0, \infty[$, $x, y \in \mathbb{R}^d$, $|x|, |y| \leqslant R$*

$$2\langle x - y, b(t, x) - b(t, y) \rangle + \|\sigma(t, x) - \sigma(t, y)\|^2$$
$$\leqslant K_t(R)|x - y|^2 \qquad \text{(local weak monotonicity)} \tag{3.3}$$

and

$$2\langle x, b(t, x) \rangle + \|\sigma(t, x)\|^2 \leqslant K_t(1)(1 + |x|^2), \quad \text{(weak coercivity)} \tag{3.4}$$

where for $R \in [0, \infty[$, $K_t(R)$ is an $\mathbb{R}_+$-valued $(\mathcal{F}_t)$-adapted process satisfying on Ω for all $R, T \in [0, \infty[$

$$\alpha_T(R) := \int_0^T K_t(R) \, \mathrm{d}t < \infty. \tag{3.5}$$

Then for any $\mathcal{F}_0$-measurable map $X_0 : \Omega \to \mathbb{R}^d$ there exists a (up to P-indistinguishability) unique solution to the stochastic differential equation

$$\mathrm{d}X(t) = b(t, X(t)) \, \mathrm{d}t + \sigma(t, X(t)) \, \mathrm{d}W(t). \tag{3.6}$$

Here solution means that $(X(t))_{t \geqslant 0}$ is a P-a.s. continuous $\mathbb{R}^d$-valued $(\mathcal{F}_t)$-adapted process such that P-a.s. for all $t \in [0, \infty[$

$$X(t) = X_0 + \int_0^t b(s, X(s)) \, \mathrm{d}s + \int_0^t \sigma(s, X(s)) \, \mathrm{d}W(s). \tag{3.7}$$

Furthermore, for all $t \in [0, \infty[$

$$E(|X(t)|^2 e^{-\alpha_t(1)}) \leqslant E(|X_0|^2) + 1. \tag{3.8}$$

Remark 3.1.2 We note that by (3.1) the integrals on the right-hand side of (3.7) are well-defined.

For the proof of the above theorem we need two lemmas.

Lemma 3.1.3 *Let* $Y(t)$, $t \in [0, \infty[$, *be a continuous*, $\mathbb{R}_+$-*valued*, $(\mathcal{F}_t)$-*adapted process on* $(\Omega, \mathcal{F}, P)$ *and* γ *an* $(\mathcal{F}_t)$-*stopping time, and let* $\varepsilon \in]0, \infty[$. *Set*

$$\tau_\varepsilon := \gamma \wedge \inf\{t \geq 0 | Y(t) \geq \varepsilon\}$$

(where as usual we set $\inf \emptyset = +\infty$*). Then*

$$P(\{\sup_{t \in [0,\gamma]} Y(t) \geq \varepsilon, \ \gamma < \infty\}) \leq \frac{1}{\varepsilon} E(1_{\{\gamma < \infty\}} Y(\tau_\varepsilon)).$$

Proof We have

$$\{\sup_{t \in [0,\gamma]} Y(t) \geq \varepsilon\} \cap \{\gamma < \infty\} = \{Y(\tau_\varepsilon) \geq \varepsilon\} \cap \{\gamma < \infty\}.$$

Hence the assertion follows by Chebyshev's inequality.   □

The following general "localization lemma" will be crucial.

Lemma 3.1.4 *Let* $n \in \mathbb{N}$ *and* $X^{(n)}(t)$, $t \in [0, \infty[$, *be a continuous*, $\mathbb{R}^d$-*valued*, $(\mathcal{F}_t)$-*adapted process on* $(\Omega, \mathcal{F}, P)$ *such that* $X^{(n)}(0) = X_0$ *for some* $\mathcal{F}_0$-*measurable function* $X_0 : \Omega \to \mathbb{R}^d$ *and*

$$dX^{(n)}(t) = b(t, X^{(n)}(t) + p^{(n)}(t)) \, dt + \sigma(t, X^{(n)}(t) + p^{(n)}(t)) \, dW(t), \quad t \in [0, \infty[$$

for some progressively measurable process $p^{(n)}(t)$, $t \in [0, \infty[$. *For* $n \in \mathbb{N}$ *and* $R \in [0, \infty[$ *let* $\tau^{(n)}(R)$ *be* $(\mathcal{F}_t)$-*stopping times such that*

(i)

$$|X^{(n)}(t)| + |p^{(n)}(t)| \leq R \quad if \quad t \in]0, \tau^{(n)}(R)] \quad P\text{-}a.e.$$

(ii)

$$\lim_{n \to \infty} E \int_0^{T \wedge \tau^{(n)}(R)} |p^{(n)}(t)| \, dt = 0 \quad for \ all \ T \in [0, \infty[.$$

(iii) There exists a function $r : [0, \infty[\to [0, \infty[$ *such that* $\lim_{R \to \infty} r(R) = \infty$ *and*

$$\lim_{R \to \infty} \varlimsup_{n \to \infty} P\left(\{\tau^{(n)}(R) \leq T, \sup_{t \in [0, \tau^{(n)}(R)]} |X^{(n)}(t)| \leq r(R)\}\right)$$

$$= 0 \ for \ all \ T \in [0, \infty[.$$

Then for every $T \in [0, \infty[$ we have

$$\sup_{t \in [0,T]} |X^{(n)}(t) - X^{(m)}(t)| \to 0 \quad \text{in probability as } n, m \to \infty.$$

Proof By (3.1) we may assume that

$$\sup_{|x| \leq R} |b(t, x)| \leq K_t(R) \quad \text{for all } R, t \in [0, \infty[. \tag{3.9}$$

(Otherwise, we replace $K_t(R)$ by the maximum of $K_t(R)$ and the integrand in (3.1).)
Fix $R \in [0, \infty[$ and define the $(\mathcal{F}_t)$-stopping times

$$\tau(R, u) := \inf\{t \geq 0 | \alpha_t(R) > u\}, \ u \in [0, \infty[.$$

Since $t \mapsto \alpha_t(R)$ is locally bounded, we have that $\tau(R, u) \uparrow \infty$ as $u \to \infty$. In particular, there exists a $u(R) \in [0, \infty[$ such that

$$P(\{\tau(R, u(R)) \leq R\}) \leq \frac{1}{R}.$$

Setting $\tau(R) := \tau(R, u(R))$ we have $\tau(R) \to \infty$ in probability as $R \to \infty$ and $\alpha_{t \wedge \tau(R)}(R) \leq u(R)$ for all $t, R \in [0, \infty[$.

Furthermore, if we replace $\tau^{(n)}(R)$ by $\tau^{(n)}(R) \wedge \tau(R)$ for $n \in \mathbb{N}, R \in [0, \infty[$, then clearly assumptions (i) and (ii) above still hold. But

$$P\left(\left\{\tau^{(n)}(R) \wedge \tau(R) \leq T, \sup_{t \in [0, \tau^{(n)}(R) \wedge \tau(R)]} |X^{(n)}(t)| \leq r(R)\right\}\right)$$

$$\leq P\left(\left\{\tau^{(n)}(R) \leq T, \sup_{t \in [0, \tau^{(n)}(R)]} |X^{(n)}(t)| \leq r(R), \tau^{(n)}(R) \leq \tau(R)\right\}\right)$$

$$+ P(\{\tau(R) \leq T, \tau^{(n)}(R) > \tau(R)\})$$

and $\lim_{R \to \infty} P(\{\tau(R) \leq T\}) = 0$. So, assumption (iii) also holds when $\tau^{(n)}(R)$ is replaced by $\tau^{(n)}(R) \wedge \tau(R)$. We may thus assume that $\tau^{(n)}(R) \leq \tau(R)$, hence

$$\alpha_{t \wedge \tau^{(n)}(R)}(R) \leq u(R) \quad \text{for all } t, R \in [0, \infty[, n \in \mathbb{N}. \tag{3.10}$$

Fix $R \in [0, \infty[$ and define

$$\lambda_t^{(n)}(R) := \int_0^t |p^{(n)}(s)| K_s(R) \, ds, \quad t \in]0, \infty[, n \in \mathbb{N}. \tag{3.11}$$

By (3.10) it follows that

$$\lim_{n \to \infty} E\left(\lambda^{(n)}_{T \wedge \tau^{(n)}(R)}(R)\right) = 0 \text{ for all } R, T \in [0, \infty[. \tag{3.12}$$

Indeed, for all $m, n \in \mathbb{N}$

$$\int_0^{T \wedge \tau^{(n)}(R)} |p^{(n)}(t)| \, K_t(R) \, dt$$

$$\leqslant m \int_0^{T \wedge \tau^{(n)}(R)} |p^{(n)}(t)| \, dt + R \int_0^{T \wedge \tau(R)} 1_{]m,\infty[}(K_t(R)) \, K_t(R) \, dt.$$

By assumption (ii) we know that as $n \to \infty$ this converges in $L^1(\Omega, \mathcal{F}, P)$ to

$$R \int_0^{T \wedge \tau(R)} 1_{]m,\infty[}(K_t(R)) \, K_t(R) \, dt,$$

which in turn is dominated by $R \, \alpha_{T \wedge \tau(R)} \leqslant R \, u(R)$ and converges P-a.e. to zero as $m \to \infty$ by (3.5). So, (3.12) follows by Lebesgue's dominated convergence theorem. Let $n, m \in \mathbb{N}$ and set

$$\psi_t(R) := \exp(-2\alpha_t(R) - |X_0|), \quad t \in [0, \infty[. \tag{3.13}$$

Then by Itô's formula we have P-a.e. for all $t \in [0, \infty[$

$$|X^{(n)}(t) - X^{(m)}(t)|^2 \psi_t(R)$$

$$= \int_0^t \psi_s(R) \Big[2\langle X^{(n)}(s) - X^{(m)}(s), b(s, X^{(n)}(s) + p^{(n)}(s))$$

$$- b(s, X^{(m)}(s) + p^{(m)}(s)) \rangle$$

$$+ \| \sigma(s, X^{(n)}(s) + p^{(n)}(s)) - \sigma(s, X^{(m)}(s) + p^{(m)}(s)) \|^2$$

$$- 2K_s(R) |X^{(n)}(s) - X^{(m)}(s)|^2 \Big] ds + M_R^{(n,m)}(t), \tag{3.14}$$

where $M_R^{(n,m)}(t), \ t \in [0, \infty[$, is a continuous local $(\mathcal{F}_t)$-martingale with $M_R^{(n,m)}(0) = 0$. Writing

$$X^{(n)}(s) - X^{(m)}(s) = (X^{(n)}(s) + p^{(n)}(s)) - (X^{(m)}(s) + p^{(m)}(s)) - p^{(n)}(s) + p^{(m)}(s)$$

and by the weak monotonicity assumption (3.3), for $t \in [0, \tau^n(R) \wedge \tau^m(R)]$ the right-hand side of (3.14) is P-a.e. dominated by

$$
\int_0^t \psi_s(R) \Big[2\langle p^{(m)}(s) - p^{(n)}(s), b(s, X^{(n)}(s) + p^{(n)}(s))
$$

$$
- b(s, X^{(m)}(s) + p^{(m)}(s)) \rangle
$$

$$
+ K_s(R)|(X^{(n)}(s) - X^{(m)}(s)) + (p^{(n)}(s) - p^{(m)}(s))|^2
$$

$$
- 2K_s(R)|X^{(n)}(s) - X^{(m)}(s)|^2 \Big] \, ds + M_R^{(n,m)}(t)
$$

$$
\leq 2 \int_0^t \psi_s(R) \, K_s(R) \left(2|p^{(m)}(s) - p^{(n)}(s)| + |p^{(m)}(s) - p^{(n)}(s)|^2 \right) \, ds
$$

$$
+ M_R^{(n,m)}(t),
$$

where we used (3.9) and assumption (i) in the last step. Since $\psi_s(R) \leq 1$ for all $s \in [0, \infty[$ and since for $s \in]0, \tau^{(n)}(R) \wedge \tau^{(m)}(R)]$

$$
|p^{(m)}(s) - p^{(n)}(s)|^2 \leq 2R(|p^{(m)}(s)| + |p^{(n)}(s)|) \quad \text{P-a.e.,}
$$

the above implies that for $T \in [0, \infty[$ fixed and $\gamma^{(n,m)}(R) := T \wedge \tau^{(n)}(R) \wedge \tau^{(m)}(R)$ we have P-a.e. for $t \in [0, \gamma^{(n,m)}(R)]$

$$
|X^{(n)}(t) - X^{(m)}(t)|^2 \psi_t(R) \leq 4(1 + R)(\lambda_t^{(n)}(R) + \lambda_t^{(m)}(R)) + M_R^{(n,m)}(t). \qquad (3.15)
$$

Hence for any $(\mathcal{F}_t)$-stopping time $\tau \leq \gamma^{(n,m)}(R)$ and $(\mathcal{F}_t)$-stopping times $\sigma_k \uparrow \infty$ as $k \to \infty$ so that $M_R^{(n,m)}(t \wedge \sigma_k)$, $t \in [0, \infty[$, is a martingale for all $k \in \mathbb{N}$, we have

$$
E(|X^{(n)}(\tau \wedge \sigma_k) - X^{(m)}(\tau \wedge \sigma_k)|^2 \psi_{\tau \wedge \sigma_k}(R))
$$

$$
\leq 4(1 + R)E(\lambda_{T \wedge \tau^{(n)}(R)}^{(n)}(R) + \lambda_{T \wedge \tau^{(m)}(R)}^{(m)}(R)).
$$

First letting $k \to \infty$ and applying Fatou's lemma, and then using Lemma 3.1.3 we obtain that for every $\varepsilon \in]0, \infty[$

$$
P(\{ \sup_{t \in [0, \gamma^{(n,m)}(R)]} (|X^{(n)}(t) - X^{(m)}(t)|^2 \psi_t(R)) > \varepsilon \})
$$

$$
\leq \frac{4(1 + R)}{\varepsilon} E(\lambda_{T \wedge \tau^{(n)}(R)}^{(n)}(R) + \lambda_{T \wedge \tau^{(m)}(R)}^{(m)}(R)).
$$

Since $[0, \infty[\ni t \mapsto \psi_t(R)(\omega)$ is strictly positive, independent of $n, m \in \mathbb{N}$, and continuous, the above inequality and (3.12) imply that

$$\sup_{t \in [0, \gamma^{(n,m)}(R)]} |X^{(n)}(t) - X^{(m)}(t)| \to 0 \quad \text{as } n, m \to \infty$$

in P-measure. So, to prove the assertion it remains to show that given $T \in [0, \infty[$,

$$\lim_{R \to \infty} \overline{\lim_{n \to \infty}} \, P(\{\tau^{(n)}(R) \leqslant T\}) = 0. \tag{3.16}$$

We first observe that replacing $K_t(R)$ by $\max(K_t(R), K_t(1))$ we may assume that

$$K_t(1) \leqslant K_t(R) \quad \text{for all } t \in [0, \infty[, \, R \in [1, \infty[. \tag{3.17}$$

Now we proceed similarly as above, but use the assumption of weak coercivity (3.4) instead of weak monotonicity (3.3). Let $n \in \mathbb{N}$ and $R \in [1, \infty[$. Then by Itô's formula P-a.e. for all $t \in [0, \infty[$ we have

$$|X^{(n)}(t)|^2 \psi_t(1)$$

$$= |X_0|^2 e^{-|X_0|} + \int_0^t \psi_s(1) \big[2 \langle X^{(n)}(s), b(s, X^{(n)}(s) + p^{(n)}(s)) \rangle$$

$$+ \|\sigma(s, X^{(n)}(s) + p^{(n)}(s))\|^2 - 2K_s(1)|X^{(n)}(s)|^2 \big] \, ds + M_R^{(n)}(t), \tag{3.18}$$

where $M_R^{(n)}(t), t \in [0, \infty[$, is a continuous local $(\mathcal{F}_t)$-martingale with $M_R^{(n)}(0) = 0$. By (3.4) and (3.9) and since $\psi_s(1) \leqslant 1$ for all $s \in [0, \infty[$ the second summand of the right-hand side of (3.18) is P-a.e. for all $t \in [0, T \wedge \tau^{(n)}(R)]$ dominated by

$$\int_0^t \psi_s(1) \big[2 \langle -p^{(n)}(s), b(s, X^{(n)}(s) + p^{(n)}(s)) \rangle$$

$$+ K_s(1)|X^{(n)}(s) + p^{(n)}(s)|^2 + K_s(1) - 2K_s(1)|X^{(n)}(s)|^2 \big] \, ds$$

$$\leqslant 2 \int_0^t K_s(R) |p^{(n)}(s)|(1 + |p^{(n)}(s)|) \, ds + \int_0^t e^{-2\alpha_s(1)} K_s(1) \, ds$$

$$\leqslant 2(1 + R)\lambda_t^{(n)}(R) + \int_0^{\alpha_t(1)} e^{-2s} \, ds, \tag{3.19}$$

where we used (3.17) and assumption (i).

Again localizing $M_R^{(n)}(t)$, $t \in [0, \infty[$, from (3.18) and (3.19) we deduce that for every $(\mathcal{F}_t)$-stopping time $\tau \leqslant T \wedge \tau^{(n)}(R)$

$$E(|X^{(n)}(\tau)|^2 \psi_\tau(1)) \leqslant E(|X_0|^2 e^{-|X_0|}) + \frac{1}{2} + 2(1 + R)E(\lambda_{T \wedge \tau^{(n)}(R)}^{(n)}(R)).$$

Hence by Lemma 3.1.3 and (3.12) we obtain

$$\lim_{c \to \infty} \sup_{R \in [0,\infty[} \overline{\lim_{n \to \infty}} P(\{ \sup_{t \in [0, T \wedge \tau^{(n)}(R)]} (|X^{(n)}(t)|^2 \psi_t(1)) \geqslant c\}) = 0.$$

Since $[0, \infty[\ni t \mapsto \psi_t(1)$ is strictly positive, independent of $n \in \mathbb{N}$ and continuous, and since $r(R) \to \infty$ as $R \to \infty$, we conclude that

$$\overline{\lim_{R \to \infty}} \, \overline{\lim_{n \to \infty}} \, P(\{ \sup_{t \in [0, \tau^{(n)}(R)]} |X^{(n)}(t)| \geqslant r(R), \ \tau^{(n)}(R) \leqslant T\})$$

$$\leqslant \lim_{R \to \infty} \sup_{\tilde{R} \in [0,\infty[} \overline{\lim_{n \to \infty}} P(\{ \sup_{t \in [0, T \wedge \tau^{(n)}(\tilde{R})]} |X^{(n)}(t)| \geqslant r(R)\}) = 0.$$

Hence (3.16) follows from assumption (iii). $\qquad\qquad\qquad\qquad\qquad\qquad\qquad\square$

Remark 3.1.5 In our application of Lemma 3.1.4 below, assumption (iii) will be fulfilled, since the event under P will be empty for all $n \in \mathbb{N}$, $R \in [0, \infty[$. For a case where assumption (iii) is more difficult to check, we refer to [52, Sect. 1].

3.2   Proof of Existence and Uniqueness

Proof of Theorem 3.1.1 The proof is based on Euler's method. Fix $n \in \mathbb{N}$ and define the processes $X^{(n)}(t)$, $t \in [0, \infty[$, iteratively by setting

$$X^{(n)}(0) := X_0$$

and for $k \in \mathbb{N} \cup \{0\}$ and $t \in \left]\frac{k}{n}, \frac{k+1}{n}\right]$ by

$$X^{(n)}(t)$$
$$= X^{(n)}\left(\frac{k}{n}\right) + \int_{\frac{k}{n}}^{t} b\left(s, X^{(n)}\left(\frac{k}{n}\right)\right) ds + \int_{\frac{k}{n}}^{t} \sigma\left(s, X^{(n)}\left(\frac{k}{n}\right)\right) dW(s).$$

This is equivalent to

$$X^{(n)}(t) = X_0 + \int_0^t b(s, X^{(n)}(\kappa(n,s)))\, ds$$

$$+ \int_0^t \sigma(s, X^{(n)}(\kappa(n,s)))\, dW(s), \; t \in [0, \infty[, \tag{3.20}$$

where $\kappa(n,t) := [tn]/n (\leq t)$, and also to

$$X^{(n)}(t) = X_0 + \int_0^t b(s, X^{(n)}(s) + p^{(n)}(s))\, ds$$

$$+ \int_0^t \sigma(s, X^{(n)}(s) + p^{(n)}(s))\, dW(s), \; t \in [0, \infty[,$$

where

$$p^{(n)}(t) := X^{(n)}(\kappa(n,t)) - X^{(n)}(t)$$

$$= - \int_{\kappa(n,t)}^t b(s, X^{(n)}(\kappa(n,s)))\, ds$$

$$- \int_{\kappa(n,t)}^t \sigma(s, X^{(n)}(\kappa(n,s)))\, dW(s), \; t \in [0, \infty[.$$

Now fix $R \in [0, \infty[$ and define

$$\tau^{(n)}(R) := \inf\left\{ t \geq 0 \,\big|\, |X^{(n)}(t)| > \frac{R}{3} \right\}$$

and

$$r(R) := \frac{R}{4}.$$

Then clearly,

$$|p^{(n)}(t)| \leq \frac{2R}{3} \text{ and } |X^{(n)}(t)| \leq \frac{R}{3} \text{ if } t \in \,]0, \tau^{(n)}(R)].$$

In particular, condition (i) in Lemma 3.1.4 holds. Since $X(\tau^{(n)}(R)) \geq \frac{R}{3}$ on $\{\tau^{(n)}(R) < \infty\}$, the event in Lemma 3.1.4(iii) is empty for all $n \in \mathbb{N}$, $R \in [0, \infty[$, so this condition is also satisfied. Let e_i, $1 \leq i \leq d$, be the canonical basis of $\mathbb{R}^d$ and

$T \in [0, \infty[$. Since for $t \in [0, T]$

$$- \langle e_i, p^{(n)}(t) \rangle$$

$$= \int_{\kappa(n,t)}^{t} \langle e_i, b(s, X^{(n)}(\kappa(n, s))) \rangle \, ds + \int_{\kappa(n,t)}^{t} \langle e_i, \sigma(s, X^{(n)}(\kappa(n, s))) \, dW(s) \rangle,$$

it follows that for $\varepsilon \in]0, \infty[$ and $1 \leqslant i \leqslant d$, $t \in [0, \infty[$

$$P(\{|\langle e_i, p^{(n)}(t) \rangle| \geqslant 2\varepsilon, \ t \leqslant \tau^{(n)}(R)\})$$

$$\leqslant P\left(\left\{ \int_{\kappa(n,t)}^{t} \sup_{|x| \leqslant R} |b(s, x)| \, ds \geqslant \varepsilon \right\} \right)$$

$$+ P\left(\left\{ \sup_{\tilde{t} \in [0,t]} \left| \int_{0}^{\tilde{t} \wedge \tau^{(n)}(R)} 1_{[\kappa(n,t),T]}(s) \right. \right. \right.$$

$$\left. \left. \left. \langle e_i, \sigma(s, X^{(n)}(\kappa(n, s))) \, dW(s) \rangle \right| \geqslant \varepsilon \right\} \right)$$

and by Corollary D.0.2 the second summand is bounded by

$$\frac{3\delta}{\varepsilon} + P\left(\left\{ \int_{\kappa(n,t)}^{t} \sup_{|x| \leqslant R} \|\sigma(s, x)\|^2 \, ds > \delta^2 \right\} \right).$$

Altogether, first letting $n \to \infty$ and using (3.1), and then letting $\delta \to 0$ we obtain that for all $t \in [0, \infty[$

$$1_{[0, \tau_n(R)]}(t) \, p^{(n)}(t) \to 0 \text{ as } n \to \infty$$

in P-measure. Since

$$1_{[0, \tau_n(R)]}(t) \, |p^{(n)}(t)| \leqslant \frac{2R}{3}, \ t \in [0, \infty[,$$

it follows by Lebesgue's dominated convergence theorem and Fubini's theorem that condition (ii) in Lemma 3.1.4 is also fulfilled. Now Lemma 3.1.4 and the fact that the space of continuous processes is complete with respect to locally (in $t \in [0, \infty[$) uniform convergence in probability imply that there exists a continuous, $(\mathcal{F}_t)$-adapted, $\mathbb{R}^d$-valued process $X(t)$, $t \in [0, \infty[$, such that for all $T \in [0, \infty[$

$$\sup_{t \in [0,T]} |X^{(n)}(t) - X(t)| \to 0 \text{ in } P\text{-measure as } n \to \infty. \tag{3.21}$$

To prove that X satisfies (3.6) we are going to take the limit in (3.20). To this end, fix $T \in [0, \infty[$ and $t \in [0, T]$. By (3.21) and because of the path continuity we only have to show that the right-hand side of (3.20) converges in P-measure to

$$X_0 + \int_0^t b(s, X(s)) \, ds + \int_0^t \sigma(s, X(s)) \, dW(s).$$

Since the convergence in (3.21) is uniform on $[0, T]$, by equicontinuity we also have that

$$\sup_{t \in [0,T]} |X^{(n)}(\kappa(n, t)) - X(t)| \to 0 \text{ in } P\text{-measure as } n \to \infty.$$

Hence for $Y^{(n)}(t) := X^{(n)}(\kappa(n, t))$ and a subsequence $(n_k)_{k \in \mathbb{N}}$

$$\sup_{t \in [0,T]} |Y^{(n_k)}(t) - X(t)| \to 0 \; P\text{-a.e. as } k \to \infty.$$

In particular, for $S(t) := \sup_{k \in \mathbb{N}} |Y^{(n_k)}(t)|$

$$\sup_{t \in [0,T]} S(t) < \infty \qquad P\text{-a.e.} \tag{3.22}$$

For $R \in [0, \infty[$ define the $(\mathcal{F}_t)$-stopping time

$$\tau(R) := \inf\{t \in [0, T] | S(t) > R\} \wedge T.$$

By the continuity of b in $x \in \mathbb{R}^d$ and by (3.1)

$$\lim_{k \to \infty} \int_0^t b(s, X^{(n_k)}(\kappa(n_k, s))) \, ds = \int_0^t b(s, X(s)) \, ds \qquad P\text{-a.e. on } \{t \leq \tau(R)\}. \tag{3.23}$$

To handle the stochastic integrals we need another sequence of stopping times. For $R, N \in [0, \infty[$ define the $(\mathcal{F}_t)$-stopping time

$$\tau_N(R) := \inf\{t \in [0, T] | \int_0^t \sup_{|x| \leq R} \|\sigma(s, x)\|^2 \, ds > N\} \wedge \tau(R).$$

Then by the continuity of σ in $x \in \mathbb{R}^d$, (3.1), and Lebesgue's dominated convergence theorem

$$\lim_{k \to \infty} E\left(\int_0^{\tau_N(R)} \|\sigma(s, X^{(n_k)}(\kappa(n_k, s))) - \sigma(s, X(s))\|^2 \, ds \right) = 0,$$

hence

$$\int_0^t \sigma(s, X^{(n_k)}(\kappa(n_k, s))) \, dW(s) \to \int_0^t \sigma(s, X(s)) \, dW(s) \tag{3.24}$$

in P-measure on $\{t \leqslant \tau_N(R)\}$ as $k \to \infty$. By (3.1), for every $\omega \in \Omega$ there exists an $N(\omega) \in [0, \infty[$ such that $\tau_N(R) = \tau(R)$ for all $N \geqslant N(\omega)$, so

$$\bigcup_{N \in \mathbb{N}} \{t \leqslant \tau_N(R)\} = \{t \leqslant \tau(R)\}.$$

Therefore, (3.24) holds in P-measure on $\{t \leqslant \tau(R)\}$. But by (3.22) for P-a.e. $\omega \in \Omega$ there exists an $R(\omega) \in [0, \infty[$ such that $\tau(R) = T$ for all $R \geqslant R(\omega)$. So, as above we conclude that (3.23) and (3.24) hold in P-measure on Ω. This completes the proof of existence.

The uniqueness is a special case of the next proposition. So, let us prove the final statement. We have by Itô's formula for our solution X that P-a.e. for all $t \in [0, \infty[$

$$|X(t)|^2 e^{-\alpha_t(1)} = |X_0|^2 + \int_0^t e^{-\alpha_s(1)} \big[2\langle X(s), b(s, X(s)) \rangle + \|\sigma(s, X(s))\|^2$$
$$- K_s(1)|X(s)|^2 \big] \, ds + M(t),$$

where $M(t)$, $t \in [0, \infty[$, is a continuous local martingale with $M(0) = 0$. By the weak coercivity assumption (3.4) the right-hand side of the above equation is dominated by

$$|X_0|^2 + \int_0^{\alpha_t(1)} e^{-s} \, ds + M(t).$$

So, again by localizing $M(t)$, $t \in [0, \infty[$, and Fatou's lemma we get

$$E(|X(t)|^2 e^{-\alpha_t(1)}) \leqslant E(|X_0|^2) + 1, \ t \in [0, \infty[.$$

$\square$

Proposition 3.2.1 *Let the assumptions of Theorem 3.1.1 apart from (3.4) be satisfied. Let $X_0, X_0^{(n)} : \Omega \to \mathbb{R}^d$, $n \in \mathbb{N}$, be $\mathcal{F}_0$-measurable such that*

$$P - \lim_{n \to \infty} X_0^{(n)} = X_0.$$

Let $T \in [0, \infty[$ and assume that $X(t), X^{(n)}(t), t \in [0, T], n \in \mathbb{N}$, be solutions of (3.6) (up to time T) such that $X(0) = X_0$ and $X^{(n)}(0) = X_0^{(n)}$ P-a.e. for all $n \in \mathbb{N}$. Then

$$P - \lim_{n \to \infty} \sup_{t \in [0,T]} |X^{(n)}(t) - X(t)| = 0. \tag{3.25}$$

Proof By the characterization of convergence in P-measure in terms of P-a.e. convergent subsequences (cf. e.g. [5]), we may assume that $X_0^{(n)} \to X_0$ as $n \to \infty$ P-a.e.

Fix $R \in [0, \infty[$ and define

$$\phi_t(R) := \exp(-\alpha_t(R) - \sup_n |X_0^{(n)}|), \ t \in [0, \infty[.$$

We note that since $|X_0| < \infty$, we have $\phi_t(R) > 0$ P-a.e. for all $t \in [0, \infty[$. Define

$$\gamma^{(n)}(R) := \inf\{t \geq 0 | |X^{(n)}(t)| + |X(t)| > R\} \wedge T.$$

Analogously to deriving (3.15) in the proof of Lemma 3.1.4 using the weak monotonicity assumption (3.3), we obtain that P-a.e. for all $t \in [0, T]$ and all $n \in \mathbb{N}$

$$|X^{(n)}(t \wedge \gamma^{(n)}(R)) - X(t \wedge \gamma^{(n)}(R))|^2 \phi_{t \wedge \gamma^{(n)}(R)}(R)$$

$$\leq |X_0^{(n)} - X_0|^2 e^{-\sup_n |X_0^{(n)}|} + m_R^{(n)}(t),$$

where $m_R^{(n)}(t), t \in [0, T]$, are continuous local $(\mathcal{F}_t)$-martingales such that $m_R^{(n)}(0) = 0$. Hence localizing $m_R^{(n)}(t), t \in [0, T]$, for any $(\mathcal{F}_t)$-stopping time $\tau \leq \gamma^{(n)}(R)$ we obtain that

$$E(|X^{(n)}(\tau) - X(\tau)|^2 \phi_\tau(R)) \leq E(|X_0^{(n)} - X_0|^2 e^{-\sup_n |X_0^{(n)}|}). \tag{3.26}$$

Since the right-hand side of (3.26) converges to zero, by Lemma 3.1.3 we conclude that

$$P - \lim_{n \to \infty} \sup_{t \in [0,T]} \left(|X^n)(t \wedge \gamma^{(n)}(R)) - X(t \wedge \gamma^{(n)}(R))|^2 \phi_{t \wedge \gamma^{(n)}(R)}(R) \right) = 0. \tag{3.27}$$

Since P-a.e. the function $[0, \infty[\ni t \mapsto \phi_t(R)$ is continuous and strictly positive, (3.27) implies

$$P - \lim_{n \to \infty} \sup_{t \in [0,T]} |X^{(n)}(t \wedge \gamma^{(n)}(R)) - X(t \wedge \gamma^{(n)}(R))| = 0. \tag{3.28}$$

But

$$P(\{\gamma^{(n)}(R) < T\})$$

$$\leq P(\{\sup_{t \in [0.T]} (|X^{(n)}(t \wedge \gamma^{(n)}(R))| + |X(t \wedge \gamma^{(n)}(R))|) \geq R\})$$

$$\leq P(\{\sup_{t \in [0.T]} (|X^{(n)}(t \wedge \gamma^{(n)}(R)) - X(t \wedge \gamma^{(n)}(R))|) \geq 1\})$$

$$+ P(\{2 \sup_{t \in [0.T]} |X(t)| \geq R - 1\}).$$

This together with (3.28) implies that

$$\lim_{R \to \infty} \overline{\lim_{n \to \infty}} P(\{\gamma^{(n)}(R) < T\}) = 0. \tag{3.29}$$

(3.28) and (3.29) imply (3.25).   □

Chapter 4
SDEs in Infinite Dimensions and Applications to SPDEs

In this chapter we will present one specific method to solve stochastic differential equations in infinite dimensional spaces, known as the *variational approach*. The main criterion for this approach to work is that the coefficients satisfy certain monotonicity assumptions. As the main references for Sect. 4.2 we mention [54, 69], but one should also check the references therein, in particular the pioneering work [65]. Section 4.1 is devoted to formulating the necessary conditions and a number of key applications. In the last section we study the Markov property and invariant measures.

4.1 Gelfand Triples, Conditions on the Coefficients and Examples

Let H be a separable Hilbert space with inner product $\langle \, , \, \rangle_H$ and H^* its dual. Let V be a reflexive Banach space, such that $V \subset H$ continuously and densely. Then for its dual space V^* it follows that $H^* \subset V^*$ continuously and densely. Identifying H and H^* via the Riesz isomorphism we have that

$$V \subset H \subset V^* \tag{4.1}$$

continuously and densely and if $_{V^*}\langle \, , \, \rangle_V$ denotes the dualization between V^* and V (i.e. $_{V^*}\langle z, v \rangle_V := z(v)$ for $z \in V^*, v \in V$), it follows that

$$_{V^*}\langle z, v \rangle_V = \langle z, v \rangle_H \quad \text{for all } z \in H, v \in V. \tag{4.2}$$

(V, H, V^*) is called a *Gelfand triple*. Note that since $H \subset V^*$ continuously and densely, V^* is also separable, hence so is V. Furthermore, $\mathcal{B}(V)$ is generated by

© Springer International Publishing Switzerland 2015
W. Liu, M. Röckner, *Stochastic Partial Differential Equations: An Introduction*,
Universitext, DOI 10.1007/978-3-319-22354-4_4

V^* and $\mathcal{B}(H)$ by H^*. We also have by Kuratowski's theorem that $V \in \mathcal{B}(H)$, $H \in \mathcal{B}(V^*)$ and $\mathcal{B}(V) = \mathcal{B}(H) \cap V$, $\mathcal{B}(H) = \mathcal{B}(V^*) \cap H$.

Below we want to study stochastic differential equations on H of type

$$dX(t) = A(t, X(t))\, dt + B(t, X(t))\, dW(t) \qquad (4.3)$$

with $W(t)$, $t \in [0, T]$, a cylindrical Q-Wiener process with $Q = I$ on another separable Hilbert space $(U, \langle\ ,\ \rangle_U)$ and with B taking values in $L_2(U, H)$ as in Chap. 2, but with A taking values in the larger space V^*.

The solution X will, however, take values in H again. In this section we give precise conditions on A and B.

Let $T \in [0, \infty[$ be fixed and let $(\Omega, \mathcal{F}, P)$ be a complete probability space with normal filtration $\mathcal{F}_t$, $t \in [0, \infty[$. Let

$$A : [0, T] \times V \times \Omega \to V^*, \quad B : [0, T] \times V \times \Omega \to L_2(U, H)$$

be *progressively measurable*, i.e. for every $t \in [0, T]$, these maps restricted to $[0, t] \times V \times \Omega$ are $\mathcal{B}([0, t]) \otimes \mathcal{B}(V) \otimes \mathcal{F}_t$-measurable. As usual by writing $A(t, v)$ we mean the map $\omega \mapsto A(t, v, \omega)$. Analogously for $B(t, v)$. We impose the following conditions on A and B:

(H1) *(Hemicontinuity)* For all $u, v, w \in V$, $\omega \in \Omega$ and $t \in [0, T]$ the map

$$\mathbb{R} \ni \lambda \mapsto {}_{V^*}\langle A(t, u + \lambda v, \omega), w \rangle_V$$

   is continuous.

(H2) *(Weak monotonicity)* There exists a $c \in \mathbb{R}$ such that for all $u, v \in V$

$$2\, {}_{V^*}\langle A(\cdot, u) - A(\cdot, v), u - v \rangle_V + \|B(\cdot, u) - B(\cdot, v)\|^2_{L_2(U,H)}$$

$$\leqslant c\|u - v\|^2_H \text{ on } [0, T] \times \Omega.$$

(H3) *(Coercivity)* There exist $\alpha \in]1, \infty[$, $c_1 \in \mathbb{R}$, $c_2 \in]0, \infty[$ and an $(\mathcal{F}_t)$-adapted process $f \in L^1([0, T] \times \Omega,\ dt \otimes P)$ such that for all $v \in V, t \in [0, T]$

$$2\, {}_{V^*}\langle A(t, v), v \rangle_V + \|B(t, v)\|^2_{L_2(U,H)} \leqslant c_1\|v\|^2_H - c_2\|v\|^\alpha_V + f(t) \quad \text{on } \Omega.$$

(H4) *(Boundedness)* There exist $c_3 \in [0, \infty[$ and an $(\mathcal{F}_t)$-adapted process $g \in L^{\frac{\alpha}{\alpha-1}}([0, T] \times \Omega,\ dt \otimes P)$ such that for all $v \in V, t \in [0, T]$

$$\|A(t, v)\|_{V^*} \leqslant g(t) + c_3\|v\|^{\alpha-1}_V \quad \text{on } \Omega,$$

   where α is as in (H3).

Remark 4.1.1

1. By (H3) and (H4) it follows that for all $v \in V$, $t \in [0, T]$

$$\|B(t, v)\|^2_{L_2(U,H)} \leq c_1 \|v\|^2_H + f(t) + 2\|v\|_V \, g(t) + (2c_3 - c_2)\|v\|^\alpha_V \quad \text{on } \Omega.$$

2. Fix $(t, \omega) \in [0, T] \times \Omega$ and set for $u \in V$

$$A(u) := A(t, u, \omega).$$

Analogously to the finite dimensional case (see 3.1.3) we introduce:

($H2_{loc}$) For every $R \in {]0, \infty[}$ there exists a C_R such that for all $u, v \in V$
with $\|u\|_V, \|v\|_V \leq R$

$$2 \, {}_{V^*}\langle A(u) - A(v), u - v \rangle_V \leq C_R \|u - v\|^2_H.$$

Then the following holds:
(H1) and ($H2_{loc}$) imply that A is *demicontinuous*, i.e.

$$u_n \to u \text{ as } n \to \infty \text{ (strongly) in } V$$

implies

$$A(u_n) \to A(u) \text{ as } n \to \infty \text{ weakly in } V^*$$

(cf. [82, Proposition 26.4]).

In particular, if $H = \mathbb{R}^d$, $d \in \mathbb{N}$, hence $V = V^* = \mathbb{R}^d$, then (H1) and ($H2_{loc}$) imply that $u \mapsto A(t, u, \omega)$ is continuous from $\mathbb{R}^d$ to $\mathbb{R}^d$.

Proof Set for $u \in V, R \in {]0, \infty[}$

$$A_R(u) := A(u) - C_R \, u.$$

The proof will be done in four steps.   □

Claim 1: A is locally bounded, i.e. for all $u \in V$ there exists a neighborhood $U(u)$ such that $A(U(u))$ is a bounded subset of V^*.

Proof of Claim 1 First consider $u = 0$. Suppose $A(U(0))$ is unbounded for all neighborhoods $U(0)$ of 0. Then there exist $u_n \in V$, $n \in \mathbb{N}$, such that

$$\|u_n\|_V \to 0 \text{ and } \|A(u_n)\|_{V^*} \to \infty \text{ as } n \to \infty.$$

Then there exists an $R \in {]0, \infty[}$ such that $\|u_n\|_V \leq R$ for all $n \in \mathbb{N}$. Set

$$a_n := (1 + \|A_R(u_n)\|_{V^*} \|u_n\|_V)^{-1}.$$

Then by $(H2_{loc})$ for all $v \in V \setminus \{0\}$ and $v_R := R \frac{v}{\|v\|_V}$

$$a_n \,_{V^*}\langle A_R(u_n), u_n - (\pm v_R) \rangle_V - a_n \,_{V^*}\langle A_R(\pm v_R), u_n - (\pm v_R) \rangle_V \leqslant 0,$$

hence

$$\mp a_n \,_{V^*}\langle A_R(u_n), v_R \rangle_V \leqslant -a_n \,_{V^*}\langle A_R(u_n), u_n \rangle_V + a_n \,_{V^*}\langle A_R(\pm v_R), u_n \mp v_R \rangle_V$$
$$\leqslant a_n \|A_R(u_n)\|_{V^*} \|u_n\|_V + \|A_R(\pm v_R)\|_{V^*} \|u_n \mp v_R\|_V$$
$$\leqslant 1 + \|A_R(\pm v_R)\|_{V^*} (R + \|v_R\|_V).$$

Consequently,

$$\sup_n |\,_{V^*}\langle a_n A_R(u_n), v \rangle_V | < \infty \text{ for all } v \in V.$$

Therefore, by the Banach–Steinhaus theorem

$$N := \sup_n \|a_n A_R(u_n)\|_{V^*} < \infty,$$

and thus for $n_0 \in \mathbb{N}$ so large that $\|u_n\| \leqslant \frac{1}{2N}$ for all $n \geqslant n_0$ we obtain

$$\|A_R(u_n)\|_{V^*} \leqslant a_n^{-1} N \leqslant N + \frac{1}{2} \|A_R(u_n)\|_{V^*},$$

i.e.

$$\|A_R(u_n)\|_{V^*} \leqslant 2N \text{ for all } n \geqslant n_0.$$

Hence also $\sup_n \|A(u_n)\|_{V^*} < \infty$, which is a contradiction. So, $A(U(0))$ is bounded for some neighborhood $U(0)$ of 0.

For arbitrary $u \in V$ we apply the above argument to the operator

$$A_u(v) := A(u + v), \; v \in V$$

which obviously also satisfies (H1) and $(H2_{loc})$. So, Claim 1 is proved.   □

Claim 2:   Let $u \in V$, $b \in V^*$ such that there exist $R \in [\|u\|_V, \infty[, \gamma \in]0, \infty[$ such that

$$_{V^*}\langle b - A(v), u - v \rangle_V \leqslant \gamma \|u - v\|_H^2 \text{ for all } v \in V \text{ with } \|v\|_V \leqslant R + 1.$$

Then $A(u) = b$.

Proof of Claim 2 Let $w \in V$, $\|w\|_V \leqslant 1, t \in {]0, 1[}$ and set $v := u - tw$. Then by assumption

$$_{V^*}\langle b - A(u - tw), tw \rangle_V = {}_{V^*}\langle b - A(v), u - v \rangle_V \leqslant \gamma t^2 \|w\|_H^2.$$

Dividing first by t and then letting $t \to 0$, by (H1) we obtain

$$_{V^*}\langle b - A(u), w \rangle_V \leqslant 0 \text{ for all } w \in V.$$

So, replacing w by $-w$, $w \in V$, we get

$$_{V^*}\langle b - A(u), w \rangle_V = 0 \text{ for all } w \in V,$$

hence $A(u) = b$. $\square$

Claim 3: ("monotonicity trick"). Let u_n, $u \in V$, $n \in \mathbb{N}$, and $b \in V^*$ such that

$$u_n \to u \quad \text{as} \quad n \to \infty \text{ weakly in } V,$$

$$u_n \to u \quad \text{as} \quad n \to \infty \text{ strongly in } H,$$

$$A(u_n) \to b \quad \text{as} \quad n \to \infty \text{ weakly in } V^*$$

and

$$\limsup_{n \to \infty} {}_{V^*}\langle A(u_n), u_n \rangle_V \geqslant {}_{V^*}\langle b, u \rangle_V.$$

Then $A(u) = b$.

Proof of Claim 3 Let $R := \|u\|_V + \sup_n \|u_n\|_V$. We have for all $v \in V$ with $\|v\|_V \leqslant R + 1$

$$_{V^*}\langle A(u_n), u_n \rangle_V - {}_{V^*}\langle A(v), u_n \rangle_V - {}_{V^*}\langle A(u_n) - A(v), v \rangle_V$$
$$= {}_{V^*}\langle A(u_n) - A(v), u_n - v \rangle_V \leqslant C_{R+1} \|u_n - v\|_H^2$$

by $(H2_{loc})$. Taking $\limsup\limits_{n \to \infty}$ we obtain

$$_{V^*}\langle b, u \rangle_V - {}_{V^*}\langle A(v), u \rangle_V - {}_{V^*}\langle b - A(v), v \rangle_V \leqslant C_{R+1} \|u - v\|_H^2,$$

so

$$_{V^*}\langle b - A(v), u - v \rangle_V \leqslant C_{R+1} \|u - v\|_H^2 \text{ for all } v \in V \text{ with } \|v\|_V \leqslant R + 1.$$

Hence Claim 2 implies that $A(u) = b$. $\square$

Claim 4:   Let $u_n, u \in V$, $n \in \mathbb{N}$, such that

$$u_n \to u \text{ as } n \to \infty \text{ (strongly) in } V.$$

Then

$$A(u_n) \to A(u) \text{ as } n \to \infty \text{ weakly in } V^*.$$

Proof of Claim 4 Since $u_n \to u$ as $n \to \infty$ in V, by Claim 1 $\{A(u_n)|n \in \mathbb{N}\}$ is bounded in V^*. Since V is separable, closed bounded sets in V^* are weakly sequentially compact. Hence there exists a subsequence $(n_k)_{k \in \mathbb{N}}$ and $b \in V^*$ such that $A(u_{n_k}) \to b$ as $k \to \infty$ weakly in V^*. Since $u_{n_k} \to u$ strongly in V as $k \to \infty$, we get

$$\lim_{k \to \infty} {}_{V^*}\langle A(u_{n_k}), u_{n_k} \rangle_V = {}_{V^*}\langle b, u \rangle_V .$$

Therefore, since $V \subset H$ continuously, all conditions in Claim 3 are fulfilled and we can conclude that $A(u) = b$. So, for all such subsequences their weak limit is $A(u)$, hence $A(u_n) \to A(u)$ as $n \to \infty$ weakly in V^*.   □

Let us now discuss the above conditions. We shall solely concentrate on A and take $B \equiv 0$. The latter we do because of the following:

Exercise 4.1.2

1. Suppose A, B satisfy (H2), (H3) above and $\tilde{A}$ is another map satisfying (H2), (H3). Then $A + \tilde{A}$, B satisfy (H2),(H3). Likewise, if A and $\tilde{A}$ both satisfy (H1), (H4) then so does $A + \tilde{A}$.
2. If A satisfies (H2), (H3) (with $B \equiv 0$) and for all $t \in [0, T]$, $\omega \in \Omega$, the map $u \mapsto B(t, u, \omega)$ is Lipschitz with respect to $\| \ \|_H$ with Lipschitz constant independent of $t \in [0, T]$, $\omega \in \Omega$, then A, B satisfy (H2), (H3).

Below, we only look at A independent of $t \in [0, T]$, $\omega \in \Omega$. From here examples for A dependent on (t, ω) are then immediate.

Example 4.1.3 $V = H = V^*$ (which includes the case $H = \mathbb{R}^d$).
Clearly, since for all $v \in V$

$$2 \, {}_{V^*}\langle A(v), v \rangle_V \leqslant 2 \, {}_{V^*}\langle A(v) - A(0), v \rangle_V + \|A(0)\|_{V^*}^2 + \|v\|_V^2$$

in the present case where $V = H = V^*$, (H2) implies (H3) with $c_1 > c_2$ and $\alpha := 2$. Furthermore, obviously, if A is Lipschitz in u, then (H1)–(H4) are immediately satisfied. But for (H1)–(H3) to hold, conditions (with respect to u) on A, which can be checked locally, can be sufficient, as the following proposition shows.

Proposition 4.1.4 *Suppose $A : H \to H$ is Fréchet differentiable such that for some $c \in [0, \infty[$ the operator $DA(x) - cI$ $(\in L(H))$ is negative definite for all $x \in H$. Then A satisfies (H1)–(H3) (with $B \equiv 0$).*

Proof Since A is Fréchet differentiable, it is continuous, so, in particular, (H1) holds. Furthermore, for $x, y \in H$ we have

$$
\begin{aligned}
A(x) - A(y) &= \int_0^1 \frac{\mathrm{d}}{\mathrm{d}s} A(y + s(x - y))\, \mathrm{d}s \\
&= \int_0^1 DA(y + s(x - y))(x - y)\, \mathrm{d}s.
\end{aligned}
$$

Hence by assumption

$$
\begin{aligned}
\langle A(x) - A(y), x - y \rangle_H &= \int_0^1 \langle DA(y + s(x - y))(x - y), x - y \rangle_H\, \mathrm{d}s \\
&\leqslant c \int_0^1 \langle x - y, x - y \rangle_H\, \mathrm{d}s \\
&= c \|x - y\|_H^2,
\end{aligned}
$$

and so (H2) holds and hence (H3), as shown above.   □

We again note that Proposition 4.1.4 shows that locally checkable conditions on A can already imply (H1)–(H3), if ($V = H = V^*$ and) $\alpha = 2$. However, the global condition (H4) then requires that A is of at most linear growth since $\alpha - 1 = 1$ if $\alpha = 2$. We also note that for $H = \mathbb{R}^1$ the conditions in Proposition 4.1.4 just mean that A is differentiable and decreasing.

If H is a space of functions, a possible and easy choice for A would be, for example, $Au = -u^3$. But then we cannot choose $H = L^2$ because A would not leave L^2 invariant. This is one motivation to look at triples $V \subset H \subset V^*$ because then we can take $V = L^p$ and $H = L^2$ and define A from V to $V^* = L^{p/(p-1)}$. Let us look at this case more precisely.

Example 4.1.5 ($L^p \subset L^2 \subset L^{p/(p-1)}$ and $A(u) := -u|u|^{p-2}$) Hence the stochastic differential equation (4.3) becomes

$$
\mathrm{d}X(t) = -X(t)|X(t)|^{p-2}\, \mathrm{d}t + B(t, X(t))\, \mathrm{d}W(t).
$$

Let $p \in [2, \infty[$, $\Lambda \subset \mathbb{R}^d$, Λ open. Let

$$
V := L^p(\Lambda) := L^p(\Lambda,\ \mathrm{d}\xi),
$$

equipped with its usual norm $\| \ \|_p$, and

$$
H := L^2(\Lambda) := L^2(\Lambda,\ \mathrm{d}\xi),
$$

where $d\xi$ denotes Lebesgue measure on Λ. Then

$$V^* = L^{p/(p-1)}(\Lambda).$$

If $p > 2$ we assume that

$$|\Lambda| := \int_{\mathbb{R}^d} \mathbb{I}_\Lambda(\xi)\, d\xi < \infty. \tag{4.4}$$

Then

$$V \subset H \subset V^*,$$

or concretely

$$L^p(\Lambda) \subset L^2(\Lambda) \subset L^{p/(p-1)}(\Lambda)$$

continuously and densely. Recall that since $p > 1$, $L^p(\Lambda)$ is reflexive.

Define $A : V \to V^*$ by

$$Au := -u|u|^{p-2}, \ u \in V = L^p(\Lambda).$$

Indeed, A takes values in $V^* = L^{p/(p-1)}(\Lambda)$, since

$$\int |Au(\xi)|^{p/(p-1)}\, d\xi = \int |u(\xi)|^p\, d\xi < \infty$$

for all $u \in L^p(\Lambda)$.

Claim A satisfies (H1)–(H4).

Proof Let $u, v, x \in V$. Then for $\lambda \in \mathbb{R}$

$$_{V^*}\langle A(u + \lambda v) - A(u), x\rangle_V$$

$$= \int (u(\xi)|u(\xi)|^{p-2} - (u(\xi) + \lambda v(\xi))|u(\xi) + \lambda v(\xi)|^{p-2})x(\xi)\, d\xi$$

$$\leqslant \left\| u|u|^{p-2} - (u + \lambda v)|u + \lambda v|^{p-2} \right\|_{V^*} \|x\|_V$$

which converges to zero as $\lambda \to 0$ by Lebesgue's dominated convergence theorem. So, (H1) holds.

Furthermore,

$$_{V^*}\langle A(u) - A(v), u - v\rangle_V$$

$$= \int (v(\xi)|v(\xi)|^{p-2} - u(\xi)|u(\xi)|^{p-2})(u(\xi) - v(\xi))\, d\xi \leqslant 0,$$

since the map $s \mapsto s|s|^{p-2}$ is increasing on $\mathbb{R}$. Thus (H2) holds, with $c := 0$.
We also have that

$$_{V^*}\langle A(v), v \rangle_V = -\int |v(\xi)|^p \, d\xi = -\|v\|_V^p,$$

so (H3) holds with $\alpha := p$. In addition,

$$\|A(v)\|_{V^*} = \left(\int |v(\xi)|^p \, d\xi \right)^{\frac{p-1}{p}} = \|v\|_V^{p-1}$$

so (H4) holds with $\alpha := p$ as required.   □

Remark 4.1.6 In the example above we may take $A : V := L^p(\Lambda) \to L^{\frac{p}{p-1}}(\Lambda) = V^*$ defined by

$$A(v) := -\Psi(v), v \in L^p(\Lambda),$$

where $\Psi : \mathbb{R} \to \mathbb{R}$ is a fixed function satisfying properties $(\Psi 1) - (\Psi 4)$ specified in Example 4.1.11 below.

Now we turn to cases where A is given by a (possibly nonlinear) partial differential operator. We shall start with the linear case; more concretely, A will be given by the classical Laplace operator

$$\Delta = \sum_{i=1}^{d} \frac{\partial^2}{\partial \xi_i^2}$$

with initial domain given by $C_0^\infty(\Lambda)$. We want to take A to be an extension of Δ to a properly chosen Banach space V so that $A : V \to V^*$ is (defined on all of V and) continuous with respect to $\| \ \|_V$ and $\| \ \|_{V^*}$. The right choice for V is the classical Sobolev space $H_0^{1,p}(\Lambda)$ for $p \in [2, \infty[$ with Dirichlet boundary conditions. So, as a preparation we need to introduce (first-order) Sobolev spaces.

Again let $\Lambda \subset \mathbb{R}^d$, Λ open, and let $C_0^\infty(\Lambda)$ denote the set of all infinitely differentiable real-valued functions on Λ with compact support. Let $p \in [1, \infty[$ and for $u \in C_0^\infty(\Lambda)$ define

$$\|u\|_{1,p} := \left(\int (|u(\xi)|^p + |\nabla u(\xi)|^p) \, d\xi \right)^{1/p}. \tag{4.5}$$

Then define

$$H_0^{1,p}(\Lambda) := \text{ completion of } C_0^\infty(\Lambda) \text{ with respect to } \| \ \|_{1,p}. \tag{4.6}$$

At this stage $H_0^{1,p}(\Lambda)$, called the *Sobolev space* of order 1 in $L^p(\Lambda)$ with *Dirichlet boundary conditions*, just consists of abstract objects, namely equivalence classes of $\| \ \|_{1,p}$-Cauchy sequences. The main point is to show that

$$H_0^{1,p}(\Lambda) \subset L^p(\Lambda), \tag{4.7}$$

i.e. that the unique continuous extension

$$\bar{i} : H_0^{1,p}(\Lambda) \to L^p(\Lambda)$$

of the embedding

$$i : C_0^\infty(\Lambda) \hookrightarrow L^p(\Lambda)$$

is one-to-one. To this end it suffices (in fact it is equivalent) to show that if $u_n \in C_0^\infty(\Lambda)$, $n \in \mathbb{N}$, such that

$$u_n \to 0 \quad \text{in } L^p(\Lambda)$$

and

$$\int |\nabla(u_n - u_m)(\xi)|^p \, d\xi \to 0 \text{ as } n, m \to \infty,$$

then

$$\int |\nabla(u_n(\xi))|^p \, d\xi \to 0 \text{ as } n \to \infty. \tag{4.8}$$

But by the completeness of $L^p(\Lambda; \mathbb{R}^d)$ there exists an

$$F = (F_1, \ldots, F_d) \in L^p(\Lambda; \mathbb{R}^d)$$

such that $\nabla u_n \to F$ as $n \to \infty$ in $L^p(\Lambda; \mathbb{R}^d)$. Let $v \in C_0^\infty(\Lambda)$. Then for $1 \leqslant i \leqslant d$, integrating by parts we obtain that

$$\int v(\xi) F_i(\xi) \, d\xi = \lim_{n \to \infty} \int v(\xi) \frac{\partial}{\partial \xi_i} u_n(\xi) \, d\xi$$

$$= - \lim_{n \to \infty} \int \frac{\partial}{\partial \xi_i} v(\xi) u_n(\xi) \, d\xi$$

$$= 0.$$

Hence $F_i = 0$ $d\xi$-a.e. for all $1 \leqslant i \leqslant d$, so (4.8) holds.

Consider the operator

$$\nabla : C_0^\infty(\Lambda) \subset L^p(\Lambda) \to L^p(\Lambda; \mathbb{R}^d).$$

By what we have shown above, we can extend ∇ to all of $H_0^{1,p}(\Lambda)$ as follows. Let $u \in H_0^{1,p}(\Lambda)$ and let $u_n \in C_0^\infty(\Lambda)$ such that $\lim_{n\to\infty} \|u - u_n\|_{1,p} = 0$. In particular, $(\nabla u_n)_{n\in\mathbb{N}}$ is a Cauchy sequence in $L^p(\Lambda; \mathbb{R}^d)$, hence has a limit there. So, define

$$\nabla u := \lim_{n\to\infty} \nabla u_n \quad \text{in } L^p(\Lambda; \mathbb{R}^d). \tag{4.9}$$

By what we have shown above this limit only depends on u and not on the chosen sequence. We recall the fact that $H_0^{1,p}(\Lambda)$ is reflexive for all $p \in\,]1, \infty[$ (cf. [82]).

Example 4.1.7 ($H_0^{1,2} \subset L^2 \subset (H_0^{1,2})^*$, $A = \Delta$) Though later we shall see that to have (H3) we have to take $p = 2$, we shall first consider $p \in [2, \infty[$ and define

$$V := H_0^{1,p}(\Lambda), H := L^2(\Lambda),$$

so

$$V^* := H_0^{1,p}(\Lambda)^*.$$

Again we assume (4.4) to hold if $p > 2$. Since then $V \subset L^p(\Lambda) \subset H$, continuously and densely, identifying H with its dual we obtain the continuous and dense embeddings

$$V \subset H \subset V^*$$

or concretely

$$H_0^{1,p}(\Lambda) \subset L^2(\Lambda) \subset H_0^{1,p}(\Lambda)^*. \tag{4.10}$$

Now we are going to extend Δ with initial domain $C_0^\infty(\Lambda)$ to a bounded linear operator $A : V \to V^*$. First of all we can consider Δ as an operator taking values in V^* since

$$\Delta : C_0^\infty(\Lambda) \to C_0^\infty(\Lambda) \subset L^2(\Lambda) \subset V^*.$$

Furthermore, for $u, v \in C_0^\infty(\Lambda)$ again integrating by parts we obtain

$$\left|_{V^*}\langle \Delta u, v \rangle_V\right| = |\langle \Delta u, v \rangle_H|$$

$$= \left|-\int \langle \nabla u(\xi), \nabla v(\xi) \rangle \, d\xi\right|$$

$$\leqslant \left(\int |\nabla u(\xi)|^{\frac{p}{p-1}} \, d\xi \right)^{\frac{p-1}{p}} \left(\int |\nabla v(\xi)|^{p} \, d\xi \right)^{\frac{1}{p}}$$

$$\leqslant \left(\int |\nabla u(\xi)|^{\frac{p}{p-1}} \, d\xi \right)^{\frac{p-1}{p}} \|v\|_{1,p}.$$

Hence for all $u \in C_0^\infty(\Lambda)$

$$\|\Delta u\|_{V^*} \leqslant \| |\nabla u| \|_{\frac{p}{p-1}}. \tag{4.11}$$

So, by (4.4) and since $\frac{p}{p-1} \leqslant 2 \leqslant p$, we get by Hölder's inequality

$$\|\Delta u\|_{V^*} \leqslant |\Lambda|^{\frac{p-2}{p}} \|u\|_{1,p} \quad \text{for all } u \in C_0^\infty(\Lambda), \tag{4.12}$$

where for $p = 2$ the factor on the right is just equal to 1.

So, Δ with domain $C_0^\infty(\Lambda)$ extends (uniquely) to a bounded linear operator $A : V \to V^*$ (with domain all of V), also denoted by Δ.

Now let us check (H1)–(H4) for A.

Claim

$$A(= \Delta) : H_0^{1,p}(\Lambda) \to \left(H_0^{1,p}(\Lambda) \right)^*$$

satisfies (H1),(H2),(H4) and provided $p = 2$, also (H3).

Proof Since $A : V \to V^*$ is linear, (H1) is obviously satisfied. Furthermore, if $u, v \in V$ then there exist $u_n, v_n \in C_0^\infty(\Lambda)$, $n \in \mathbb{N}$, such that $u_n \to u$, $v_n \to v$ as $n \to \infty$ in V. Hence, integrating by parts, we get

$$\begin{aligned}
{}_{V^*}\langle A(u) - A(v), u - v \rangle_V &= \lim_{n \to \infty} {}_{V^*}\langle \Delta u_n - \Delta v_n, u_n - v_n \rangle_V \\
&= \lim_{n \to \infty} \langle \Delta(u_n - v_n), u_n - v_n \rangle_H \\
&= -\lim_{n \to \infty} \int |\nabla(u_n - v_n)(\xi)|^2 \, d\xi \leqslant 0.
\end{aligned}$$

So (H2) is satisfied. Furthermore,

$$\begin{aligned}
2 \, {}_{V^*}\langle A(v), v \rangle_V &= \lim_{n \to \infty} 2\langle \Delta v_n, v_n \rangle_H \\
&= -\lim_{n \to \infty} 2 \int |\nabla v_n(\xi)|^2 \, d\xi \\
&= -2 \int |\nabla v(\xi)|^2 \, d\xi \\
&= 2 \left(\|v\|_H^2 - \|v\|_{1,2}^2 \right).
\end{aligned}$$

So (H3) is satisfied if $p = 2$ with $\alpha = 2$. Furthermore, (H4), with $\alpha = 2$, is clear by (4.12).   □

Remark 4.1.8   The corresponding SDE (4.3) then reads

$$dX(t) = \Delta X(t)\, dt + B(t, X(t))\, dW(t).$$

If $B \equiv 0$, this is just the classical *heat equation*. If $B \not\equiv 0$, but constant, the solution is an *Ornstein–Uhlenbeck process* on H.

Example 4.1.9 ($H_0^{1,p} \subset L^2 \subset (H_0^{1,p})^*$, $A = p$-*Laplacian*)   Hence the stochastic differential equation (4.3) becomes

$$dX(t) = \operatorname{div}\left(|\nabla X(t)|^{p-2}\nabla X(t)\right)\, dt + B(t, X(t))\, dW(t).$$

Again we take $p \in [2, \infty[$, $\Lambda \subset \mathbb{R}^d$, Λ open and bounded, and $V := H_0^{1,p}(\Lambda)$, $H := L^2(\Lambda)$, so $V^* = (H_0^{1,p}(\Lambda))^*$. Define $A : H_0^{1,p}(\Lambda) \to H_0^{1,p}(\Lambda)^*$ by

$$A(u) := \operatorname{div}(|\nabla u|^{p-2}\nabla u), \quad u \in H_0^{1,p}(\Lambda);$$

more precisely, given $u \in H_0^{1,p}(\Lambda)$ for all $v \in H_0^{1,p}(\Lambda)$

$$_{V^*}\langle A(u), v\rangle_V := -\int |\nabla u(\xi)|^{p-2}\langle \nabla u(\xi), \nabla v(\xi)\rangle\, d\xi \quad \text{for all } v \in H_0^{1,p}(\Lambda). \tag{4.13}$$

A is called the *p-Laplacian*, also denoted by Δ_p. Note that $\Delta_2 = \Delta$. To show that $A : V \to V^*$ is well-defined we have to show that the right-hand side of (4.13) defines a linear functional in $v \in V$ which is continuous with respect to $\|\ \|_V = \|\ \|_{1,p}$. First we recall that by (4.9) $\nabla u \in L^p(\Lambda; \mathbb{R}^d)$ for all $u \in H_0^{1,p}(\Lambda)$. Hence by Hölder's inequality

$$\int |\nabla u(\xi)|^{p-1}|\nabla v(\xi)|\, d\xi \leq \left(\int |\nabla u(\xi)|^p\, d\xi\right)^{\frac{p-1}{p}} \left(\int |\nabla v(\xi)|^p\, d\xi\right)^{\frac{1}{p}}$$

$$\leq \|u\|_{1,p}^{p-1}\|v\|_{1,p}.$$

Since this dominates the absolute value of the right-hand side of (4.13) for all $u \in H_0^{1,p}(\Lambda)$ we have that $A(u)$ is a well-defined element of $(H_0^{1,p}(\Lambda))^*$ and that

$$\|A(u)\|_{V^*} \leq \|u\|_V^{p-1}. \tag{4.14}$$

Now we are going to check that A satisfies (H1)–(H4).

(H1) Let $u, v, x \in H_0^{1,p}(\Lambda)$, then by (4.13) we have to show for $\lambda \in \mathbb{R}$, $|\lambda| \leqslant 1$

$$\lim_{\lambda \to 0} \int \Big(|\nabla(u + \lambda v)(\xi)|^{p-2} \langle \nabla(u + \lambda v)(\xi), \nabla x(\xi) \rangle$$
$$- |\nabla u(\xi)|^{p-2} \langle \nabla u(\xi), \nabla x(\xi) \rangle \Big) \, d\xi = 0.$$

Since obviously the integrands converge to zero as $\lambda \to 0$ $d\xi$-a.e., we only have to find a dominating function to apply Lebesgue's dominated convergence theorem. But obviously, since $|\lambda| \leqslant 1$

$$|\nabla(u + \lambda v)(\xi)|^{p-2} |\langle \nabla(u + \lambda v)(\xi), \nabla x(\xi) \rangle|$$
$$\leqslant 2^{p-2} \left(|\nabla u(\xi)|^{p-1} + |\nabla v(\xi)|^{p-1} \right) |\nabla x(\xi)|$$

and the right-hand side is in $L^1(\Lambda)$ by Hölder's inequality as we have seen above.

(H2) Let $u, v \in H_0^{1,p}(\Lambda)$. Then by (4.13)

$$-_{V^*} \langle A(u) - A(v), u - v \rangle_V$$
$$= \int \langle |\nabla u(\xi)|^{p-2} \nabla u(\xi) - |\nabla v(\xi)|^{p-2} \nabla v(\xi), \nabla u(\xi) - \nabla v(\xi) \rangle \, d\xi$$
$$= \int \big(|\nabla u(\xi)|^p + |\nabla v(\xi)|^p - |\nabla u(\xi)|^{p-2} \langle \nabla u(\xi), \nabla v(\xi) \rangle$$
$$- |\nabla v(\xi)|^{p-2} \langle \nabla u(\xi), \nabla v(\xi) \rangle \big) \, d\xi$$
$$\geqslant \int \big(|\nabla u(\xi)|^p + |\nabla v(\xi)|^p - |\nabla u(\xi)|^{p-1} |\nabla v(\xi)|$$
$$- |\nabla v(\xi)|^{p-1} |\nabla u(\xi)| \big) \, d\xi$$
$$= \int (|\nabla u(\xi)|^{p-1} - |\nabla v(\xi)|^{p-1})(|\nabla u(\xi)| - |\nabla v(\xi)|) \, d\xi$$
$$\geqslant 0,$$

since the map $\mathbb{R}_+ \ni s \mapsto s^{p-1}$ is increasing. Hence (H2) is shown with $c = 0$.

(H3) Because Λ is bounded by Poincaré's inequality (cf. [39]) there exists a constant $c = c(p, d, |\Lambda|) \in]0, \infty[$ such that

$$\int |\nabla u(\xi)|^p \, d\xi \geqslant c \int |u(\xi)|^p \, d\xi \quad \text{for all } u \in H_0^{1,p}(\Lambda). \tag{4.15}$$

Hence by (4.13) for all $u \in H_0^{1,p}(\Lambda)$

$$_{V^*}\langle A(u), u \rangle_V = - \int |\nabla u(\xi)|^p \, d\xi \leqslant - \frac{\min(1,c)}{2} \|u\|_{1,p}^p.$$

So, (H3) holds with $\alpha = p$ and $c_1 = 0$. (We note that only for (H3) have we used that Λ is bounded.)

(H4) This condition holds for A by (4.14) with $\alpha = p$.

Before we go on to our last example, which will include the case of the porous medium equation, we would like to stress the following:

Remark 4.1.10

1. If one is given $V \subset H \subset V^*$ and $A : V \to V^*$ (e.g. as in the above examples) satisfying (H1)–(H4) (with $B \equiv 0$) one can consider a "smaller" space V_0, i.e. another reflexive separable Banach space such that

$$V_0 \subset V$$

continuously and densely, hence (by restricting the linear functionals to V_0)

$$V^* \subset V_0^*$$

continuously and densely, so altogether

$$V_0 \subset V \subset H \subset V^* \subset V_0^*.$$

Restricting A to V_0 we see that A satisfies (H1),(H2) and (H4) with respect to the Gelfand triple

$$V_0 \subset H \subset V_0^*.$$

However, since $\| \ \|_{V_0}$ is up to a multiplicative constant larger than $\| \ \|_V$, property (H3) might no longer hold. Therefore, e.g. if one considers a map A which is given by a sum of the Laplacian (cf. Example 4.1.7) and a monomial (cf. Example 4.1.5) one cannot just take any $V_0 \subset H_0^{1,2}(\Lambda) \cap L^p(\Lambda)$, since (H3) might get lost.

In the case of the p-Laplacian, $p \in [2, \infty[$, it is possible to add monomials of order p, since $H_0^{1,p}(\Lambda) \subset L^p(\Lambda)$ continuously and densely, so

$$A(u) := \operatorname{div}(|\nabla u|^{p-2}\nabla u) - u|u|^{p-2}, \ u \in H_0^{1,p}(\Lambda),$$

satisfies (H1)–(H4), if Λ is bounded, with respect to the Gelfand triple

$$H_0^{1,p}(\Lambda) \subset L^2(\Lambda) \subset (H_0^{1,p}(\Lambda))^*.$$

But generally, taking sums of A as above requires some care and is not always possible.

2. In all our analysis the space V^* is only used as a tool. Eventually, since the solutions to our SDE (4.3) will take values in H, V^* will be of no relevance. Therefore, no further information about V^* such as its explicit representation (e.g. as a space of Schwartz distributions) is necessary.

Example 4.1.11 $[L^p \subset (H_0^{1,2})^* \subset (L^p)^*, A$ = porous medium operator] The stochastic differential equation (4.3) becomes

$$dX(t) = \Delta\Psi(X(t)) \, dt + B(t, X(t)) \, dW(t).$$

As references for this example we refer to [4, 25, 69].

Let $\Lambda \subset \mathbb{R}^d$, Λ open and bounded, $p \in [2, \infty[$ and

$$V := L^p(\Lambda), \quad H := (H_0^{1,2}(\Lambda))^*.$$

Since Λ is bounded we have by Poincaré's inequality (4.15) that for some constant $c = c(2, d, |\Lambda|) > 0$

$$\|u\|_{1,2} \geq \|u\|_{H_0^{1,2}} := \left(\int |\nabla u(\xi)|^2 \, d\xi \right)^{\frac{1}{2}}$$

$$\geq \left(\frac{\min(1, c)}{2} \right)^{\frac{1}{2}} \|u\|_{1,2} \quad \text{for all } u \in H_0^{1,2}(\Lambda). \tag{4.16}$$

So, we can (and will do so below) consider $H_0^{1,2}(\Lambda)$ with norm $\| \ \|_{H_0^{1,2}}$ and corresponding scalar product

$$\langle u, v \rangle_{H_0^{1,2}} := \int \langle \nabla u(\xi), \nabla v(\xi) \rangle \, d\xi, \ u, v \in H_0^{1,2}(\Lambda).$$

Since $H_0^{1,2}(\Lambda) \subset L^2(\Lambda)$ continuously and densely, so is

$$H_0^{1,2}(\Lambda) \subset L^{\frac{p}{p-1}}(\Lambda).$$

Hence

$$L^p(\Lambda) \equiv \left(L^{\frac{p}{p-1}}(\Lambda) \right)^* \subset (H_0^{1,2}(\Lambda))^* = H,$$

continuously and densely. Now we would like to identify H with its dual $H^* = H_0^{1,2}(\Lambda)$ via the corresponding Riesz isomorphism $R : H \to H^*$ defined by $Rx := \langle x, \cdot \rangle_H$, $x \in H$. Let us calculate the latter.

Lemma 4.1.12 *The map* $\Delta : H_0^{1,2}(\Lambda) \to (H_0^{1,2}(\Lambda))^* = H$ *(defined by (4.13) for* $p = 2$*) is an isometric isomorphism. In particular,*

$$\langle \Delta u, \Delta v \rangle_H = \langle u, v \rangle_{H_0^{1,2}} \quad \text{for all } u, v \in H_0^{1,2}(\Lambda). \tag{4.17}$$

Furthermore, $(-\Delta)^{-1} : H \to H^* = H_0^{1,2}(\Lambda)$ *is the Riesz isomorphism for* H*, i.e. for every* $x \in H$

$$\langle x, \cdot \rangle_H = {}_{H_0^{1,2}}\langle (-\Delta)^{-1} x, \cdot \rangle_H . \tag{4.18}$$

Proof Let $u \in H_0^{1,2}(\Lambda)$. Since by (4.13) for all $v \in H_0^{1,2}(\Lambda)$

$$_H\langle -\Delta u, v \rangle_{H_0^{1,2}} = \int \langle \nabla u(\xi), \nabla v(\xi) \rangle \, d\xi = \langle u, v \rangle_{H_0^{1,2}}, \tag{4.19}$$

it follows that $-\Delta : H_0^{1,2}(\Lambda) \to H$ is just the Riesz isomorphism for $H_0^{1,2}(\Lambda)$ and the first part of the assertion including (4.17) follows. To prove the last part, fix $x \in H$. Then by (4.17) and (4.19) for all $y \in H$

$$\langle x, y \rangle_H = \langle (-\Delta)^{-1} x, (-\Delta)^{-1} y \rangle_{H_0^{1,2}} = {}_H\langle x, (-\Delta)^{-1} y \rangle_{H_0^{1,2}} .$$

$\square$

Now we identify H with its dual H^* by the Riesz map $(-\Delta)^{-1} : H \to H^*$, so $H \equiv H^*$ in this sense, hence

$$V = L^p(\Lambda) \subset H \subset (L^p(\Lambda))^* = V^* \tag{4.20}$$

continuously and densely.

Lemma 4.1.13 *The map*

$$\Delta : H_0^{1,2}(\Lambda) \to (L^p(\Lambda))^*$$

extends to a linear isometry

$$\Delta : L^{\frac{p}{p-1}}(\Lambda) \to (L^p(\Lambda))^* = V^*$$

and for all $u \in L^{\frac{p}{p-1}}(\Lambda)$*,* $v \in L^p(\Lambda)$

$$_{V^*}\langle -\Delta u, v \rangle_V = {}_{L^{\frac{p}{p-1}}}\langle u, v \rangle_{L^p} = \int u(\xi) v(\xi) \, d\xi. \tag{4.21}$$

Remark 4.1.14 One can prove that this isometry is in fact surjective, hence

$$(L^p(\Lambda))^* = \Delta(L^{\frac{p}{p-1}}) \neq L^{\frac{p}{p-1}}.$$

We shall not use this below, but it shows that the embedding (4.20) has to be handled with care, always taking into account that H is identified with H^* by $(-\Delta)^{-1} : H \to H^*$, giving rise to a different dualization between $L^p(\Lambda)$ and $(L^p(\Lambda))^*$. In particular, for all $x \in H$, $v \in L^p(\Lambda)$

$$_{(L^p)^*}\langle x, v \rangle_{L^p} = \langle x, v \rangle_H$$

$$\left(\neq \,_{L^{\frac{p}{p-1}}}\langle x, v \rangle_{L^p} = \int x(\xi)v(\xi) \, \mathrm{d}\xi \quad \text{provided } x \in L^{\frac{p}{p-1}} \right).$$

Proof of Lemma 4.1.13 Let $u \in H_0^{1,2}(\Lambda)$. Then since $\Delta u \in H$, by (4.2) and (4.18) we obtain that for all $v \in V$

$$_{V^*}\langle \Delta u, v \rangle_V = \langle \Delta u, v \rangle_H = -\,_{H_0^{1,2}}\langle u, v \rangle_H = -\langle u, v \rangle_{L^2} \tag{4.22}$$

since $v \in V \subset L^2(\Lambda)$. Therefore,

$$\|\Delta u\|_{V^*} \leqslant \|u\|_{\frac{p}{p-1}}.$$

So, Δ extends to a continuous linear map

$$\Delta : L^{\frac{p}{p-1}}(\Lambda) \to V^*$$

such that (4.22) holds for all $u \in L^{\frac{p}{p-1}}(\Lambda)$, i.e. (4.21) is proved.

So, applying it to $u \in L^{\frac{p}{p-1}}(\Lambda)$ and

$$v := -\|u\|_q^{-\frac{q}{p}} u|u|^{q-2} \in L^p(\Lambda),$$

where $q := \frac{p}{p-1}$, by (4.21) we obtain that

$$_{V^*}\langle \Delta u, v \rangle_V = \|u\|_{\frac{p}{p-1}}$$

and $\|v\|_p = 1$, so $\|\Delta u\|_{V^*} = \|u\|_{\frac{p}{p-1}}$ and the assertion is completely proved.

□

Now we want to define the "porous medium operator A". So, let $\Psi : \mathbb{R} \to \mathbb{R}$ be a function having the following properties:

(Ψ1) Ψ is continuous.
(Ψ2) For all $s, t \in \mathbb{R}$

$$(t - s)(\Psi(t) - \Psi(s)) \geqslant 0.$$

(Ψ3) There exist $p \in [2, \infty[$, $a \in]0, \infty[$, $c \in [0, \infty[$ such that for all $s \in \mathbb{R}$

$$s\Psi(s) \geq a|s|^p - c.$$

(Ψ4) There exist $c_3, c_4 \in]0, \infty[$ such that for all $s \in \mathbb{R}$

$$|\Psi(s)| \leq c_4 + c_3|s|^{p-1},$$

where p is as in (Ψ3).

We note that (Ψ4) implies that

$$\Psi(v) \in L^{\frac{p}{p-1}}(\Lambda) \quad \text{for all } v \in L^p(\Lambda). \tag{4.23}$$

Now we can define the *porous medium operator* $A : L^p(\Lambda) = V \to V^* = (L^p(\Lambda))^*$ by

$$A(u) := \Delta\Psi(u), \quad u \in L^p(\Lambda). \tag{4.24}$$

Note that by Lemma 4.1.13 the operator A is well-defined. Now let us check (H1)–(H4).

(H1) Let $u, v, x \in V = L^p(\Lambda)$ and $\lambda \in \mathbb{R}$. Then by (4.21)

$$_{V^*}\langle A(u + \lambda v), x\rangle_V = {}_{V^*}\langle \Delta\Psi(u + \lambda v), x\rangle_V$$

$$= -\int \Psi(u(\xi) + \lambda v(\xi))x(\xi) \, d\xi. \tag{4.25}$$

By (Ψ4) for $|\lambda| \leq 1$ the integrand in the right-hand side of (4.25) is bounded by

$$[c_4 + c_3 2^{p-2}(|u|^{p-1} + |v|^{p-1})]|x|$$

which by Hölder's inequality is in $L^1(\Lambda)$. So, (H1) follows by (Ψ1) and Lebesgue's dominated convergence theorem.

(H2) Let $u, v \in V = L^p(\Lambda)$. Then by (4.21)

$$_{V^*}\langle A(u) - A(v), u - v\rangle_V = {}_{V^*}\langle \Delta(\Psi(u) - \Psi(v)), u - v\rangle_V$$

$$= -\int [\Psi(u(\xi)) - \Psi(v(\xi))](u(\xi) - v(\xi)) \, d\xi$$

$$\leq 0,$$

where we used (Ψ2) in the last step.

(H3) Let $v \in L^p(\Lambda) = V$. Then by (4.21) and (Ψ3)

$$_{V^*}\langle A(v), v \rangle_V = - \int \Psi(v(\xi))v(\xi) \, d\xi$$

$$\leqslant \int (-a|v(\xi)|^p + c) \, d\xi.$$

Hence (H3) is satisfied with $c_1 := 0$, $c_2 := 2a$, $\alpha = p$ and $f(t) = 2c|\Lambda|$.

(H4) Let $v \in L^p(\Lambda) = V$. Then by Lemma 4.1.13 and (Ψ4)

$$\|A(v)\|_{V^*} = \|\Delta\Psi(v)\|_{V^*}$$

$$= \|\Psi(v)\|_{L^{\frac{p}{p-1}}}$$

$$\leqslant c_4 |\Lambda|^{\frac{p-1}{p}} + c_3 \left(\int |v(\xi)|^p \, d\xi \right)^{\frac{p-1}{p}}$$

$$= c_4 |\Lambda|^{\frac{p-1}{p}} + c_3 \|v\|_V^{p-1},$$

so (H4) holds with $\alpha = p$.

Remark 4.1.15

1. For $p \in [2, \infty[$ and $\Psi(s) := s|s|^{p-2}$ we have

$$A(v) = \Delta(v|v|^{p-2}), \quad v \in L^p(\Lambda),$$

which is the non-linear operator appearing in the classical porous medium equation, i.e.

$$\frac{\partial X(t)}{\partial t} = \Delta(X(t)|X(t)|^{p-2}), \quad X(0, \cdot) = X_0,$$

whose solution describes the time evolution of the density $X(t)$ of a substance in a porous medium (cf. e.g. [4]).

2. Let $\Psi : \mathbb{R} \to \mathbb{R}$ be given such that (Ψ1)–(Ψ4) are satisfied with some $p \in \,]1, \infty[$ (in (Ψ3), (Ψ4)). One can see that the above assumptions that Λ is bounded and $p \geqslant 2$, can be avoided. But p then depends on the dimension of the underlying space $\mathbb{R}^d$. Let us assume first that $d \geqslant 3$. We distinguish two cases:

Case 1. $|\Lambda| = \infty$ and $p := \frac{2d}{d+2}$, $c = c_4 = 0$, where c, c_4 are the constants in (Ψ3) and in (Ψ4) respectively.

Case 2. $|\Lambda| < \infty$ and $p \in \left[\frac{2d}{d+2}, \infty \right[$.

By the Sobolev embedding theorem (cf. [39, Theorem 7.10]) we have

$$H_0^{1,2}(\Lambda) \subset L^{\frac{2d}{d-2}}(\Lambda)$$

continuously and densely, and

$$\|u\|_{\frac{2d}{d-2}} \leq \frac{2(d-1)}{\sqrt{d(d-2)}} \|u\|_{H_0^{1,2}} \text{ for all } u \in H_0^{1,2}(\Lambda).$$

In Case 1 we have $\frac{2d}{d-2} = \frac{p}{p-1}$ and in Case 2 (hence in both cases)

$$\frac{2d}{d-2} \geq \frac{p}{p-1}$$

and thus

$$H_0^{1,2}(\Lambda) \subset L^{\frac{p}{p-1}}(\Lambda)$$

densely and for some $c_0 \in]0, \infty[$

$$\|u\|_{\frac{p}{p-1}} \leq c_0 \|u\|_{H_0^{1,2}} \text{ for all } u \in H_0^{1,2}(\Lambda).$$

Now the above arguments generalize to both Cases 1 and 2, i.e. for the Gelfand triple

$$V := L^p(\Lambda) \subset H := (H_0^{1,2}(\Lambda))^* \subset (L^p(\Lambda))^*$$

the operator

$$A : L^p(\Lambda) =: V \to V^* = (L^p(\Lambda))^*$$

defined in (4.24), satisfies (H1)–(H4).

We note that in Case 1 the norm $\| \ \|_{H_0^{1,2}}$ defined in (4.16) is in general not equivalent to $\| \ \|_{1,2}$, because the Poincaré inequality does not hold. In Case 1, if $d = 6$, or in Case 2, if $(3 \leq)d \leq 6$, we may take $p = \frac{3}{2}$ and

$$\Psi(s) := \text{sign}(s)\sqrt{|s|}, s \in \mathbb{R}.$$

In this case the equation in Remark 4.1.15 is called *fast diffusion equation*. For Λ bounded the above extends, of course, also to the case $d = 1, 2$ where even stronger Sobolev embeddings hold (cf. [39, Theorems 7.10 and 7.15]).

4.2 The Main Result and An Itô Formula

Consider the general situation described at the beginning of the previous section. So, we have a Gelfand triple

$$V \subset H \subset V^*$$

and maps

$$A : [0, T] \times V \times \Omega \to V^*, \quad B : [0, T] \times V \times \Omega \to L_2(U, H)$$

as specified there, satisfying (H1)–(H4), and consider the stochastic differential equation

$$dX(t) = A(t, X(t)) \, dt + B(t, X(t)) \, dW(t) \tag{4.26}$$

on H with $W(t)$, $t \in [0, T]$, a cylindrical Q-Wiener process with $Q := I$ taking values in another separable Hilbert space $(U, \langle \ , \ \rangle_U)$ and being defined on a complete probability space $(\Omega, \mathcal{F}, P)$ with normal filtration $\mathcal{F}_t$, $t \in [0, T]$.

Before we formulate our main existence and uniqueness result for solutions of (4.26) we have to define what we mean by "solution".

Definition 4.2.1 A continuous H-valued $(\mathcal{F}_t)$-adapted process $(X(t))_{t \in [0, T]}$ is called a *solution of* (4.26), if for its $dt \otimes P$-equivalence class $\hat{X}$ we have $\hat{X} \in L^\alpha([0, T] \times \Omega, dt \otimes P; V) \cap L^2([0, T] \times \Omega, dt \otimes P; H)$ with α as in (H3) and P-a.s.

$$X(t) = X(0) + \int_0^t A(s, \bar{X}(s)) \, ds + \int_0^t B(s, \bar{X}(s)) \, dW(s), \quad t \in [0, T], \tag{4.27}$$

where $\bar{X}$ is any V-valued progressively measurable $dt \otimes P$-version of $\hat{X}$.

Remark 4.2.2

1. The existence of the special version $\bar{X}$ above follows from Exercise 4.2.3 below. Furthermore, for technical reasons in Definition 4.2.1 and below we consider all processes initially as V^*-valued, hence by $dt \otimes P$-equivalence classes we always mean classes of V^*-valued processes.
2. The integral with respect to ds in (4.27) is initially a V^*-valued Bochner integral which turns out to be in fact H-valued.
3. Solutions in the sense of Definition 4.2.1 are often called *variational solutions* in the literature. There are various other notions of solutions for stochastic (partial) differential equations. We recall the definition of (probabilistically) weak and strong solutions in Appendix E below. The notions of analytically weak and strong solutions as well as the notion of mild solutions and their relations are recalled in Appendix G below.
4. We stress that the solution $(X(t))_{t \in [0, T]}$ from Definition 4.2.1 is in general not an H-valued semimartingale, since the first integral in the right-hand side of (4.27) is not necessarily of bounded variation in H. Therefore, the classical Itô-formula on Hilbert spaces (see Sect. 6.1 below) does not apply to $(X(t))_{t \in [0, T]}$.

Exercise 4.2.3

1. Let $B_1^{V^*}$ denote the closed unit ball in V^*. Since $B_1^{V^*} \cap H \neq \emptyset$, it has a countable subset $\{l_i | i \in \mathbb{N}\}$, which is dense in $B_1^{V^*} \cap H$ with respect to H-norm. Define

$\Theta : H \to [0, \infty]$ by

$$\Theta(h) := \sup_{i \in \mathbb{N}} |\langle l_i, h \rangle_H|, \ h \in H.$$

Then Θ is lower semicontinuous on H, hence $\mathcal{B}(H)$-measurable. Since $_{V^*}\langle l_i, v \rangle_V = \langle l_i, v \rangle_H, i \in \mathbb{N}, v \in V$, we have

$$\Theta(v) = \|v\|_V \quad \text{for all } v \in V,$$

and furthermore (by the reflexivity of V)

$$\{\Theta < \infty\} = V.$$

2. Let $X : [0, T] \times \Omega \to H$ be any progressively measurable (i.e. $\mathcal{B}([0, t]) \otimes \mathcal{F}_t/\mathcal{B}(H)$-measurable for all $t \in [0, T]$) $dt \otimes P$-version of $\hat{X} \in L^\alpha([0, T] \times \Omega, \, dt \otimes P; V)$, $\alpha \in]0, \infty[$. Then

$$\bar{X} := \mathbb{I}_{\{\Theta \circ X < \infty\}} X$$

is a V-valued progressively measurable (i.e. $\mathcal{B}([0, t]) \otimes \mathcal{F}_t/\mathcal{B}(V)$-measurable) $dt \otimes P$-version of $\hat{X}$.
3. Both $A(\cdot, \bar{X})$ and $B(\cdot, \bar{X})$ are V^*-valued respectively $L_2(U, H)$-valued progressively measurable processes.

Now the main result (cf. [54]):

Theorem 4.2.4 *Let A, B be as above satisfying (H1)–(H4) and let $X_0 \in L^2(\Omega, \mathcal{F}_0, P; H)$. Then there exists a unique solution X to (4.26) in the sense of Definition 4.2.1. Moreover,*

$$E(\sup_{t \in [0,T]} \|X(t)\|_H^2) < \infty. \tag{4.28}$$

The proof of Theorem 4.2.4 strongly depends on the following Itô formula, from [54, Theorem I.3.1], which we shall prove here first. The presentation of its proof and that of Theorem 4.2.4 is an extended adaptation of those in [69].

Theorem 4.2.5 *Let $\alpha \in]1, \infty[$, $X_0 \in L^2(\Omega, \mathcal{F}_0, P; H)$ and $Y \in L^{\frac{\alpha}{\alpha-1}}([0, T] \times \Omega, \, dt \otimes P; V^*)$, $Z \in L^2([0, T] \times \Omega, \, dt \otimes P; L_2(U, H))$, both progressively measurable. Define the continuous V^*-valued process*

$$X(t) := X_0 + \int_0^t Y(s) \, ds + \int_0^t Z(s) \, dW(s), \ t \in [0, T].$$

If for its $dt \otimes P$-equivalence class $\hat{X}$ we have $\hat{X} \in L^\alpha([0, T] \times \Omega, \, dt \otimes P; V)$ and if $E(\|X(t)\|_H^2) < \infty$ for dt-a.e. $t \in [0, T]$ (which is automatically the case if $\alpha \geq 2$), *then X is a continuous H-valued ($\mathcal{F}_t$)-adapted process,*

$$E\left(\sup_{t\in[0,T]} \|X(t)\|_H^2\right) \leq 4\left[E\left(\|X(0)\|_H^2\right) + \|Y\|_{K^*}\left(\|X\|_K + 1\right)\right.$$

$$\left. + 10\,E\left(\int_0^T \|Z(s)\|_{L_2(U,H)}^2\,ds\right)\right] \tag{4.29}$$

and the following Itô-formula holds for the square of its H-norm P-a.s.

$$\|X(t)\|_H^2 = \|X_0\|_H^2 + \int_0^t \left(2\,_{V^*}\langle Y(s),\bar{X}(s)\rangle_V + \|Z(s)\|_{L_2(U,H)}^2\right)\,ds$$

$$+ 2\int_0^t \langle X(s),Z(s)\,dW(s)\rangle_H \quad \text{for all } t \in [0,T] \tag{4.30}$$

for any V-valued progressively measurable $dt \otimes P$-version $\bar{X}$ of $\hat{X}$.

As in [54] for the proof of Theorem 4.2.5 we need the following lemma about piecewise constant approximations based on an argument due to [29]. For abbreviation below we set

$$K := L^\alpha([0,T] \times \Omega,\ dt \otimes P; V). \tag{4.31}$$

Lemma 4.2.6 *Let $X : [0,T] \times \Omega \to V^*$ be $\mathcal{B}([0,T]) \otimes \mathcal{F}/\mathcal{B}(V^*)$-measurable such that for its $dt \otimes P$-equivalence class $\hat{X}$ we have $\hat{X} \in K$. Then there exists a sequence of partitions $I_l := \{0 = t_0^l < t_1^l < \cdots < t_{k_l}^l = T\}$ such that $I_l \subset I_{l+1}$ and $\delta(I_l) := \max_i(t_i^l - t_{i-1}^l) \to 0$ as $l \to \infty$, $X(t_i^l) \in V$ P-a.e. for all $l \in \mathbb{N}$, $1 \leq i \leq k_l - 1$, and for*

$$\bar{X}^l := \sum_{i=2}^{k_l} 1_{]t_{i-1}^l,t_i^l]} X(t_{i-1}^l), \quad \tilde{X}^l := \sum_{i=1}^{k_l-1} 1_{]t_{i-1}^l,t_i^l]} X(t_i^l), \quad l \in \mathbb{N},$$

we have that $\bar{X}^l, \tilde{X}^l$ are ($dt \otimes P$-versions of elements) in K such that

$$\lim_{l\to\infty} \left\{\|\hat{X} - \bar{X}^l\|_K + \|\hat{X} - \tilde{X}^l\|_K\right\} = 0.$$

Proof For simplicity we assume that $T = 1$ and let $\bar{X} : [0,1] \times \Omega \to V$ be a $dt \otimes P$-version of $\hat{X}$ such that $\int_0^1 \|\bar{X}(t)\|_V^\alpha\,dt < \infty$ P-a.s. We extend $\bar{X}$ to $\mathbb{R} \times \Omega$ by setting $\bar{X} = 0$ on $[0,1]^c \times \Omega$. There exists an $\Omega' \in \mathcal{F}$ with full probability such that for every $\omega \in \Omega'$ there exists a sequence $(f_n)_{n\in\mathbb{N}} \subset C(\mathbb{R};V)$ with compact support such that

$$\int_{\mathbb{R}} \|f_n(s) - \bar{X}(s,\omega)\|_V^\alpha\,ds \leq \frac{1}{2n}, \quad n \in \mathbb{N}.$$

Thus, for every $n \in \mathbb{N}$,

$$\limsup_{\delta \to 0} \int_{\mathbb{R}} \|\bar{X}(\delta + s, \omega) - \bar{X}(s, \omega)\|_V^\alpha \, ds$$

$$\leq 3^{\alpha-1} \limsup_{\delta \to 0} \int_{\mathbb{R}} \left[\|\bar{X}(\delta + s, \omega) - f_n(\delta + s)\|_V^\alpha + \|\bar{X}(s, \omega) - f_n(s)\|_V^\alpha \right] ds$$

$$\leq \frac{3^{\alpha-1}}{n}, \quad n \in \mathbb{N}.$$

Here we used that since each f_n is uniformly continuous, by Lebesgue's dominated convergence theorem we have that for all $n \in \mathbb{N}$

$$\lim_{\delta \to 0} \int_{\mathbb{R}} \|f_n(\delta + s) - f_n(s)\|_V^\alpha \, ds = 0.$$

Letting $n \to \infty$ we obtain

$$\lim_{\delta \to 0} \int_{\mathbb{R}} \|\bar{X}(\delta + s, \omega) - \bar{X}(s, \omega)\|_V^\alpha \, ds = 0, \quad \omega \in \Omega'. \tag{4.32}$$

Now, given $t \in \mathbb{R}$, let $[t]$ denote the largest integer $\leq t$. Let $\gamma_n(t) := 2^{-n}[2^n t]$, $n \in \mathbb{N}$, that is, $\gamma_n(t)$ is the largest number of the form $\frac{k}{2^n}$, $k \in \mathbb{Z}$, less or equal to t. Shifting the integral in (4.32) by t and taking $\delta = \gamma_n(t) - t$ we obtain

$$\lim_{n \to \infty} \int_{\mathbb{R}} \|\bar{X}(\gamma_n(t) + s) - \bar{X}(t + s)\|_V^\alpha \, ds = 0 \text{ on } \Omega'.$$

Moreover, since $\bar{X}(r) = 0$ for all $r \in \mathbb{R} \setminus [0, 1]$,

$$\int_0^1 \|\bar{X}(\gamma_n(t) + s) - \bar{X}(t + s)\|_V^\alpha \, ds$$

$$\leq 1_{[-2,2]}(t) 2^{\alpha-1} \int_{\mathbb{R}} \left[\|\bar{X}(\gamma_n(t) + s)\|_V^\alpha + \|\bar{X}(t + s)\|_V^\alpha \right] ds$$

$$= 2^\alpha 1_{[-2,2]}(t) \int_0^1 \|\bar{X}(s)\|_V^\alpha \, ds \text{ on } \Omega'.$$

So, by Lebesgue's dominated convergence theorem, we obtain that

$$0 = \lim_{n \to \infty} E \int_{\mathbb{R}} dt \int_0^1 \|\bar{X}(\gamma_n(t) + s) - \bar{X}(t + s)\|_V^\alpha \, ds$$

$$\geq \lim_{n \to \infty} E \int_0^1 ds \int_0^1 \|\bar{X}(\gamma_n(t - s) + s) - \bar{X}(t)\|_V^\alpha \, dt. \tag{4.33}$$

Given $s \in [0, 1[$ and $n \in \mathbb{N}$, let the partition $I_n(s)$ be defined by

$$t_0^n(s) := 0, \ t_i^n(s) := \left(s - \frac{[2^n s]}{2^n}\right) + \frac{i-1}{2^n}, \ 1 \leq i \leq 2^n, \ t_{2^n+1}^n(s) := 1.$$

Then, for $t \in [t_{i-1}^n(s), t_i^n(s)[, \ 2 \leq i \leq 2^n + 1$, one has $t - s \in [2^{-n}(i - [2^n s] - 2), 2^{-n}(i - [2^n s] - 1)[$ and hence,

$$\gamma_n(t - s) + s = 2^{-n}(i - [2^n s] - 2) + s = t_{i-1}^n(s), \ \ 2 \leq i \leq 2^n + 1.$$

Therefore, (4.33) implies

$$\lim_{n \to \infty} E \int_0^1 ds \int_0^1 \|\bar{X}(t) - \bar{X}^{n,s}(t)\|_V^\alpha \, dt = 0,$$

where $\bar{X}^{n,s}$ is the process defined as $\bar{X}^l$ for the partition $I_n(s)$ but with $X(t_{i-1}^l(s))$ replaced by $\bar{X}(t_{i-1}^l(s))$. Similarly, the same holds for $\tilde{X}^{n,s}$ in place of $\bar{X}^{n,s}$ by using $\tilde{\gamma}_n := \gamma_n + 2^{-n}$ instead of γ_n, where $\tilde{X}^{n,s}$ is defined as $\tilde{X}^l$ for the partition $I_n(s)$ but with $X(t_i^l(s))$ replaced by $\bar{X}(t_i^l(s))$. Hence, there exist a subsequence $n_k \to \infty$ and a ds-zero set $N_1 \in \mathcal{B}([0, 1])$ such that

$$\lim_{k \to \infty} E \int_0^1 \left[\|\bar{X}(t) - \bar{X}^{n_k,s}(t)\|_V^\alpha + \|\bar{X}(t) - \tilde{X}^{n_k,s}(t)\|_V^\alpha\right] dt = 0, \quad s \in [0, 1] \setminus N_1.$$

In particular, $\bar{X}^{n_k,s}$ and $\tilde{X}^{n_k,s}$ are ($dt \otimes P$-versions of elements) in K. Since for $1 \leq i \leq 2^n$ the maps $s \mapsto t_i^n(s)$ are piecewise C^1-diffeomorphisms, the image measures of ds under these maps are absolutely continuous with respect to ds. Therefore, since $\bar{X} = X$ $ds \otimes P$-a.e., there exists a ds-zero set $N_2 \in \mathcal{B}([0, 1])$ such that

$$\bar{X}(t_i^n(s)) = X(t_i^n(s)) \ P\text{-a.e. for all } s \in [0, 1] \setminus N_2, 1 \leq i \leq 2^n.$$

Therefore, fixing $s \in [0, 1[\setminus(N_1 \cup N_2)$, the sequence of the corresponding partitions $I_{n_l}(s), l \geq 1$, has all properties of the assertion. $\square$

Remark 4.2.7

(i) As follows from the above proof all the partition points $t_i^l, l \geq 1, 1 \leq i \leq k_l - 1$, in the assertion of Lemma 4.2.6 can be chosen outside an a priori given Lebesgue zero set in $[0, T]$ instead of N_2 above.

(ii) Tracing through the proof of Lemma 4.2.6 one sees that it also holds if we replace K by the following space

$$K_0 := L_W^\alpha([0, T] \times \Omega, dt \otimes P; V) \cap L_W^2([0, T] \times \Omega, dt \otimes P; H),$$

equipped with the sum of the two respective norms, where we adopt the standard notation of a subscript W for the subspace of those elements which possess an $(\mathcal{F}_t)$-adapted $dt \otimes P$-version.

Proof of Theorem 4.2.5 Since $M(t) := \int_0^t Z(s) \, dW(s)$, $t \in [0, T]$, is already a continuous martingale on H and since $Y \in K^* = L^{\alpha/(\alpha-1)}([0, T] \times \Omega \to V^*; \, dt \otimes P)$ is progressively measurable, $\int_0^t Y(s) \, ds$ is a continuous adapted process on V^*. Thus, X is a continuous adapted process on V^*, hence $\mathcal{B}([0, T]) \otimes \mathcal{F}/\mathcal{B}(V^*)$-measurable.

Claim (a):

$$\|X(t)\|_H^2 = \|X(s)\|_H^2 + 2 \int_s^t {}_{V^*}\langle Y(r), X(t)\rangle_V \, dr + 2\langle X(s), M(t) - M(s)\rangle_H$$
$$+ \|M(t) - M(s)\|_H^2 - \|X(t) - X(s) - M(t) + M(s)\|_H^2 \qquad (4.34)$$

holds for all $t > s$ such that $X(t), X(s) \in V$.
Indeed, this follows immediately by noting that

$$\|M(t) - M(s)\|_H^2 - \|X(t) - X(s) - M(t) + M(s)\|_H^2$$
$$+ 2\langle X(s), M(t) - M(s)\rangle_H$$
$$= 2\langle X(t), M(t) - M(s)\rangle_H - \|X(t) - X(s)\|_H^2$$
$$= 2\langle X(t), X(t) - X(s)\rangle_H - 2 \int_s^t {}_{V^*}\langle Y(r), X(t)\rangle_V \, dr$$
$$- \|X(t)\|_H^2 - \|X(s)\|_H^2 + 2\langle X(t), X(s)\rangle_H$$
$$= \|X(t)\|_H^2 - \|X(s)\|_H^2 - 2 \int_s^t {}_{V^*}\langle Y(r), X(t)\rangle_V \, dr.$$

Claim (b): We have

$$E\left(\sup_{t \in [0, T]} \|X(t)\|_H^2 \right) < \infty. \qquad (4.35)$$

To prove the claim let I_l, $l \in \mathbb{N}$, be a sequence of partitions as in Lemma 4.2.6. By Remark 4.2.7(i) and the assumption that $E(\|X(t)\|_H^2) < \infty$ for dt-a.e. $t \in [0, T]$ we may choose I_l such that

$$E(\|X(t)\|_H^2) < \infty \text{ for all } t \in \bigcup_{l \in \mathbb{N}} I_l. \qquad (4.36)$$

Then by (4.34), for any $t = t_i^l \in I_l \setminus \{0, T\}$

$$\|X(t)\|_H^2 - \|X_0\|_H^2$$

$$= \sum_{j=0}^{i-1} (\|X(t_{j+1}^l)\|_H^2 - \|X(t_j^l)\|_H^2)$$

$$= 2 \int_0^t {}_{V^*}\langle Y(s), \tilde{X}^l(s)\rangle_V \, ds$$

$$+ 2 \int_0^t \langle \bar{X}^l(s), Z(s) \, dW(s)\rangle_H + 2\langle X(0), \int_0^{t_1^l} Z(s) \, dW(s)\rangle_H$$

$$+ \sum_{j=0}^{i-1} \left(\|M(t_{j+1}^l) - M(t_j^l)\|_H^2 - \|X(t_{j+1}^l) - X(t_j^l) - M(t_{j+1}^l) + M(t_j^l)\|_H^2 \right),$$

$$(4.37)$$

where $\bar{X}^l$ and $\tilde{X}^l$ are defined as in Lemma 4.2.6. We note that since $\bar{X}^l$ is $(\mathcal{F}_t)$-adapted and pathwise bounded the stochastic integral involving $\bar{X}^l$ above is well-defined. By Lemma 4.2.6

$$E \left(\int_0^T |\, {}_{V^*}\langle Y(s), \tilde{X}^l(s)\rangle_V \,| \, ds \right) \le \|Y\|_{K^*} \|\tilde{X}^l\|_K \le c_1 \qquad (4.38)$$

for some constant $c_1 > 0$ independent of l. Moreover, by the Burkholder–Davis inequality (cf. Proposition D.0.1), Lemmas 2.4.2 and 2.4.4,

$$E \left(\sup_{t \in [0,T]} \left| \int_0^t \langle \bar{X}^l(s), Z(s) \, dW(s)\rangle_H \right| \right)$$

$$\le 3E \left(\left[\int_0^T \|Z(s)^* \bar{X}^l(s)\|_U^2 \, ds \right]^{1/2} \right)$$

$$\le 3E \left(\left[\int_0^T \|\bar{X}^l(s)\|_H^2 \|Z(s)\|_{L_2(U,H)}^2 \, ds \right]^{1/2} \right)$$

$$= 3E \left(\left[\int_0^T \|\bar{X}^l(s)\|_H^2 \, d\langle M\rangle_s \right]^{1/2} \right)$$

$$\le \frac{1}{4} E \left(\sup_{1 \le j \le k_l - 1} \|X(t_j^l)\|_H^2 \right) + 9E(\langle M\rangle_T), \qquad (4.39)$$

since $\langle M\rangle_t = \int_0^t \|Z(s)\|_{L_2(U,H)}^2 \, ds$ and we used that

$$ab \le \frac{1}{12}a^2 + 3b^2, \, a, b > 0.$$

Here we note that by (4.36)

$$E\left(\sup_{1\le j\le k_l-1}\|X(t_j^l)\|_H^2\right) < \infty.$$

Finally, by Lemma 2.4.4

$$E\left(\sum_{j=0}^{i-1}\|M(t_{j+1}^l) - M(t_j^l)\|_H^2\right) = \sum_{j=0}^{i-1} E\left(\int_{t_j^l}^{t_{j+1}^l}\|Z(s)\|_{L_2(U,H)}^2\,ds\right)$$

$$= E\left(\int_0^{t_i^l}\|Z(s)\|_{L_2(U,H)}^2\,ds\right)$$

$$= E\left(\langle M\rangle_{t_i^l}\right). \tag{4.40}$$

Combining (4.37)–(4.40), we obtain

$$E\left(\sup_{t\in I_l\setminus\{T\}}\|X(t)\|_H^2\right) \le c_2$$

with $c_2 := 4\left[E(\|X(0)\|_H^2) + \|Y\|_{K^*}(\|X\|_K + 1) + 10E(\int_0^T\|Z(s)\|_{L_2(U,H)}^2\,ds)\right]$.
Therefore, letting $l\uparrow\infty$ and setting $I := \cup_{l\ge1}I_l\setminus\{T\}$, with I_l as in Lemma 4.2.6, we obtain

$$E\left(\sup_{t\in I}\|X(t)\|_H^2\right) \le c_2,$$

since $I_l\subset I_{l+1}$ for all $l\in\mathbb{N}$. Since for all $t\in[0,T]$

$$\sum_{j=1}^N {}_{V^*}\langle X(t), e_j\rangle_V^2 \uparrow \|X(t)\|_H^2 \text{ as } N\uparrow\infty,$$

where $\{e_j|j\in\mathbb{N}\}\subset V$ is an orthonormal basis of H and as usual for $x\in V^*\setminus H$ we set $\|x\|_H := \infty$, it follows that $t\mapsto\|X(t)\|_H$ is lower semicontinuous P-a.s. Since I is dense in $[0,T]$, we have $\sup_{t\in[0,T]}\|X(t)\|_H^2 = \sup_{t\in I}\|X(t)\|_H^2$. Thus, (4.35) holds.

Claim (c):

$$\lim_{l\to\infty}\sup_{t\in[0,T]}\left|\int_0^t\langle X(s) - \bar{X}^l(s), Z(s)\,dW(s)\rangle_H\right| = 0 \text{ in probability.} \tag{4.41}$$

We first note that because of (b) X is H-valued P-a.s. and by (b) and its continuity in V^* the process X is weakly continuous in H P-a.s. and, therefore, since $\mathcal{B}(H)$ is generated by H^*, progressively measurable as an H-valued process. Hence, for any $n \in \mathbb{N}$ the process $P_n X$ is continuous in H so that

$$\lim_{l \to \infty} \int_0^T \|P_n(X(s) - \bar{X}^l(s))\|_H^2 \, d\langle M \rangle_s = 0, \quad P\text{-a.s.}$$

Here P_n denotes the orthogonal projection onto $\text{span}\{e_1, \ldots, e_n\}$ in H. Therefore, applying Corollary D.0.2 we see that it suffices to show that for any $\varepsilon > 0$,

$$\lim_{n \to \infty} \sup_{l \in \mathbb{N}} P\left(\sup_{t \in [0,T]} \left| \int_0^t \langle (1 - P_n)\bar{X}^l(s), Z(s) \, dW(s) \rangle_H \right| > \varepsilon \right) = 0,$$

$$\lim_{n \to \infty} P\left(\sup_{t \in [0,T]} \left| \int_0^t \langle (1 - P_n)X(s), Z(s) \, dW(s) \rangle_H \right| > \varepsilon \right) = 0. \tag{4.42}$$

For any $n \in \mathbb{N}$, $\delta \in (0,1)$ and $N > 1$ by Lemma 2.4.2 and Corollary D.0.2 we have that

$$P\left(\sup_{t \in [0,T]} \left| \int_0^t \langle (1 - P_n)\bar{X}^l(s), Z(s) \, dW(s) \rangle_H \right| > \varepsilon \right)$$

$$\leq \frac{3\delta}{\varepsilon} + P\left(\int_0^T \|Z(s)^*(1 - P_n)(\bar{X}^l(s))\|_U^2 \, ds > \delta^2 \right)$$

$$\leq \frac{3\delta}{\varepsilon} + P\left(\sup_{t \in [0,T]} \|X(t)\|_H > N \right) + \frac{N^2}{\delta^2} \int_0^T \|(1 - P_n)Z(s)\|_{L_2(U,H)}^2 \, ds.$$

By first letting $n \to \infty$, and using Lebesgue's dominated convergence theorem, and then letting $N \to \infty$, and using Claim (b), and finally letting $\delta \to 0$, we prove the first equality in (4.42). Similarly, the second equality is proved.

Claim (d):   (4.30) holds for $t \in I$.

Fix $t \in I$. We may assume that $t \neq 0$. In this case for each sufficiently large $l \in \mathbb{N}$ there exists a unique $0 < i < k_l$ such that $t = t_i^l$. We have $X(t_j^l) \in V$ a.s. for all j. By Lemma 4.2.6 and (4.41) the sum of the first three terms in the right-hand side of (4.37) converges in probability to $2 \int_0^t {}_{V^*}\langle Y(s), \bar{X}(s) \rangle_V \, ds + 2 \int_0^t \langle X(s), Z(s) \, dW(s) \rangle_H$, as $l \to \infty$. Hence by Lemma 2.4.4

$$\|X(t)\|_H^2 - \|X(0)\|_H^2$$

$$= 2 \int_0^t {}_{V^*}\langle Y(s), \bar{X}(s) \rangle_V \, ds + 2 \int_0^t \langle X(s), Z(s) \, dW(s) \rangle_H + \langle M \rangle_t - \varepsilon_0,$$

where

$$\varepsilon_0 := P - \lim_{l \to \infty} \sum_{j=0}^{i-1} \|X(t_{j+1}^l) - X(t_j^l) - M(t_{j+1}^l) + M(t_j^l)\|_H^2$$

exists and "$P - \lim$" denotes limit in probability. So, to prove (4.30) for t as above, it suffices to show that $\varepsilon_0 = 0$. Since for any $\varphi \in V$,

$$\langle X(t_{j+1}^l) - X(t_j^l) - M(t_{j+1}^l) + M(t_j^l), \varphi \rangle_H = \int_{t_j^l}^{t_{j+1}^l} {}_{V^*}\langle Y(s), \varphi \rangle_V \, ds,$$

letting $\tilde{M}^l$ and $\bar{M}^l$ be defined as $\tilde{X}^l$ and $\bar{X}^l$ respectively, for M replacing X, we obtain for every $n \in \mathbb{N}$

$$\varepsilon_0 = P - \lim_{l \to \infty} \left(\int_0^t {}_{V^*}\langle Y(s), \tilde{X}^l(s) - \bar{X}^l(s) - P_n(\tilde{M}^l(s) - \bar{M}^l(s)) \rangle_V \, ds \right.$$

$$+ \langle \int_0^{t_1^l} Y(s) \, ds, \, X(0) \rangle_H - \sum_{j=0}^{i-1} \langle X(t_{j+1}^l) - X(t_j^l) - M(t_{j+1}^l) $$

$$\left. + M(t_j^l), (1 - P_n)(M(t_{j+1}^l) - M(t_j^l)) \rangle_H \right).$$

The weak continuity of X in H implies that $t \mapsto \int_0^t Y(s) ds$ is weakly continuous in H, since M is continuous in H. Hence the second term converges to zero as $l \to \infty$. Lemma 4.2.6 implies that $\int_0^t {}_{V^*}\langle Y(s), \tilde{X}^l(s) - \bar{X}^l(s) \rangle_V \, ds \to 0$ in probability as $l \to \infty$. Moreover, since $P_n M(s)$ is a continuous process in V, $\int_0^t {}_{V^*}\langle Y(s), P_n(\tilde{M}^l(s) - \bar{M}^l(s)) \rangle_V \, ds \to 0$ P-a.s. as $l \to \infty$. Thus, by the Cauchy–Schwarz inequality, Lemmas 2.4.1 and 2.4.4

$$\varepsilon_0 \leq \text{P-} \lim_{l \to \infty} \left(\sum_{j=0}^{i-1} \|X(t_{j+1}^l) - X(t_j^l) - M(t_{j+1}^l) + M(t_j^l)\|_H^2 \right)^{\frac{1}{2}}$$

$$\cdot \left(\sum_{j=0}^{i-1} \|(1 - P_n)(M(t_{j+1}^l) - M(t_j^l))\|_H^2 \right)^{\frac{1}{2}}$$

$$= \varepsilon_0^{1/2} \int_0^t \|(1 - P_n)Z(s)\|_{L^2(U,H)}^2 \, ds,$$

which goes to zero as $n \to \infty$ again by Lebesgue's dominated convergence theorem. Therefore, $\varepsilon_0 = 0$.

Claim (e):   (4.30) holds for all $t \in [0, T] \backslash I$.

Take $\Omega' \in \mathcal{F}$ with full probability such that the limit in (4.41) is a pointwise limit on Ω' for some subsequence (denoted again by $l \to \infty$) and (4.30) holds for all $t \in I$ on Ω'. If $t \notin I$, for any $l \in \mathbb{N}$ there exists a unique $j(l) < k_l$ such that $t \in]t^l_{j(l)}, t^l_{j(l)+1}]$. Letting $t(l) := t^l_{j(l)}$, we have $t(l) \uparrow t$ as $l \uparrow \infty$. Using that $\|u - v\|^2_H = \|u\|^2_H - \|v\|^2_H - 2\,_{V^*}\langle u - v,\ v \rangle_V$ for all $u, v \in V$, by (4.30) for $t \in I$, for all $l > m$ we have on Ω'

$$
\begin{aligned}
\|X(t(l)) - X(t(m))\|^2_H &= \|X(t(l))\|^2_H - \|X(t(m))\|^2_H \\
&\quad - 2 \,_{V^*}\Big\langle \int_{t(m)}^{t(l)} Y(s)\,\mathrm{d}s + \int_{t(m)}^{t(l)} Z(s)\,\mathrm{d}W(s),\ X(t(m)) \Big\rangle_V \\
&= 2 \int_{t(m)}^{t(l)} {}_{V^*}\langle Y(s), \bar{X}(s) - X(t(m)) \rangle_V \ \mathrm{d}s \\
&\quad + 2 \int_{t(m)}^{t(l)} \langle X(s) - X(t(m)), Z(s)\,\mathrm{d}W(s) \rangle_H + \langle M \rangle_{t(l)} - \langle M \rangle_{t(m)} \\
&= 2 \int_{t(m)}^{t(l)} {}_{V^*}\langle Y(s), \bar{X}(s) - \bar{X}^m(s) \rangle_V \ \mathrm{d}s \\
&\quad + 2 \int_{t(m)}^{t(l)} \langle X(s) - \bar{X}^m(s), Z(s)\,\mathrm{d}W(s) \rangle_H + \langle M \rangle_{t(l)} - \langle M \rangle_{t(m)},
\end{aligned}
$$
$$(4.43)$$

where in the second equality we tacitly assumed that m is so big that $t(m) > 0$. The second summand is dominated by

$$
4 \sup_{t \in [0,T]} \left| \int_0^t \langle X(s) - \bar{X}^m(s), Z(s)\,\mathrm{d}W(s) \rangle_H \right|.
$$

Thus, by the continuity of $\langle M \rangle_t$ and (4.41) (holding pointwise on Ω'), we have that

$$
\lim_{m \to \infty} \sup_{l > m} \left\{ 2\left| \int_0^T \mathbb{1}_{[t(m), t(l)]}(s) \langle X(s) - \bar{X}^m(s), Z(s)\,\mathrm{d}W(s) \rangle_H \right| \right.
$$
$$
\left. + |\langle M \rangle_{t(l)} - \langle M \rangle_{t(m)}| \right\} = 0 \tag{4.44}
$$

holding on Ω'. Furthermore, by Lemma 4.2.6, selecting another subsequence if necessary, we have for some $\Omega'' \in \mathcal{F}$ with full probability and $\Omega'' \subset \Omega'$, that on Ω''

$$
\lim_{m \to \infty} \int_0^T |\,_{V^*}\langle Y(s), \bar{X}(s) - \bar{X}^m(s) \rangle_V|\ \mathrm{d}s = 0.
$$

Since

$$\sup_{l>m} \int_{t(m)}^{t(l)} |_{V^*}\langle Y(s), \bar{X}(s) - \bar{X}^m(s)\rangle_V| \, ds$$

$$\le \int_0^T |_{V^*}\langle Y(s), \bar{X}(s) - \bar{X}^m(s)\rangle_V| \, ds,$$

we have that on Ω''

$$\lim_{m\to\infty} \sup_{l>m} \int_{t(m)}^{t(l)} {}_{V^*}\langle Y(s), \bar{X}(s) - \bar{X}^m(s)\rangle_V \, ds = 0.$$

Combining this with (4.43) and (4.44), we conclude that

$$\lim_{m\to\infty} \sup_{l\ge m} \|X(t(l)) - X(t(m))\|_H^2 = 0$$

holds on Ω''. Thus, $(X(t(l)))_{l\in\mathbb{N}}$ converges in H on Ω''. Since we know that $X(t(l)) \to X(t)$ in V^*, it converges to $X(t)$ strongly in H on Ω''. Therefore, since (4.30) holds on Ω'' for $t(l)$, letting $l \to \infty$, we obtain (4.30) on Ω'' also for all $t \notin I$.

Claim (f): X is strongly continuous in H.

Since the right-hand side of (4.30) is on Ω'' continuous in $t \in [0, T]$, so must be its left-hand side, i.e. $t \mapsto \|X(t)\|_H$ is continuous on $[0, T]$. Therefore, the weak continuity of $X(t)$ in H implies its strong continuity in H.   □

Remark 4.2.8

(i) In the situation of Theorem 4.2.5 we have

$$E(\|X(t)\|_H^2)$$

$$=E(\|X_0\|_H^2) + \int_0^t E(2\,_{V^*}\langle Y(s), \bar{X}(s)\rangle_V + \|Z(s)\|_{L_2(U,H)}^2) \, ds, \quad t \in [0, T].$$

$$(4.45)$$

Proof Let $M(t)$, $t \in [0, T]$, denote the real valued local martingale in (4.30) and let τ_l, $l \in \mathbb{N}$, be $(\mathcal{F}_t)$-stopping times such that $M(t \wedge \tau_l)$, $t \in [0, T]$, is a martingale and $\tau_l \uparrow T$ as $l \to \infty$. Then for all $l \in \mathbb{N}$, $t \in [0, T]$, we have

$$E(\|X(t \wedge \tau_l)\|_H^2)$$

$$=E(\|X_0\|_H^2) + \int_0^t E(1_{[0,\tau_l]}(s)[2\,_{V^*}\langle Y(s), \bar{X}(s)\rangle_V + \|Z(s)\|_{L_2(U,H)}^2]) \, ds. \quad (4.46)$$

Using Claim (b) from the proof of Theorem 4.2.5 and the fact that the integrands on the right-hand side of (4.45) are $dt \otimes P$-integrable we can apply Lebesgue's dominated convergence theorem to obtain the assertion. $\square$

(ii) We note that in (4.30) the stochastic integral is always a (global) martingale. This immediately follows from Proposition D.01(ii), since

$$E(\sup_{t \in [0,T]} \|X(t)\|_H^2) < \infty.$$

(iii) Suppose $\alpha \in (1, 2)$. Then by Remark 4.2.7(ii) and by tracing through the proof of Theorem 4.2.5 one sees that the latter remains valid if we replace K by the space K_0 defined Remark 4.2.7(ii) and K^* by K_0^*. The reason is that clearly

$$K_0^* = L_W^{\frac{\alpha}{\alpha-1}}([0, T] \times \Omega, dt \otimes P; V^*) + L_W^2([0, T] \times \Omega; dt \otimes P; H)$$

$$\subset L_W^q([0, T] \times \Omega, dt \otimes P; V^*) \text{ with } q := \min(2, \frac{\alpha}{\alpha - 1}).$$

Now we turn to the proof of Theorem 4.2.4. We first need some preparations. Let $\{e_i | i \in \mathbb{N}\} \subset V$ be an orthonormal basis of H and let $H_n := \mathrm{span}\{e_1, \ldots, e_n\}$ such that $\mathrm{span}\{e_i | i \in \mathbb{N}\}$ is dense in V. Let $P_n : V^* \to H_n$ be defined by

$$P_n y := \sum_{i=1}^{n} {}_{V^*}\langle y, e_i \rangle_V e_i, \quad y \in V^*. \tag{4.47}$$

By (4.1.2), $P_n|_H$ is just the orthogonal projection onto H_n in H. Furthermore,

$${}_{V^*}\langle z, P_n y \rangle_V = {}_{V^*}\langle y, P_n z \rangle_V \text{ for all } y, z \in V^*,$$

and

$${}_{V^*}\langle P_n y, v \rangle_V = {}_{V^*}\langle y, P_n v \rangle_V \text{ for all } y \in V^*, \ v \in V.$$

Let $\{g_i | i \in \mathbb{N}\}$ be an orthonormal basis of U and set

$$W^{(n)}(t) := \sum_{i=1}^{n} \langle W(t), g_i \rangle_U \, g_i \, .$$

Here for $g \in U$ we define

$$\langle W(t), g \rangle_U := \int_0^t \langle g, \cdot \rangle_U \, dW(s), \ t \in [0, T],$$

where the stochastic integral is well-defined by Sect. 2.5.2 with $H := \mathbb{R}$, since the map $u \mapsto \langle g, u \rangle_U, u \in U$, is in $L_2(U, \mathbb{R})$.

For each finite $n \in \mathbb{N}$ we consider the following stochastic equation on H_n:

$$dX^{(n)}(t)$$
$$= P_n A(t, X^{(n)}(t)) \, dt + P_n B(t, X^{(n)}(t)) \, dW^{(n)}(t), \tag{4.48}$$

where $X^{(n)}(0) := P_n X_0$. It is easily seen (cf. in particular Remark 4.1.1) that we are in the situation of Theorem 3.1.1 which implies that (4.48) has a unique continuous strong solution. Let

$$J := L^2([0, T] \times \Omega, \, dt \otimes P; L_2(U, H)). \tag{4.49}$$

To construct the solution to (4.26), we need the following lemma.

Lemma 4.2.9 *Under the assumptions in Theorem 4.2.4, there exists a $C \in \,]0, \infty[$ such that*

$$\|X^{(n)}\|_K + \|A(\cdot, X^{(n)})\|_{K^*} + \sup_{t \in [0,T]} E\|X^{(n)}(t)\|_H^2 \leq C \tag{4.50}$$

for all $n \in \mathbb{N}$.

Proof By the finite-dimensional Itô formula we have P-a.s.

$$\|X^{(n)}(t)\|_H^2 = \|X_0^{(n)}\|_H^2 + \int_0^t \Big(2\, {}_{V^*}\langle A(s, X^{(n)}(s)), X^{(n)}(s)\rangle_V$$
$$+ \|Z^{(n)}(s)\|_{L_2(U_n, H)}^2 \Big) \, ds + M^{(n)}(t), \ t \in [0, T],$$

where $Z^{(n)}(s) := P_n B(s, X^{(n)}(s))$, $U_n := \text{span}\{e_1, \dots, e_n\}$ and

$$M^{(n)}(t) := 2 \int_0^t \langle X^{(n)}(s), P_n B(s, X^{(n)}(s)) \, dW^{(n)}(s)\rangle_H, \ t \in [0, T],$$

is a local martingale. Let τ_l, $l \in \mathbb{N}$, be $(\mathcal{F}_t)$-stopping times such that $\|X^{(n)}(t \wedge \tau_l)(\omega)\|_V$ is bounded uniformly in $(t, \omega) \in [0, T] \times \Omega$, $M^{(n)}(t \wedge \tau_l)$, $t \in [0, T]$, is a martingale for each $l \in \mathbb{N}$ and $\tau_l \uparrow T$ as $l \to \infty$. Then for all $l \in \mathbb{N}$, $t \in [0, T]$

$$E\left(\|X^{(n)}(t \wedge \tau_l)\|_H^2\right)$$
$$= E(\|X_0^{(n)}\|_H^2) + \int_0^t E\Big(1_{[0,\tau_l]}(s)(2\, {}_{V^*}\langle A(s, X^{(n)}(s)), X^{(n)}(s)\rangle_V$$
$$+ \|Z^{(n)}(s)\|_{L_2(U_n, H)}^2)\Big) \, ds.$$

Hence using the product rule we obtain

$$E(e^{-c_1 t}\|X^{(n)}(t \wedge \tau_l)\|_H^2)$$

$$= E(\|X_0^{(n)}\|_H^2) + \int_0^t E(\|X^{(n)}(s \wedge \tau_l)\|_H^2)\, d(e^{-c_1 s})$$

$$+ \int_0^t e^{-c_1 s}\, d(E(\|X^{(n)}(s \wedge \tau_l)\|_H^2))$$

$$= E(\|X_0^{(n)}\|_H^2) - \int_0^t c_1 E(\|X^{(n)}(s \wedge \tau_l)\|_H^2)e^{-c_1 s}\, ds$$

$$+ \int_0^t e^{-c_1 s} E\Big(1_{[0,\tau_l]}(s)(2\,_{V^*}\langle A(s, X^{(n)}(s)), X^{(n)}(s)\rangle_V$$

$$+ \|Z^{(n)}(s)\|_{L_2(U_n, H)}^2)\Big)\, ds. \tag{4.51}$$

Applying (H3) we arrive at

$$E(e^{-c_1 t}\|X^{(n)}(t \wedge \tau_l)\|_H^2) + \int_0^t c_1 E(\|X^{(n)}(s \wedge \tau_l)\|_H^2)e^{-c_1 s}\, ds$$

$$+ c_2 \int_0^t E(1_{[0,\tau_l]}(s)\|X^{(n)}(s \wedge \tau_l)\|_V^\alpha)e^{-c_1 s}\, ds$$

$$\leqslant E(\|X_0^{(n)}\|_H^2) + \int_0^t c_1 E(\|X^{(n)}(s)\|_H^2)e^{-c_1 s}\, ds + \int_0^T E(|f(s)|)\, ds,$$

where by the definition of τ_l, $l \in \mathbb{N}$, all terms are finite. Now taking $l \to \infty$ and applying Fatou's lemma we get

$$E(e^{-c_1 t}\|X^{(n)}(t)\|_H^2) + c_2 E\left(\int_0^t \|X^{(n)}(s)\|_V^\alpha e^{-c_1 s}\, ds \right)$$

$$\leqslant E(\|X_0^{(n)}\|_H^2) + E\left(\int_0^T |f(s)|\, ds \right)$$

for all $t \in [0, T]$. Here we used that by (3.8) the subtracted term is finite. Since $\|X_0^{(n)}\|_H \leqslant \|X_0\|_H$, now the assertion follows for the first and third summand in (4.50). For the remaining summand the assertion then follows by (H4). □

Proof of Theorem 4.2.4 By the reflexivity of K, Lemma 4.2.9 and Remark 4.1.1, part 1, we have, for a subsequence $n_k \to \infty$:

(i) $X^{(n_k)} \to \bar{X}$ weakly in K and weakly in $L^2([0, T] \times \Omega;\, dt \otimes P; H)$.
(ii) $Y^{(n_k)} := A(\cdot, X^{(n_k)}) \to Y$ weakly in K^*.

(iii) $Z^{(n_k)} := B(\cdot, X^{(n_k)}) \to Z$ weakly in J and hence

$$\int_0^{\cdot} P_{n_k} B(s, X^{(n_k)}(s)) \, dW^{(n_k)}(s) \to \int_0^{\cdot} Z(s) \, dW(s)$$

in $\mathcal{M}_T^2(H)$ and, therefore, also weakly* in $L^{\infty}([0, T], \quad dt; L^2(\Omega, P; H))$ (equipped with the supremum norm).

Here the second part in (iii) follows since also $P_{n_k} B(\cdot, X^{(n_k)}) \tilde{P}_{n_k} \to Z$ weakly in J, where $\tilde{P}_n$ is the orthogonal projection onto $\text{span}\{g_1, \cdots, g_n\}$ in U, since

$$\int_0^{\cdot} P_{n_k} B(s, X^{(n_k)}(s)) \, dW^{(n_k)}(s) = \int_0^{\cdot} P_{n_k} B(s, X^{n_k}(s)) \tilde{P}_{n_k} \, dW(s)$$

and since a bounded linear operator between two Banach spaces is trivially weakly continuous. Since the approximants are progressively measurable, so are (the $dt \otimes P$-versions) $\bar{X}, Y$ and Z.

Thus from (4.48) for all $v \in \bigcup_{n \geqslant 1} H_n (\subset V)$, $\varphi \in L^{\infty}([0, T] \times \Omega)$ by Fubini's theorem we get

$$E \left(\int_0^T {}_{V^*}\langle \bar{X}(t), \varphi(t) v \rangle_V \, dt \right)$$

$$= \lim_{k \to \infty} E \left(\int_0^T {}_{V^*}\langle X^{(n_k)}(t), \varphi(t) v \rangle_V \, dt \right)$$

$$= \lim_{k \to \infty} E \bigg(\int_0^T {}_{V^*}\langle X_0^{(n_k)}, \varphi(t) v \rangle_V \, dt$$

$$+ \int_0^T \int_0^t {}_{V^*}\langle P_{n_k} Y^{(n_k)}(s), \varphi(t) v \rangle_V \, ds \, dt$$

$$+ \int_0^T \left\langle \int_0^t P_{n_k} Z^{(n_k)}(s) \, dW^{(n_k)}(s), \varphi(t) v \right\rangle_H \, dt \bigg)$$

$$= \lim_{k \to \infty} \bigg[E \left(\langle X_0^{(n_k)}, v \rangle_H \int_0^T \varphi(t) \, dt \right)$$

$$+ E \left(\int_0^T {}_{V^*}\langle Y^{(n_k)}(s), \int_s^T \varphi(t) \, dt \, v \rangle_V \, ds \right)$$

$$+ \int_0^T E \left(\left\langle \int_0^t Z^{(n_k)}(s) \, dW^{(n_k)}(s), \varphi(t) v \right\rangle_H \right) dt \bigg]$$

$$= E \left(\int_0^T {}_{V^*}\langle X_0 + \int_0^t Y(s) \, ds + \int_0^t Z(s) \, dW(s), \varphi(t) v \rangle_V \, dt \right).$$

Therefore, defining

$$X(t) := X_0 + \int_0^t Y(s)\, ds + \int_0^t Z(s)\, dW(s), \quad t \in [0, T], \tag{4.52}$$

we have $X = \bar{X}$ $dt \otimes P$-a.e.

Now Theorem 4.2.5 applies to X in (4.52), so X is continuous in H and

$$E\left(\sup_{t \le T} \|X(t)\|_H^2 \right) < \infty.$$

Thus, it remains to verify that

$$B(\cdot, \bar{X}) = Z, \quad A(\cdot, \bar{X}) = Y, \quad dt \otimes P\text{-a.e.} \tag{4.53}$$

To this end, we first note that for any nonnegative $\psi \in L^\infty([0, T],\, dt; \mathbb{R})$ it follows from (i) that

$$E\left(\int_0^T \psi(t) \|\bar{X}(t)\|_H^2\, dt \right)$$

$$= \lim_{k \to \infty} E\left(\int_0^T \langle \psi(t)\bar{X}(t), X^{(n_k)}(t)\rangle_H\, dt \right)$$

$$\le \left(E \int_0^T \psi(t)\|\bar{X}(t)\|_H^2\, dt \right)^{1/2} \liminf_{k \to \infty} \left(E \int_0^T \psi(t)\|X^{(n_k)}(t)\|_H^2\, dt \right)^{1/2} < \infty.$$

Since $X = \bar{X}$ $dt \otimes P$-a.e., this implies

$$E\left(\int_0^T \psi(t)\|X(t)\|_H^2\, dt \right) \le \liminf_{k \to \infty} E\left(\int_0^T \psi(t)\|X^{(n_k)}(t)\|_H^2\, dt \right). \tag{4.54}$$

By (4.52) using Remark 4.2.8, the product rule and Fubini's theorem we obtain that

$$E\left(e^{-ct}\|X(t)\|_H^2 \right) - E\left(\|X_0\|_H^2 \right)$$

$$= E\left(\int_0^t e^{-cs}\left(2\, _{V^*}\langle Y(s), \bar{X}(s)\rangle_V + \|Z(s)\|_{L_2(U,H)}^2 - c\|X(s)\|_H^2 \right) ds \right). \tag{4.55}$$

Furthermore, for any $\phi \in K \cap L^2([0, T] \times \Omega,\, dt \otimes P; H)$, taking $l \to \infty$ in (4.51) with c_1 replaced by c and $t \wedge \tau_l$ replaced by t, we obtain

$$E\left(e^{-ct}\|X^{(n_k)}(t)\|_H^2 \right) - E\left(\|X_0^{(n_k)}\|_H^2 \right)$$

$$= E\left(\int_0^t e^{-cs}\left(2\, _{V^*}\langle A(s, X^{(n_k)}(s)), X^{(n_k)}(s)\rangle_V \right.$$

$$+ \|P_{n_k} B(s, X^{(n_k)}(s)) \tilde{P}_{n_k}\|^2_{L_2(U,H)} - c\|X^{(n_k)}(s)\|^2_H) \, ds \Big)$$

$$\leq E\Big(\int_0^t e^{-cs} \Big(2 \, {}_{V^*}\langle A(s, X^{(n_k)}(s)), X^{(n_k)}(s)\rangle_V$$

$$+ \|B(s, X^{(n_k)}(s))\|^2_{L_2(U,H)} - c\|X^{(n_k)}(s)\|^2_H) \, ds \Big)$$

$$= E\Big(\int_0^t e^{-cs} \Big(2 \, {}_{V^*}\langle A(s, X^{(n_k)}(s)) - A(s, \phi(s)), X^{(n_k)}(s) - \phi(s)\rangle_V$$

$$+ \|B(s, X^{(n_k)}(s)) - B(s, \phi(s))\|^2_{L_2(U,H)} - c\|X^{(n_k)}(s) - \phi(s)\|^2_H \Big) \, ds$$

$$+ E\Big(\int_0^t e^{-cs} \Big(2 \, {}_{V^*}\langle A(s, \phi(s)), X^{(n_k)}(s)\rangle_V$$

$$+ 2 \, {}_{V^*}\langle A(s, X^{(n_k)}(s)) - A(s, \phi(s)), \phi(s)\rangle_V$$

$$- \|B(s, \phi(s))\|^2_{L_2(U,H)} + 2\langle B(s, X^{(n_k)}(s)), B(s, \phi(s))\rangle_{L_2(U,H)}$$

$$- 2c\langle X^{(n_k)}(s), \phi(s)\rangle_H + c\|\phi(s)\|^2_H \Big) \, ds \Big). \tag{4.56}$$

Note that by (H2) the first of the two summands above is negative. Hence by letting $k \to \infty$ we conclude by (i)–(iii), Fubini's theorem, and (4.54) that for every nonnegative $\psi \in L^\infty([0, T], \, dt; \mathbb{R})$

$$E\left(\int_0^T \psi(t)(e^{-ct}\|X(t)\|^2_H - \|X_0\|^2_H) \, dt \right)$$

$$\leq E\Big(\int_0^T \psi(t)\Big(\int_0^t e^{-cs}\Big[2 \, {}_{V^*}\langle A(s, \phi(s)), \bar{X}(s)\rangle_V + 2 \, {}_{V^*}\langle Y(s)$$

$$- A(s, \phi(s)), \phi(s)\rangle_V - \|B(s, \phi(s))\|^2_{L_2(U,H)} + 2\langle Z(s), B(s, \phi(s))\rangle_{L_2(U,H)}$$

$$- 2c\langle X(s), \phi(s)\rangle_H + c\|\phi(s)\|^2_H \Big] \, ds \Big) \, dt \Big).$$

Inserting (4.55) for the left-hand side and rearranging we arrive at

$$0 \geq E\Big(\int_0^T \psi(t)\Big(\int_0^t e^{-cs}\Big[2 \, {}_{V^*}\langle Y(s) - A(s, \phi(s)), \bar{X}(s) - \phi(s)\rangle_V$$

$$+ \|B(s, \phi(s)) - Z(s)\|^2_{L_2(U,H)} - c\|X(s) - \phi(s)\|^2_H \Big] \, ds \Big) \, dt \Big). \tag{4.57}$$

Taking $\phi = \bar{X}$ we obtain from (4.57) that $Z = B(\cdot, \bar{X})$. Finally, first applying (4.57) to $\phi = \bar{X} - \varepsilon \tilde{\phi} \, v$ for $\varepsilon > 0$ and $\tilde{\phi} \in L^\infty([0, T] \times \Omega, \, dt \otimes P; \mathbb{R})$, $v \in V$, then dividing both sides by ε and letting $\varepsilon \to 0$, by Lebesgue's dominated convergence theorem, (H1) and (H4), we obtain

$$0 \geq E\left(\int_0^T \psi(t)\left(\int_0^t e^{-cs}\tilde{\phi}(s) \, _{V^*}\langle Y(s) - A(s, \bar{X}(s)), v\rangle_V \, \mathrm{d}s\right) \mathrm{d}t\right).$$

By the arbitrariness of ψ, $\tilde{\phi}$ and v, we conclude that $Y = A(\cdot, \bar{X})$. This completes the existence proof.

The uniqueness is a consequence of the following proposition.   □

Proposition 4.2.10 *Consider the situation of Theorem 4.2.4 and let* X, Y *be two solutions. Then for* $c \in \mathbb{R}$ *as in (H2)*

$$E(\|X(t) - Y(t)\|_H^2) \leq e^{ct} E(\|X(0) - Y(0)\|_H^2) \text{ for all } t \in [0, T]. \tag{4.58}$$

Proof We first note that by our definition of solution (cf. Definition 4.2.1) and by Remark 4.1.1, part 1 we can apply Remark 4.2.8 to $X - Y$ and the product rule to obtain for $t \in [0, T]$

$$\begin{aligned}
E(e^{-ct}\|X(t) - Y(t)\|_H^2) &= E(\|X_0 - Y_0\|_H^2) \\
&\quad + \int_0^t (E(2 \, _{V^*}\langle A(s, \bar{X}(s)) - A(s, \bar{Y}(s)), \bar{X}(s) - \bar{Y}(s)\rangle_V \\
&\quad + \|B(s, \bar{X}(s)) - B(s, \bar{Y}(s))\|_{L_2(U,H)}^2 \\
&\quad - cE(\|X(s) - Y(s)\|_H^2))e^{-cs} \, \mathrm{d}s \\
&\leq E(\|X(0) - Y(0)\|_H^2),
\end{aligned}$$

where we used (H2) in the last step. Applying Gronwall's lemma we obtain the assertion.   □

Remark 4.2.11 Let $s \in [0, T]$ and $X_s \in L^2(\Omega, \mathcal{F}_s, P; H)$. Consider the equation

$$X(t) = X_s + \int_s^t A(u, \bar{X}(u)) \, \mathrm{d}u + \int_s^t B(u, \bar{X}(u)) \, \mathrm{d}W(u), \quad t \in [s, T] \tag{4.59}$$

with underlying Wiener process $W(t) - W(s)$, $t \in [s, T]$, and filtration $(\mathcal{F}_t)_{t \geq s}$, i.e. we just start our time at s. We define the notion of solution for (4.59) analogously to Definition 4.2.1. Then all results above in the case $s = 0$ carry over to this more general case. In particular, there exists a unique solution with initial condition X_s denoted by $X(t, s, X_s)$, $t \in [s, T]$. Let $0 \leq r \leq s \leq T$. Then for $X_r \in L^2(\Omega, \mathcal{F}_r, P; H)$

$$X(t, r, X_r) = X(t, s, X(s, r, X_r)), \quad t \in [s, T] \text{ P-a.e.} \tag{4.60}$$

Indeed, we have

$$X(t, r, X_r) = X_r + \int_r^t A(u, \bar{X}(u, r, X_r)) \, du + \int_r^t B(u, \bar{X}(u, r, X_r)) \, dW(u)$$

$$= X(s, r, X_r) + \int_s^t A(u, \bar{X}(u, r, X_r)) \, du$$

$$+ \int_s^t B(u, \bar{X}(u, r, X_r)) \, dW(u), \quad t \in [s, T].$$

But by definition $X(t, s, X(s, r, X_r))$, $t \in [s, T]$, satisfies the same equation. So, (4.60) follows by uniqueness. Furthermore, if for $s \in [0, T]$, $X_s = x$ for some $x \in H$ and A and B are independent of $\omega \in \Omega$, then by construction $X(t, s, x)$ obviously is independent of $\mathcal{F}_s$ for all $t \in [s, T]$, since so are collections of increments of $W(t)$, $t \in [s, T]$.

4.3 Markov Property and Invariant Measures

Now we are going to prove some qualitative results about the solutions of (4.2.1) or (4.59) and about their transition probabilities, i.e. about

$$p_{s,t}(x, \, dy) := P \circ (X(t, s, x))^{-1}(\, dy), 0 \leqslant s \leqslant t \leqslant T, x \in H. \tag{4.61}$$

As usual we set for $\mathcal{B}(H)$-measurable $F : H \to \mathbb{R}$, and $t \in [s, T]$, $x \in H$

$$p_{s,t}F(x) := \int F(y) p_{s,t}(x, \, dy),$$

provided F is $p_{s,t}(x, \, dy)$-integrable.

Remark 4.3.1 The measures $p_{s,t}(x, \, dy)$, $0 \leqslant s \leqslant t \leqslant T$, $x \in H$, could in principle depend on the chosen Wiener process and the respective filtration. However, the construction of our solutions $X(t, s, x)$, $t \in [s, T]$, suggests that this is not the case. This can be rigorously proved in several ways. It is, for example, a consequence of the famous Yamada–Watanabe theorem which is included in Appendix E below.

Proposition 4.3.2 *Consider the situation of Theorem 4.2.4. Let $F : H \to \mathbb{R}$ be Lipschitz with*

$$\mathrm{Lip}(F) := \sup_{x,y \in H, x \neq y} \frac{|F(x) - F(y)|}{\|x - y\|_H}(< \infty)$$

denoting its Lipschitz constant. Then for all $0 \leqslant s \leqslant t \leqslant T$

$$p_{s,t}|F|(x) < \infty \text{ for all } x \in H$$

and for all $x, y \in H$

$$|p_{s,t}F(x) - p_{s,t}F(y)| \leqslant e^{\frac{c}{2}(t-s)}\mathrm{Lip}(F)\,\|x - y\|_H, \tag{4.62}$$

where c is as in (H2).

Proof Clearly, for all $x \in H$

$$|F(x)| \leqslant |F(0)| + \mathrm{Lip}(F)\,\|x\|_H,$$

and thus for all $0 \leqslant s \leqslant t \leqslant T$

$$\begin{aligned}
p_{s,t}|F|(x) &= E(|F|(X(t, s, x))) \\
&\leqslant |F(0)| + \mathrm{Lip}(F)\,E(\|X(t, s, x)\|_H) \\
&\leqslant |F(0)| + \mathrm{Lip}(F)\left(E\left(\sup_{t \in [s,T]}\|X(t, s, x)\|_H^2\right)\right)^{1/2} \\
&< \infty.
\end{aligned}$$

Furthermore, for $x, y \in H$ by (the "started at s" analogue of) (4.58)

$$\begin{aligned}
|p_{s,t}F(x) - p_{s,t}F(y)| &\leqslant E(|F(X(t, s, x)) - F(X(t, s, y)))|) \\
&\leqslant \mathrm{Lip}(F)\,E(\|X(t, s, x) - X(t, s, y)\|_H) \\
&\leqslant \mathrm{Lip}(F)\,e^{\frac{c}{2}(t-s)}\|x - y\|_H.
\end{aligned}$$

$\square$

Proposition 4.3.3 *Consider the situation of Theorem 4.2.4 and, in addition, assume that both A and B as well as f and g in (H3),(H4) respectively, are independent of $\omega \in \Omega$. Then any solution $X(t)$, $t \in [r, T]$, of (4.59) (with r replacing s) is Markov in the following sense:*
for every bounded, $\mathcal{B}(H)$-measurable $G : H \to \mathbb{R}$, and all $s, t \in [r, T]$, $s \leqslant t$

$$E(G(X(t))|\mathcal{F}_s)(\omega) = E(G(X(t, s, X(s)(\omega)))) \text{ for } P\text{-a.e. } \omega \in \Omega. \tag{4.63}$$

Proof Let $s, t \in [r, T]$, $s \leq t$. Then by Theorem E.0.8 in Appendix E there exists an $F \in \hat{\mathcal{E}}$ as in Definition E.0.5 such that for P-a.e. $\omega \in \Omega$

$$X(t)(\omega) = F_{P \circ X(s)^{-1}}(X(s)(\omega), \tilde{W}(\omega))(t)$$

with $\tilde{W} := W - W(s)$ and W is defined by (2.12) as a Q_1-Wiener process on U_1. Then for every bounded, $\mathcal{B}(H)$-measurable $G : H \to \mathbb{R}$ and P-a.e. $\omega \in \Omega$

$$E[G(X(t))|\mathcal{F}_s](\omega)$$

$$=E[G(F_{P\circ X(s)^{-1}}(X(s)(\omega),\ \tilde{W})(t))]$$

$$=E[G(F_{\delta_{X(s)(\omega)}}(X(s)(\omega),\ \tilde{W})(t))]$$

$$=E[G(X(t,s,X(s)(\omega)))],$$

where in the last equality we applied Theorem E.0.8 again and in the third equality we used the fact that by the definition of $\hat{\mathcal{E}}$, for $P \circ X(s)^{-1}$-a.e. $x \in H$

$$F_\mu(x,\ \cdot\ ) = F(x,\ \cdot\ ) = F_{\delta_x}(x,\ \cdot\ ) = P^{Q_1}\text{-a.e.}$$

□

Corollary 4.3.4 *Consider the situation of Proposition 4.3.3 and let $0 \leqslant r \leqslant s \leqslant t \leqslant T$. Then we have ("Chapman–Kolmogorov equations")*

$$p_{r,s}p_{s,t} = p_{r,t}, \tag{4.64}$$

i.e. for $F : H \to \mathbb{R}$, bounded and $\mathcal{B}(H)$-measurable, $x \in H$,

$$p_{r,s}(p_{s,t}F)(x) = p_{r,t}F(x).$$

Proof For $F : H \to \mathbb{R}$ as above and $x \in H$ by Proposition 4.3.3 we have

$$p_{r,s}(p_{s,t}F)(x) = E(p_{s,t}F(X(s,r,x))) = \int E(F(X(t,s,X(s,r,x)(\omega))))P(\,d\omega)$$

$$= \int E(F(X(t,r,x))|\mathcal{F}_s)(\omega)P(\,d\omega)$$

$$= E(F(X(t,r,x))) = p_{r,t}F(x).$$

□

Now let us assume that in the situation of Theorem 4.2.4 both A and B as well as f and g in (H3), (H4) respectively are independent of $(t,\omega) \in [0,T] \times \Omega$ (so they particularly hold for all $T \in [0,\infty[$). Then again using the notation introduced in Remark 4.2.11 for $0 \leqslant s \leqslant t < \infty$ and $x \in H$ we have

$$X(t,s,x) = X^{\tilde{W}}(t-s,0,x)\ P\text{-a.e.}, \tag{4.65}$$

where $X^{\tilde{W}}(t, 0, x)$, $t \in [0, \infty[$, is the solution of

$$X(t) = x + \int_0^t A(\bar{X}(u))\, du + \int_0^t B(\bar{X}(u))\, d\tilde{W}(u)$$

and $\tilde{W} := W(\cdot + s) - W(s)$ with filtration $\mathcal{F}_{s+u}$, $u \in [0, \infty[$, which is again a Wiener process. To show this let us express the dependence of the solution $X(t, s, x)$, $t \in [s, \infty)$ of (4.59) with $X_s := x$ on the Wiener process W by writing $X^W(t, s, x)$ instead of $X(t, s, x)$ and similarly, $p_{s,t}^W(s, dy)$ instead of $p_{s,t}(x, dy)$. Then, for all $s, u \in [0, \infty[$

$$X^W(u + s, s, x)$$

$$= x + \int_s^{u+s} A(\bar{X}^W(u', s, x))\, du' + \int_s^t B(\bar{X}^W(u', s, x))\, dW(u')$$

$$= x + \int_0^u A(\bar{X}^W(u' + s, s, x))\, du' + \int_0^u B(\bar{X}^W(u' + s, s, x))\, d\tilde{W}(u').$$

So, by uniqueness the process $X^W(u+s, s, x)$, $u \in [0, \infty[$, must P-a.e. coincide with $X^{\tilde{W}}(u, 0, x)$, $u \in [0, \infty[$, and (4.65) follows with $u := t - s$. In particular, it follows that

$$p_{s,t}^W(x,\ dy) = P \circ (X^{\tilde{W}}(t-s, 0, x))^{-1}(\ dy) = p_{0,t-s}^{\tilde{W}}(x,\ dy) = p_{0,t-s}^W(x,\ dy) \qquad (4.66)$$

("time homogeneity"), where we used Remark 4.3.1 for the last equality.

Defining

$$p_t := p_{0,t}^W, \quad t \in [0, \infty[,$$

equality (4.64) for $r = 0$ and $s + t$ replacing t by (4.66) turns into

$$p_{s+t} = p_s p_t \text{ for } s, t \in [0, \infty[. \qquad (4.67)$$

For $x \in H$ we define

$$P_x := P \circ (X(\cdot, 0, x))^{-1}, \qquad (4.68)$$

i.e. P_x is the distribution of the solution to (4.2.1) with initial condition $x \in H$, defined as a measure on $C([0, \infty[, H)$. We equip $C([0, \infty[, H)$ with the σ-algebra

$$\mathcal{G} := \sigma(\pi_s | s \in [0, \infty[)$$

and filtration

$$\mathcal{G}_t := \sigma(\pi_s | s \in [0, t]), \ t \in [0, \infty[,$$

where $\pi_t(w) := w(t)$ for $w \in C([0,\infty[,H)$, $t \in [0,\infty[$.

Proposition 4.3.5 *Consider the situation of Theorem 4.2.4 and, in addition, assume that both A and B as well as f and g in (H3),(H4) respectively, are independent of $(t,\omega) \in [0,T] \times \Omega$ (so they particularly hold for all $T \in [0,\infty[$). Then the following assertions hold:*

1. $P_x, x \in H$, *form a time-homogenous Markov process on $C([0,\infty),H)$ with respect to the filtration $\mathcal{G}_t, t \in [0,\infty[$, i.e. for all $s,t \in [0,\infty[$, and all bounded, $\mathcal{B}(H)$-measurable $F : H \to \mathbb{R}$*

$$E_x(F(\pi_{t+s})|\mathcal{G}_s) = E_{\pi_s}(F(\pi_t)) \quad P_x - a.e., \tag{4.69}$$

where E_x and $E_x(\cdot|\mathcal{G}_s)$ denote expectation and conditional expectation with respect to P_x, respectively.
2. *Suppose $\dim H < \infty$. If there exist $\eta, f \in]0,\infty[$ such that*

$$2\,_{V^*}\langle A(v), v\rangle_V + \|B(v)\|^2_{L_2(U,H)} \leqslant -\eta\|v\|^2_H + f \quad for\ all\ v \in V, \tag{4.70}$$

("strict coercivity") then there exists an invariant measure μ for $(p_t)_{t\geqslant 0}$, i.e. μ is a probability measure on $(H, \mathcal{B}(H))$ such that

$$\int p_t F\,d\mu = \int F\,d\mu \quad for\ all\ t \in [0,\infty[\tag{4.71}$$

and all bounded, $\mathcal{B}(H)$-measurable $F : H \to \mathbb{R}$.

Proof

1. The right-hand side of (4.69) is $\mathcal{G}_s$-measurable by Proposition 4.3.2 and a monotone class argument. So, let $0 \leqslant t_1 < t_2 < \ldots < t_n \leqslant s$ and let $G : H^n \to \mathbb{R}$ be bounded and $\otimes_{i=1}^n \mathcal{B}(H)$-measurable. Then by (4.63) and (4.66)

$$E_x(G(\pi_{t_1},\ldots,\pi_{t_n})F(\pi_{t+s}))$$
$$= E(G(X(t_1,0,x),\ldots,X(t_n,0,x))F(X(t+s,0,x))$$
$$= E(G(X(t_1,0,x),\ldots,X(t_n,0,x))E(F(X(t+s,0,x))|\mathcal{F}_s))$$
$$= \int G(X(t_1,0,x)(\omega),\ldots,X(t_n,0,x)(\omega))$$
$$\quad E(F(X(t+s,s,X(s,0,x)(\omega))))P(d\omega)$$
$$= \int G(X(t_1,0,x)(\omega),\ldots,X(t_n,0,x)(\omega))$$
$$\quad E(F(X(t,0,X(s,0,x)(\omega))))P(d\omega)$$

$$= \int G(\pi_{t_1}(\omega), \ldots, \pi_{t_n}(\omega)) E(F(X(t, 0, \pi_s(\omega)))) P_x(\,d\omega)$$

$$= \int G(\pi_{t_1}(\omega), \ldots, \pi_{t_n}(\omega)) E_{\pi_s(\omega)}(F(\pi_t)) P_x(\,d\omega).$$

Since the functions $G(\pi_{t_1}, \ldots, \pi_{t_n})$ considered above generate $\mathcal{F}_s$, equality (4.69) follows.

2. Let δ_0 be the Dirac measure in $0 \in H$ considered as a measure on $(H, \mathcal{B}(H))$ and for $n \in \mathbb{N}$ define the Krylov–Bogoliubov measure

$$\mu_n := \frac{1}{n} \int_0^n \delta_0 p_t \, dt,$$

i.e. for $\mathcal{B}(H)$-measurable $F : H \to [0, \infty[$

$$\int F \, d\mu_n = \frac{1}{n} \int_0^n p_t F(0) \, dt.$$

Clearly, each μ_n is a probability measure. We first prove that $\{\mu_n | n \in \mathbb{N}\}$ is tight. By Remark 4.2.8 for any solution X to (4.2.1) applying the product rule and using (4.70) we get that

$$E(e^{\eta t} \|X(t)\|_H^2) = E(\|X(0)\|_H^2) + E\left(\int_0^t e^{\eta s} \left(2 \,_{V^*}\langle A(\bar{X}(s)), \bar{X}(s)\rangle_V \right.\right.$$

$$\left.\left. + \|B(\bar{X}(s))\|_{L_2(U,H)}^2 + \eta\|\bar{X}(s)\|_H^2\right) \, ds \right)$$

$$\leq E(\|X(0)\|_H^2) + f \int_0^t e^{\eta s} \, ds, \quad t \in [0, \infty[.$$

Therefore,

$$E(\|X(t)\|_H^2) \leq e^{-\eta t} E(\|X(0)\|_H^2) + \frac{f}{\eta}, \quad t \in [0, \infty[, \tag{4.72}$$

which in turn implies that

$$\int \|x\|_H^2 \mu_n(dx) = \frac{1}{n} \int_0^n E(\|X(t, 0, 0)\|_H^2) \, dt \leq \frac{f}{\eta} \quad \text{for all } n \in \mathbb{N}. \tag{4.73}$$

Hence by Chebychev's inequality

$$\sup_{n \in \mathbb{N}} \mu_n(\{\| \; \|_H^2 > R\}) \leq \frac{1}{R} \frac{f}{\eta} \to 0 \text{ as } R \to \infty. \tag{4.74}$$

Since $\dim H < \infty$, the closed balls $\{\| \ \|_H^2 \leq R\}$, $R \in {]}0, \infty[$, are compact. Hence by Prohorov's theorem there exists a probability measure μ and a subsequence $(\mu_{n_k})_{k\in\mathbb{N}}$ such that $\mu_{n_k} \to \mu$ weakly as $k \to \infty$.

Now let us prove that μ is invariant for $(p_t)_{t\geq 0}$. So, let $t \in [0, \infty[$ and let $F : H \to \mathbb{R}$ be bounded and $\mathcal{B}(H)$-measurable. By a monotone class argument we may assume that F is Lipschitz continuous. Then $p_t F$ is bounded and (Lipschitz) continuous by Proposition 4.3.2. Hence using (4.67) for the third equality below, we obtain

$$\int p_t F \, d\mu$$

$$= \lim_{k\to\infty} \int p_t F \, d\mu_{n_k}$$

$$= \lim_{k\to\infty} \frac{1}{n_k} \int_0^{n_k} p_s(p_t F)(0) \, ds$$

$$= \lim_{k\to\infty} \frac{1}{n_k} \int_0^{n_k} p_{s+t} F(0) \, ds$$

$$= \lim_{k\to\infty} \int F \, d\mu_{n_k} + \lim_{k\to\infty} \frac{1}{n_k} \int_{n_k}^{n_k+t} p_s F(0) \, ds - \lim_{k\to\infty} \frac{1}{n_k} \int_0^t p_s F(0) \, ds$$

$$= \int F \, d\mu, \tag{4.75}$$

since $|p_s F(0)| \leq \sup_{x\in H} |F(x)|$, so the second and third limits above are equal to zero. $\qquad\Box$

Remark 4.3.6 If $\dim H = \infty$, the above proof of Proposition 4.3.5, part 2 works up to and including (4.74). However, since closed balls are no longer compact, one can apply Prohorov's theorem only on a Hilbert space H_1 into which H is compactly embedded. So, let H_1 be a separable Hilbert space such that $H \subset H_1$ compactly and densely (e.g. take H_1 to be the completion of H in the norm

$$\|x\|_1 := \left[\sum_{i=1}^{\infty} \alpha_i \langle x, e_i \rangle_H^2 \right]^{1/2}, \ x \in H,$$

where $\alpha_i \in {]}0, \infty[$, $\sum_{i=1}^{\infty} \alpha_i < \infty$, and $\{e_i | i \in \mathbb{N}\}$ is an orthonormal basis of H); extending the measures μ_n by zero to $\mathcal{B}(H_1)$ we obtain that $\{\mu_n | n \in \mathbb{N}\}$ is tight on H_1. This extension is possible, since by Kuratowski's theorem $H \in \mathcal{B}(H_1)$ and $\mathcal{B}(H_1) \cap H = \mathcal{B}(H)$. Hence by Prohorov's theorem there exists a probability measure $\bar{\mu}$ on $(H_1, \mathcal{B}(H_1))$ and a subsequence $(\mu_{n_k})_{k\in\mathbb{N}}$ such that $\mu_{n_k} \to \bar{\mu}$ weakly on H_1 as $k \to \infty$. As in Exercise 4.2.3, part 1 one constructs a lower semicontinuous function $\Theta : H_1 \to [0, \infty]$ such that

$$\Theta := \begin{cases} \| \ \|_H & \text{on } H \\ +\infty & \text{on } H_1 \backslash H. \end{cases}$$

Then (4.73) implies that for $l_i, i \in \mathbb{N}$, as in Example 4.2.3, part 1,

$$
\begin{aligned}
\int_{H_1} \Theta^2(x) \bar{\mu}(\, dx) &= \lim_{N \to \infty} \lim_{M \to \infty} \int \sup_{i \leq N} \langle l_i, x \rangle_{H_1}^2 \wedge M \, \bar{\mu}(\, dx) \\
&= \sup_{M,N \in \mathbb{N}} \lim_{k \to \infty} \int \sup_{i \leq N} \langle l_i, x \rangle_{H_1}^2 \wedge M \, \mu_{n_k}(\, dx) \\
&\leq \liminf_{k \to \infty} \sup_{N,M \in \mathbb{N}} \int \sup_{i \leq N} \langle l_i, x \rangle_{H_1}^2 \wedge M \, \mu_{n_k}(\, dx) \\
&= \liminf_{k \to \infty} \int_H \|x\|_H^2 \, \mu_{n_k}(\, dx) \\
&\leq \frac{f}{\eta}.
\end{aligned}
$$

Hence $\Theta < \infty$ $\bar{\mu}$-a.e., so $\bar{\mu}(H) = 1$. Therefore, $\mu := \bar{\mu}\big|_{\mathcal{B}(H)}$ is a probability measure on $(H, \mathcal{B}(H))$.

Unfortunately, the last part of the proof of Proposition 4.3.5, which shows that μ is invariant, does not work. More precisely, for the first equality in (4.75) we need that $p_t F$ is continuous with respect to the same topology with respect to which $(\mu_{n_k})_{k \in \mathbb{N}}$ converges weakly, i.e. the topology on H_1. This is, however, weaker than that on H. So, unless we can construct H_1 in such a way that $p_t F$ has a continuous extension to H_1, the first equality in (4.75) may not hold.

So far, we have taken a positive time s as the starting time for our SDE (see Remark 4.2.11). In the case of coefficients independent of t and ω, it is also possible and convenient to consider negative starting times. For this, however, we need a Wiener process with negative time. To this end we recall that we can run a cylindrical Wiener process $W(t), t \in [0, \infty[$ on H (with positive time) backwards in time and again get a Wiener process. More precisely, for fixed $T \in [0, \infty[$ we have that $W(T - t) - W(T), t \in [0, T]$ is again a cylindrical Wiener process with respect to the filtration $\sigma(\{W(T - s) - W(T) | s \in [0, t]\}), t \in [0, T]$, and also with respect to the filtration $\sigma(\{W(r_2) - W(r_1) | r_1, r_2 \in [T - t, \infty[, r_2 \leq r_1\}), t \in [0, T]$, where the latter will be more convenient for us.

So, let A, B be independent of $(t, \omega) \in [0, T] \times \Omega$ and let $W^{(1)}(t), t \in [0, \infty[$, be another cylindrical Wiener process on $(\Omega, \mathcal{F}, P)$ with covariance operator $Q = I$, independent of $W(t), t \in [0, \infty[$. Define

$$\bar{W}(t) := \begin{cases} W(t), & \text{if } t \in [0, \infty[, \\ W^{(1)}(-t), & \text{if } t \in] -\infty, 0] \end{cases} \qquad (4.76)$$

with filtration

$$\bar{\mathcal{F}}_t := \bigcap_{s>t} \bar{\mathcal{F}}_s^{\circ}, \quad t \in \mathbb{R}, \tag{4.77}$$

where $\bar{\mathcal{F}}_s^{\circ} := \sigma(\{\bar{W}(r_2) - \bar{W}(r_1)|r_1, r_2 \in] - \infty, s], r_2 \geqslant r_1\}, \mathcal{N})$ and $\mathcal{N} := \{A \in \mathcal{F}|P(A) = 0\}$. As in the proof of Proposition 2.1.13 one shows that if $-\infty < s < t < \infty$, then $\bar{W}(t) - \bar{W}(s)$ is independent of $\bar{\mathcal{F}}_s$. Now for $s \in \mathbb{R}$ fixed consider the SDE

$$dX(t) = A(X(t))\,dt + B(X(t))\,d\bar{W}(t), \ t \in [s, \infty[. \tag{4.78}$$

Remark 4.3.7 Let $s \in \mathbb{R}$ and $X_s \in L^2(\Omega, \bar{\mathcal{F}}_s, P; H)$ and consider the integral version of (4.78)

$$X(t) = X_s + \int_s^t A(\bar{X}(u))\,du + \int_s^t B(\bar{X}(u))\,d\bar{W}(u), \quad t \in [s, \infty[, \tag{4.79}$$

with underlying Wiener process $\bar{W}(t) - \bar{W}(s)$, $t \in [s, \infty[$ and filtration $(\bar{\mathcal{F}}_t)_{t \geqslant s}$ (cf. Remark 4.2.11). We define the notion of solution for (4.79) analogously to Definition 4.2.1. Then again all results above for $s = 0$ (respectively for $s \in [0, \infty[$, see Remark 4.2.11) carry over to this more general case. In particular, we have the analogue of (4.64), namely

$$p_{r,s}p_{s,t} = p_{r,t} \quad \text{for all } -\infty < r \leqslant s \leqslant t < \infty, \tag{4.80}$$

where for $s, t \in \mathbb{R}$, $s \leqslant t, x \in H$

$$p_{s,t}(x, dy) := P \circ (X(t, s, x))^{-1}(\,dy),$$

and analogously to (4.66) one shows that

$$p_{s,t}(x, \ dy) = p_{0,t-s}(x, \ dy).$$

In particular, for $t = 0$ we have

$$p_{-s,0}(x, \ dy) = p_{0,s}(x, \ dy) \quad \text{for all } x \in H, \ s \in [0, \infty[. \tag{4.81}$$

Furthermore, for every $s \in \mathbb{R}$ there exists a unique solution with initial condition X_s denoted by $X(t, s, X_s)$, $t \in [s, \infty[$, and (4.60) as well as the final part of Remark 4.2.11 also hold in this case.

Our next main aim (cf. Theorem 4.3.9 below) is to prove the existence of a unique invariant measure for (4.78) if the constant c in (H2) is strictly negative ("strict monotonicity"). The method of the proof is an adaptation from [27, Sect. 6.3.1]. We shall need the following:

Lemma 4.3.8 *Suppose (H3), (H4) hold and that (H2) holds for $c := -\lambda$ for some $\lambda \in]0, \infty[$. Let $\eta \in]0, \lambda[$. Then there exists a $\delta_\eta \in]0, \infty[$ such that for all $v \in V$*

$$2 \,_{V^*}\langle A(v), v\rangle_V + \|B(v)\|^2_{L_2(U,H)} \leq -\eta \|v\|^2_H + \delta_\eta. \tag{4.82}$$

Proof Let $v \in V$ and $\varepsilon \in]0, 1[$. Then first using (H2) (with $c = -\lambda$ according to our assumption), then Remark 4.1.1, part 1, and finally (H3) we obtain

$$2 \,_{V^*}\langle A(v), v\rangle_V + \|B(v)\|^2_{L^2(U,H)}$$

$$= 2 \,_{V^*}\langle A(v) - A(0), v\rangle_V + 2 \,_{V^*}\langle A(0), v\rangle_V + \|B(v) - B(0)\|^2_{L_2(U,H)}$$

$$\quad - \|B(0)\|^2_{L_2(U,H)} + 2\langle B(v), B(0)\rangle_{L_2(U,H)}$$

$$\leq -\lambda \|v\|^2_H + 2\varepsilon \|v\|^\alpha_V + 2\varepsilon^{-\frac{1}{\alpha-1}}(\alpha - 1)\alpha^{\frac{-\alpha}{\alpha-1}} \|A(0)\|^{\frac{\alpha}{\alpha-1}}_{V^*} + \varepsilon^{-1}\|B(0)\|^2_{L_2(U,H)}$$

$$\quad + \varepsilon \|B(v)\|^2_{L_2(U,H)}$$

$$\leq -\lambda \|v\|^2_H + 2\varepsilon \|v\|^\alpha_V + \beta_\varepsilon$$

$$\quad + \varepsilon \left(c_1 \|v\|^2_H + f + \frac{2}{\alpha}\|v\|^\alpha_V + 2\frac{\alpha-1}{\alpha}g^{\frac{\alpha}{\alpha-1}} + 2c_3\|v\|^\alpha_V \right)$$

$$\leq \left[-\lambda + \varepsilon c_1\left(1 + \frac{2}{c_2}(1 + \alpha^{-1} + c_3)\right) \right] \|v\|^2_H + \tilde{\beta}_\varepsilon + \frac{2}{c_2}\varepsilon(1 + \alpha^{-1} + c_3)f$$

$$\quad - \frac{2}{c_2}\varepsilon(1 + \alpha^{-1} + c_3)(2 \,_{V^*}\langle A(v), v\rangle_V + \|B(v)\|^2_{L_2(U,H)})$$

with $\beta_\varepsilon, \tilde{\beta}_\varepsilon \in]0, \infty[$ independent of v and where we applied Young's inequality in the form

$$ab = [(\alpha\varepsilon)^{-1/\alpha}a][(\alpha\varepsilon)^{1/\alpha}b] \leq \frac{(\alpha\varepsilon)^{-1/(\alpha-1)}}{\alpha/(\alpha-1)}a^{\alpha/(\alpha-1)} + \varepsilon b^\alpha,$$

$a, b \in [0, \infty[$ in the second step. Hence taking ε small enough we can find $\delta_\eta \in]0, \infty[$ such that for all $v \in V$

$$2 \,_{V^*}\langle A(v), v\rangle_V + \|B(v)\|^2_{L_2(U,H)} \leq -\eta \|v\|^2_H + \delta_\eta.$$

$\square$

Theorem 4.3.9 *Consider the situation of Proposition 4.3.5 and, in addition, assume that $c \in \mathbb{R}$ in (H2) is strictly negative, i.e. $c = -\lambda$, $\lambda \in]0, \infty[$ ("strict monotonicity"). Then there exists an invariant measure μ for $(p_t)_{t \geq 0}$ such that*

$$\int \|y\|^2_H \, \mu(\,dy) < \infty.$$

Moreover, for $F : H \to \mathbb{R}$ Lipschitz, $x \in H$ and any invariant measure μ for $(p_t)_{t \geq 0}$

$$\left| p_t F(x) - \int F \, d\mu \right| \leq e^{-\frac{\lambda}{2}t} \mathrm{Lip}(F) \int \|x - y\|_H \, \mu(\, dy) \quad \text{for all } t \in [0, \infty[. \quad (4.83)$$

In particular, there exists exactly one invariant measure for $(p_t)_{t \geq 0}$.

Remark 4.3.10 (4.83) is referred to as "exponential convergence of $(p_t)_{t \geq 0}$ to equilibrium" (uniformly with respect to x in balls in H).

For the proof of Theorem 4.3.9 we need one lemma.

Lemma 4.3.11 *Consider the situation of Theorem 4.3.9. Let $t \in \mathbb{R}$. Then there exists an $\eta_t \in L^2(\Omega, \mathcal{F}, P; H)$, such that for all $x \in H$*

$$\lim_{s \to -\infty} X(t, s, x) = \eta_t \text{ in } L^2(\Omega, \mathcal{F}, P; H).$$

Moreover, there exists a $C \in [0, \infty[$ such that for all $s \in \,]-\infty, t]$

$$E(\|X(t, s, x) - \eta_t\|_H^2) \leq Ce^{\lambda(s-t)}(1 + \|x\|_H^2).$$

Proof For $s_1, s_2 \in \,]-\infty, t]$, $s_1 \leq s_2$, and $x \in H$

$$X(t, s_1, x) - X(t, s_2, x)$$

$$= \int_{s_2}^{t} [A(\bar{X}(u, s_1, x)) - A(\bar{X}(u, s_2, x))] \, ds$$

$$+ \int_{s_2}^{t} [B(\bar{X}(u, s_1, x)) - B(\bar{X}(u, s_2, x))] \, d\bar{W}(u) + X(s_2, s_1, x) - x,$$

since

$$X(s_2, s_1, x) = x + \int_{s_1}^{s_2} A(\bar{X}(u, s_1, x)) \, du + \int_{s_1}^{s_2} B(\bar{X}(u, s_1, x)) \, d\bar{W}(u). \quad (4.84)$$

Since Remark 4.2.8 extends to our present case we can use the product rule and (H2) with $c = -\lambda$ to obtain

$$E(e^{\lambda t} \|X(t, s_1, x) - X(t, s_2, x)\|_H^2) = E(e^{\lambda s_2} \|X(s_2, s_1, x) - x\|_H^2)$$

$$+ \int_{s_2}^{t} e^{\lambda u} E\Big(2 \,_{V^*}\langle A(\bar{X}(u, s_1, x)) - A(\bar{X}(u, s_2, x)), \bar{X}(u, s_1, x) - \bar{X}(u, s_2, x)\rangle_V$$

$$+ \|B(\bar{X}(u, s_1, x)) - B(\bar{X}(u, s_2, x))\|_{L_2(U, H)}^2\Big) \, du$$

$$+ \int_{s_2}^{t} e^{\lambda u} \lambda E \left(\| X(u, s_1, x) - X(u, s_2, x) \|_H^2 \right) \, du$$

$$\leqslant 2 e^{\lambda s_2} [E \left(\| X(s_2, s_1, x) \|_H^2 \right) + \| x \|_H^2]. \tag{4.85}$$

But again by Remark 4.2.8 extended to the present case, the product rule and (4.82) imply

$$E(e^{\eta s_2} \| X(s_2, s_1, x) \|_H^2)$$

$$= e^{s_1 \eta} \| x \|_H^2 + \int_{s_1}^{s_2} e^{\eta u} E \Big(2 \,_{V^*} \langle A(\bar{X}(u, s_1, x)), \bar{X}(u, s_1, x) \rangle_V$$

$$+ \| B(\bar{X}(u, s_1, x)) \|_{L_2(U, H)}^2 \Big) \, du + \int_{s_1}^{s_2} e^{\eta u} \eta E(\| X(u, s_1, x) \|_H^2) \, du$$

$$\leqslant e^{s_1 \eta} \| x \|_H^2 + \delta_\eta \int_{s_1}^{s_2} e^{\eta u} \, du \leqslant e^{s_1 \eta} \| x \|_H^2 + \frac{\delta_\eta}{\eta} e^{s_2 \eta}. \tag{4.86}$$

Combining (4.85) and (4.86) we obtain

$$E(\| X(t, s_1, x) - X(t, s_2, x) \|_H^2) \leqslant 2 \left(\frac{\delta_\eta}{\eta} + 2 \| x \|_H^2 \right) e^{\lambda(s_2 - t)}. \tag{4.87}$$

Letting s_2 (hence s_1) tend to $-\infty$, it follows that there exists an $\eta_t(x) \in L^2(\Omega, \mathcal{F}, P; H)$ such that

$$\lim_{s \to -\infty} X(t, s, x) = \eta_t(x) \text{ in } L^2(\Omega, \mathcal{F}, P; H),$$

and letting $s_1 \to -\infty$ in (4.87) the last part of the assertion also follows, provided we can prove that $\eta_t(x)$ is independent of $x \in H$. To this end let $x, y \in H$ and $s \in]-\infty, t]$. Then

$$X(t, s, x) - X(t, s, y)$$

$$= x - y + \int_s^t (A(\bar{X}(u, s, x)) - A(\bar{X}(u, s, y))) \, du$$

$$+ \int_s^t (B(\bar{X}(u, s, x)) - B(\bar{X}(u, s, y))) \, d\bar{W}(u).$$

Hence by the same arguments to derive (4.85) we get

$$E(e^{\lambda t} \| X(t, s, x) - X(t, s, y) \|_H^2) \leqslant e^{\lambda s} \| x - y \|_H^2,$$

so

$$\lim_{s \to -\infty} (X(t, s, x) - X(t, s, y)) = 0 \text{ in } L^2(\Omega, \mathcal{F}, P; H).$$

Hence both assertions are completely proved. □

Proof of Theorem 4.3.9 Define

$$\mu := P \circ \eta_0^{-1}$$

with η_0 as in Lemma 4.3.11. Since $\eta_0 \in L^2(\Omega, \mathcal{F}, P; H)$ we have that

$$\int \|y\|_H^2 \mu(\mathrm{d}y) < \infty.$$

Let $t \in [0, \infty[$. We note that by (4.80) and (4.81) for all $s \in [0, \infty[$

$$p_{-s,0} p_{0,t} = p_{-s,t} = p_{0,t+s} = p_{-(t+s),0}. \tag{4.88}$$

Let $F : H \to \mathbb{R}$, F bounded and Lipschitz. Then by Proposition 4.3.2 we have that $p_{0,t} F$ is (bounded and) Lipschitz. Furthermore, by Lemma 4.3.11 for all $x \in H$

$$p_{-s,0}(x, \mathrm{d}y) \to \mu \text{ weakly as } s \to \infty.$$

Hence by (4.88) for all $x \in H$

$$\int p_{0,t} F \, \mathrm{d}\mu = \lim_{s \to \infty} p_{-s,0}(p_{0,t} F)(x) = \lim_{s \to \infty} p_{-(t+s),0} F(x) = \int F \, \mathrm{d}\mu.$$

Recalling that by definition $p_t = p_{0,t}$, it follows that μ is an invariant measure for $(p_t)_{t \geq 0}$. Furthermore, if μ is an invariant measure for $(p_t)_{t \geq 0}$, then by Proposition 4.3.2 for all $t \in [0, \infty[$

$$\left| p_t F(x) - \int F \, \mathrm{d}\mu \right| = \left| \int (p_t F(x) - p_t F(y)) \mu(\mathrm{d}y) \right|$$

$$\leq \int (e^{-\frac{\lambda}{2} t} \mathrm{Lip}(F) \|x - y\|) \wedge 2 \|F\|_\infty \, \mu(\mathrm{d}y),$$

which implies (4.83) and by letting $t \to \infty$ also that all such invariant measures μ coincide. □

Chapter 5
SPDEs with Locally Monotone Coefficients

In this chapter we will present more general results on the existence and uniqueness of solutions to SPDEs. More precisely, we will replace the standard monotonicity condition and coercivity condition in Chap. 4 by a local monotonicity condition and generalized coercivity condition respectively. The main references for this chapter are [57–59].

In the applications we make a slight change of notation by writing for the standard L^p-norms $\|\cdot\|_{L^p}$ instead of $\|\cdot\|_p$, $p \in [1, \infty]$, as in previous chapters to be more consistent with the notation for norms in Sobolev spaces, used below.

5.1 Local Monotonicity

5.1.1 Main Result

Let

$$V \subset H \equiv H^* \subset V^*$$

be a Gelfand triple and for $T > 0$ let $W(t)$, $t \in [0, T]$, be a cylindrical Wiener process in a separable Hilbert space U on a probability space $(\Omega, \mathcal{F}, P)$ with normal filtration $\mathcal{F}_t$, $t \in [0, T]$. We consider the following stochastic differential equation on H

$$dX(t) = A(t, X(t)) \, dt + B(t, X(t)) \, dW(t), \qquad (5.1)$$

where for some fixed time $T > 0$

$$A : [0, T] \times V \times \Omega \to V^*; \quad B : [0, T] \times V \times \Omega \to L_2(U, H)$$

are progressively measurable.

© Springer International Publishing Switzerland 2015
W. Liu, M. Röckner, *Stochastic Partial Differential Equations: An Introduction*,
Universitext, DOI 10.1007/978-3-319-22354-4_5

In Chap. 4 we have shown that (5.1) has a unique solution if A, B satisfy the classical monotonicity and coercivity conditions. The main aim of this section is to provide a more general framework for the variational approach, being conceptually not more complicated than the classical one (cf. Chap. 4 or [54]), but including a large number of new applications, for example, fundamental SPDEs such as the stochastic 2-D Navier–Stokes equation and stochastic Burgers type equations. The main changes consist of localizing the monotonicity condition and relaxing the growth condition. Let us now state the precise conditions on the coefficients of (5.1):

Suppose that there exist constants $\alpha \in]1, \infty[$, $\beta \in [0, \infty[$, $\theta \in]0, \infty[$, $C_0 \in \mathbb{R}$ and a nonnegative adapted process $f \in L^1([0, T] \times \Omega; \, dt \otimes P)$ such that the following conditions hold for all $u, v, w \in V$ and $(t, \omega) \in [0, T] \times \Omega$:

(H1) (Hemicontinuity) The map $\lambda \mapsto {}_{V*}\langle A(t, u + \lambda v), w \rangle_V$ is continuous on $\mathbb{R}$.

(H2′) (Local monotonicity)

$$2_{V*}\langle A(t, u) - A(t, v), u - v \rangle_V + \|B(t, u) - B(t, v)\|^2_{L_2(U, H)}$$

$$\leq (f(t) + \rho(v)) \|u - v\|^2_H,$$

   where $\rho : V \to [0, +\infty[$ is a measurable hemicontinuous function and locally bounded in V.

(H3) (Coercivity)

$$2_{V*}\langle A(t, v), v \rangle_V + \|B(t, v)\|^2_{L_2(U, H)} \leq C_0 \|v\|^2_H - \theta \|v\|^\alpha_V + f(t).$$

(H4′) (Growth)

$$\|A(t, v)\|^{\frac{\alpha}{\alpha-1}}_{V*} \leq (f(t) + C_0 \|v\|^\alpha_V)(1 + \|v\|^\beta_H).$$

Remark 5.1.1

(1) (H2′) is significantly weaker than the standard monotonicity condition (H2) (i.e. $\rho \equiv 0$) in Chap. 4. One typical choice of ρ in applications is

$$\rho(v) = C\|v\|^\gamma_V,$$

   where C and γ are some nonnegative constants.

   One important example is the stochastic 2-D Navier–Stokes equation on a bounded or unbounded domain. It satisfies (H2′), but not (H2) (see Sect. 5.1.3 below).

(2) (H4′) is also weaker than the standard growth condition (H4) in Chap. 4. The advantage of (H4′) is, e.g. that it allows the inclusion of many semilinear type equations with nonlinear perturbation terms. For example, if we consider a reaction-diffusion type equation, i.e. $A(u) = \Delta u + F(u)$, then for (H3) to hold we need $\alpha = 2$. Hence (H4) would imply that F has at most linear

growth. However, for the weaker condition $(H4')$ we can allow F to have some polynomial growth. We refer to Sect. 5.1.3 for more details.

(3) By $(H3)$, $(H4')$ and Young's inequality it easily follows that

$$\|B(t,v)\|_{L_2(U,H)}^2 \le f(t) + C_0\|v\|_H^2 + \frac{2(\alpha-1)}{\alpha}f(t)\left(1 + \|v\|_H^\beta\right)$$
$$+ \frac{2}{\alpha}\left[C_0(\alpha-1)(1 + \|v\|_H^\beta) + 1 - \frac{\alpha}{2}\theta\right]\|v\|_V^\alpha.$$

(4) Since ρ is locally bounded, $(H2')$ implies that for fixed $\omega \in \Omega, t \in [0,T]$, $A(t,\cdot,\omega)$ satisfies $(H2_{loc})$ in Remark 4.1.1,2. Hence $(H1)$ and $(H2')$ imply that $A(t,\cdot,\omega) : V \to V^*$ is demicontinuous for all $(t,\omega) \in [0,T] \times \Omega$.

Definition 5.1.2 A continuous H-valued $(\mathcal{F}_t)$-adapted process $(X(t))_{t\in[0,T]}$ is called a solution of (5.1), if for its $dt \otimes P$-equivalence class $\hat{X}$ we have

$$\hat{X} \in L^\alpha([0,T] \times \Omega, \, dt \otimes P; V) \cap L^2([0,T] \times \Omega, \, dt \otimes P; H)$$

with α in $(H3)$ and P-a.s.

$$X(t) = X(0) + \int_0^t A(s,\bar{X}(s))\,ds + \int_0^t B(s,\bar{X}(s))\,dW(s), \; t \in [0,T],$$

where $\bar{X}$ is any V-valued progressively measurable $dt \otimes P$-version of $\hat{X}$ (which exists by Exercise 4.2.3).

Theorem 5.1.3 *Suppose* $(H1), (H2'), (H3), (H4')$ *hold for some* $f \in L^{p/2}([0,T] \times \Omega; \, dt \otimes P)$ *with some* $p \ge \beta + 2$, *and there exists a constant C such that*

$$\|B(t,v)\|_{L^2(U,H)}^2 \le C(f(t) + \|v\|_H^2), \; t \in [0,T], v \in V;$$

$$\rho(v) \le C(1 + \|v\|_V^\alpha)(1 + \|v\|_H^\beta), \; v \in V. \tag{5.2}$$

Then for any $X_0 \in L^p(\Omega, \mathcal{F}_0, P; H)$, *(5.1) has a unique solution* $(X(t))_{t\in[0,T]}$ *such that* $X(0) = X_0$ *and it satisfies*

$$E\left(\sup_{t\in[0,T]} \|X(t)\|_H^p\right) < \infty.$$

Moreover, if $A(t,\cdot)(\omega), B(t,\cdot)(\omega)$ *are independent of* $t \in [0,T]$ *and* $\omega \in \Omega$, *then the solution* $(X(t))_{t\in[0,T]}$ *of (5.1) is a Markov process.*

5.1.2 Proof of the Main Theorem

The first step of the proof is mainly based on the Galerkin approximation. Let

$$\{e_1, e_2, \cdots\} \subset V$$

be an orthonormal basis of H and let $H_n := span\{e_1, \cdots, e_n\}$ such that $span\{e_1, e_2, \cdots\}$ is dense in V. Let $P_n : V^* \to H_n$ be defined by

$$P_n y := \sum_{i=1}^{n} {}_{V^*}\langle y, e_i \rangle_V e_i, \ y \in V^*.$$

Recall that $P_n \restriction_H$ is just the orthogonal projection onto H_n in H and we have

$${}_{V^*}\langle P_n A(t, u), v \rangle_V = \langle P_n A(t, u), v \rangle_H = {}_{V^*}\langle A(t, u), v \rangle_V, \ u \in V, v \in H_n.$$

Let $\{g_1, g_2, \cdots\}$ be an orthonormal basis of U and

$$W^{(n)}(t) := \sum_{i=1}^{n} \langle W(t), g_i \rangle_U g_i = \tilde{P}_n W(t),$$

where $\tilde{P}_n$ is the orthogonal projection onto $span\{g_1, \cdots, g_n\}$ in U.

Then for each finite $n \in \mathbb{N}$ we consider the following stochastic equation on H_n

$$dX^{(n)}(t) = P_n A(t, X^{(n)}(t)) \, dt + P_n B(t, X^{(n)}(t)) \, dW^{(n)}(t), \ X^{(n)}(0) = P_n X_0. \quad (5.3)$$

By Remark 5.1.1, (3) and (4) it is easy to check that all assumptions of Theorem 3.1.1 hold, which hence implies that (5.3) has a unique strong solution.

In order to construct the solution of (5.1), we need some a priori estimates for $X^{(n)}$. As in Chap. 4 we use the following notations:

$$K := L^\alpha([0, T] \times \Omega, \, dt \otimes P; V);$$

$$J := L^2([0, T] \times \Omega, \, dt \otimes P; L_2(U, H)).$$

Lemma 5.1.4 *Under the assumptions in Theorem 5.1.3, there exists a $C > 0$ such that for all $n \in \mathbb{N}$*

$$\|X^{(n)}\|_K + \sup_{t \in [0, T]} E\|X^{(n)}(t)\|_H^2 \leq C.$$

Proof The assertion follows from (H3) as in the proof of Lemma 4.2.9. □

Lemma 5.1.5 *Under the assumptions in Theorem 5.1.3, there exists a $C > 0$ such that for all $n \in \mathbb{N}$ we have*

$$E\left(\sup_{t\in[0,T]} \|X^{(n)}(t)\|_H^p\right) + E\int_0^T \|X^{(n)}(t)\|_H^{p-2}\|X^{(n)}(t)\|_V^\alpha \, dt$$

$$\leq C\left(E\|X_0\|_H^p + E\int_0^T f^{p/2}(t) \, dt\right). \tag{5.4}$$

In particular, there exists a $C > 0$ such that for all $n \in \mathbb{N}$

$$\|A(\cdot, X^{(n)})\|_{K^*} \leq C.$$

Proof By Itô's formula (for $\mathbb{R}^n$-valued semimartingales)

$$\|X^{(n)}(t)\|_H^p$$

$$= \|X^{(n)}(0)\|_H^p + p(p-2)\int_0^t \|X^{(n)}(s)\|_H^{p-4}\|(P_n B(s, X^{(n)}(s))\tilde{P}_n)^* X^{(n)}(s)\|_H^2 \, ds$$

$$+ \frac{p}{2}\int_0^t \|X^{(n)}(s)\|_H^{p-2}\left(2_{V^*}\langle A(s, X^{(n)}(s)), X^{(n)}(s)\rangle_V + \|P_n B(s, X^{(n)}(s))\tilde{P}_n\|_{L_2(U,H^n)}^2\right) ds$$

$$+ p\int_0^t \|X^{(n)}(s)\|_H^{p-2}\langle X^{(n)}(s), P_n B(s, X^{(n)}(s)) \, dW^{(n)}(s)\rangle_H.$$

Hence by (H3), (5.1.3) and Young's inequality

$$\|X^{(n)}(t)\|_H^p + \frac{p\theta}{2}\int_0^t \|X^{(n)}(s)\|_H^{p-2}\|X^{(n)}(s)\|_V^\alpha \, ds$$

$$\leq \|X^{(n)}(0)\|_H + C\int_0^t \left(\|X^{(n)}(s)\|_H^p + f(s)\cdot\|X^{(n)}(s)\|_H^{p-2}\right) ds$$

$$+ p\int_0^t \|X^{(n)}(s)\|_H^{p-2}\langle X^{(n)}(s), P_n B(s, X^{(n)}(s)) \, dW^{(n)}(s)\rangle_H$$

$$\leq \|X_0\|_H^p + C\int_0^t \left(\|X^{(n)}(s)\|_H^p + f^{p/2}(s)\right) ds$$

$$+ p\int_0^t \|X^{(n)}(s)\|_H^{p-2}\langle X^{(n)}(s), P_n B(s, X^{(n)}(s)) \, dW^{(n)}(s)\rangle_H, \quad t \in [0, T], \tag{5.5}$$

where C is a constant (independent of n) and may change from line to line.

For $n \in \mathbb{N}$ we define the stopping time

$$\tau_R^{(n)} := \inf\{t \in [0, T] : \|X^{(n)}(t)\|_H > R\} \wedge T, \ R > 0.$$

Here, as usual, $\inf \emptyset := \infty$. It is obvious that

$$\lim_{R \to \infty} \tau_R^{(n)} = T, \ P\text{-}a.s., \ n \in \mathbb{N}.$$

Then by Proposition D.0.1(i), (5.2) and Young's inequality we have for all $t \in [0, T]$

$$E \sup_{r \in [0, \tau_R^{(n)} \wedge t]} \left| \int_0^r \|X^{(n)}(s)\|_H^{p-2} \langle X^{(n)}(s), P_n B(s, X^{(n)}(s)) \, dW^{(n)}(s) \rangle_H \right|$$

$$\leq 3E \left(\int_0^{\tau_R^{(n)} \wedge t} \|X^{(n)}(s)\|_H^{2p-2} \|B(s, X^{(n)}(s))\|_{L_2(U,H)}^2 \, ds \right)^{1/2}$$

$$\leq 3E \left(\sup_{s \in [0, \tau_R^{(n)} \wedge t]} \|X(s)^{(n)}\|_H^{2p-2} \cdot C \int_0^{\tau_R^{(n)} \wedge t} \left(\|X(s)^{(n)}\|_H^2 + f(s) \right) \, ds \right)^{1/2}$$

$$\leq 3E \left[\varepsilon \sup_{s \in [0, \tau_R^{(n)} \wedge t]} \|X^{(n)}(s)\|_H^p + C_\varepsilon \left(\int_0^t \left(\sup_{r \in [0, \tau_R^{(n)} \wedge t]} \|X^{(n)}(r)\|_H^2 + f(s) \right) \, ds \right)^{p/2} \right]$$

$$\leq 3\varepsilon E \left(\sup_{s \in [0, \tau_R^{(n)} \wedge t]} \|X^{(n)}(s)\|_H^p \right) + 3 \cdot (2T)^{p/2-1} C_\varepsilon \int_0^t E \sup_{r \in [0, \tau_R^{(n)} \wedge t]} \left(\|X^{(n)}(r)\|_H^2 + f^{p/2}(s) \right) \, ds,$$

$$(5.6)$$

where $\varepsilon > 0$ is a small constant.

Then by (5.5), (5.6) and Gronwall's lemma we have

$$E \left(\sup_{t \in [0, \tau_R^{(n)}]} \|X^{(n)}(t)\|_H^p \right) + E \int_0^{\tau_R^{(n)}} \|X^{(n)}(s)\|_H^{p-2} \|X^{(n)}(s)\|_V^\alpha \, ds$$

$$\leq C \left(E \|X_0\|_H^p + E \int_0^T f^{p/2}(s) \, ds \right), \ n \in \mathbb{N},$$

where C is a constant independent of n.

Letting $R \to \infty$, (5.4) follows from the monotone convergence theorem.

Moreover, by $(H4')$ and because $p \geq \beta + 2$, we deduce from (5.4) and Lemma 5.1.4 that

$$\|A(\cdot, X^{(n)})\|_{K^*} \leq C, \ n \geq 1,$$

where C is a constant independent of n. $\qquad\qquad\qquad\qquad\qquad\qquad\qquad\qquad\quad \square$

Proof of Theorem 5.1.3

(1) Existence: By Lemmas 5.1.4, 5.1.5 and (5.2) there exists a subsequence $n_k \to \infty$ such that

  (i) $X^{(n_k)} \to \bar{X}$ weakly in K and weakly* in $L^p(\Omega; L^\infty([0,T];H))$.
  (ii) $Y^{(n_k)} := A(\cdot, X^{(n_k)}) \to Y$ weakly in K^*.
  (iii) $Z^{(n_k)} := P_{n_k} B(\cdot, X^{(n_k)}) \to Z$ weakly in J and hence

$$\int_0^{\cdot} P_{n_k} B(s, X^{(n_k)}(s))\, dW^{(n_k)}(s) \to \int_0^{\cdot} Z(s)\, dW(s)$$

  weakly in $\mathcal{M}_T^2(H)$ and, therefore, also weakly* in $L^\infty([0,T], \, dt; L^2(\Omega, P; H))$.

Now we define

$$X(t) := X_0 + \int_0^t Y(s)\, ds + \int_0^t Z(s)\, dW(s), \ t \in [0,T]. \tag{5.7}$$

As in the proof of Theorem 4.2.4 one shows that $X = \bar{X}$ $dt \otimes P$-a.e.

Then by Theorem 4.2.5 we know that X is an H-valued continuous $(\mathcal{F}_t)$-adapted process. Since $X^{(n_k)}$, $k \in \mathbb{N}$, converges weakly to X in $L^p(\Omega; L^\infty([0,T]; H))$, it follows by Lemma 5.1.5 that

$$E\left(\sup_{t \in [0,T]} \|X(t)\|_H^p\right) < \infty. \tag{5.8}$$

Therefore, it remains to verify that

$$A(\cdot, \bar{X}) = Y, \ B(\cdot, \bar{X}) = Z \ dt \otimes P\text{-}a.e.$$

Define

$$\mathcal{M} := K \cap L^p(\Omega; L^\infty([0,T]; H)).$$

Let ϕ be a V-valued progressively measurable version of an element in $\mathcal{M}$ and let $\tau^\phi : \Omega \to [0,T]$ be a stopping time such that

$$c_\phi := P\text{-ess} \sup_{\Omega} \int_0^{\tau^\phi} \left(f(s) + \|\phi(s)\|_V^\alpha\right) ds < \infty.$$

Applying the Itô-formula (4.30) together with Itô's product rule we obtain for $t \in [0, T]$

$$E\left(e^{-\int_0^{t \wedge \tau^\phi} (f(s) + \rho(\phi(s))) \, ds} \|X^{(n_k)}(t \wedge \tau^\phi)\|_H^2\right) - E\left(\|X^{(n_k)}(0)\|_H^2\right)$$

$$= E\left[\int_0^{t \wedge \tau^\phi} e^{-\int_0^s (f(r) + \rho(\phi(r))) \, dr} \left(2 \, _{V^*}\langle A(s, X^{(n_k)}(s)), X^{(n_k)}(s)\rangle_V \right.\right.$$

$$\left.\left. + \|P_{n_k} B(s, X^{(n_k)}(s)) \tilde{P}_{n_k}\|_{L_2(U, H^n)}^2 - (f(s) + \rho(\phi(s)))\|X^{(n_k)}(s)\|_H^2\right) ds\right]$$

$$\leq E\left[\int_0^{t \wedge \tau^\phi} e^{-\int_0^s (f(r) + \rho(\phi(r))) \, dr} \left(2 \, _{V^*}\langle A(s, X^{(n_k)}(s)), X^{(n_k)}(s)\rangle_V \right.\right.$$

$$\left.\left. + \|B(s, X^{(n_k)}(s))\|_{L_2(U, H)}^2 - (f(s) + \rho(\phi(s)))\|X^{(n_k)}(s)\|_H^2\right) ds\right]$$

$$= E\left[\int_0^{t \wedge \tau^\phi} e^{-\int_0^s (f(r) + \rho(\phi(r))) \, dr} \left(2 \, _{V^*}\langle A(s, X^{(n_k)}(s)) - A(s, \phi_s), X^{(n_k)}(s) - \phi(s)\rangle_V \right.\right.$$

$$+ \|B(s, X^{(n_k)}(s)) - B(s, \phi(s))\|_{L_2(U, H)}^2$$

$$\left.\left. - (f(s) + \rho(\phi(s)))\|X^{(n_k)}(s) - \phi(s)\|_H^2\right) ds\right]$$

$$+ E\left[\int_0^{t \wedge \tau^\phi} e^{-\int_0^s (f(r) + \rho(\phi(r))) \, dr} \left(2 \, _{V^*}\langle A(s, X^{(n_k)}(s)) - A(s, \phi(s)), \phi(s)\rangle_V \right.\right.$$

$$+ 2 \, _{V^*}\langle A(s, \phi(s)), X^{(n_k)}(s)\rangle_V - \|B(s, \phi(s))\|_{L_2(U, H)}^2$$

$$+ 2\langle B(s, X^{(n_k)}(s)), B(s, \phi(s))\rangle_{L_2(U, H)}$$

$$\left.\left. - 2(f(s) + \rho(\phi(s)))\langle X^{(n_k)}(s), \phi(s)\rangle_H + (f(s) + \rho(\phi(s)))\|\phi(s)\|_H^2\right) ds\right].$$

$$(5.9)$$

Above in (5.9) we used that by Proposition D.0.1(ii), (5.2) and Lemma 5.1.5 (used for fixed n) the local martingale appearing after applying Itô's formula is a martingale, hence has expectation zero. Furthermore, by the definition of $\mathcal{M}$, the property of τ^ϕ, (5.2), (H4′) and because of Lemma 5.1.5 (used for fixed k) it is easy to check that all integrals in (5.9) are well-defined. Letting $k \to \infty$, by (H2′), (i)-(iii) and the same argument as in (4.53) we have for every nonnegative

$\psi \in L^\infty([0, T]; \, dt)$,

$$E\left[\int_0^T \psi(t)\left(e^{-\int_0^{t \wedge \tau^\phi}(f(s)+\rho(\phi(s)))\,ds}\|X(t)\|_H^2 - \|X_0\|_H^2\right)\,dt\right]$$

$$\leq \liminf_{k \to \infty} E\left[\int_0^T \psi(t)\left(e^{-\int_0^{t \wedge \tau^\phi}(f(s)+\rho(\phi(s)))\,ds}\|X^{(n_k)}(t)\|_H^2 - \|X^{(n_k)}(0)\|_H^2\right)\,dt\right]$$

$$\leq E\left[\int_0^T \psi(t)\left(\int_0^{t \wedge \tau^\phi} e^{-\int_0^s(f(r)+\rho(\phi(r)))\,dr}\Big(2_{V^*}\langle Y(s) - A(s, \phi(s)), \phi(s)\rangle_V\right.\right.$$

$$+ 2_{V^*}\langle A(s, \phi(s)), \bar{X}(s)\rangle_V - \|B(s, \phi(s))\|_{L_2(U,H)}^2 + 2\langle Z(s), B(s, \phi(s))\rangle_{L_2(U,H)}$$

$$\left.\left.- 2(f(s) + \rho(\phi(s)))\langle X(s), \phi(s)\rangle_H + (f(s) + \rho(\phi(s)))\|\phi(s)\|_H^2\Big)\,ds\right)\,dt\right].$$

$$(5.10)$$

By Itô's formula and Proposition D.0.1(iii) we have

$$E\left(e^{-\int_0^{t \wedge \tau^\phi}(f(s)+\rho(\phi(s)))\,ds}\|X(t)\|_H^2\right) - E\left(\|X_0\|_H^2\right)$$

$$= E\left[\int_0^{t \wedge \tau^\phi} e^{-\int_0^s(f(r)+\rho(\phi(r)))\,dr}\left(2_{V^*}\langle Y(s), \bar{X}(s)\rangle_V\right.\right.$$

$$\left.\left.+ \|Z(s)\|_{L_2(U,H)}^2 - (f(s) + \rho(\phi(s)))\|X(s)\|_H^2\right)\,ds\right].$$

$$(5.11)$$

By inserting (5.11) into (5.10) we obtain

$$0 \geq E\left[\int_0^T \psi(t)\left(\int_0^{t \wedge \tau^\phi} e^{-\int_0^s(f(r)+\rho(\phi(r)))\,dr}\left(2_{V^*}\langle Y(s) - A(s, \phi(s)), \bar{X}(s) - \phi(s)\rangle_V\right.\right.\right.$$

$$\left.\left.\left.+ \|B(s, \phi(s)) - Z(s)\|_{L_2(U,H)}^2 - (f(s) + \rho(\phi(s)))\|X(s) - \phi(s)\|_H^2\right)\,ds\right)\,dt\right].$$

$$(5.12)$$

Note that (5.2), Lemmas 5.1.4 and 5.1.5 imply that

$$\bar{X} \in \mathcal{M}.$$

Taking $\phi = \bar{X}$ and for $R > 0$

$$\tau^\phi := \tau_R^X := \inf\left\{t \in [0, T] : \int_0^t \left(f(s) + \|\bar{X}(s)\|_V^\alpha\right)\,ds > R^\alpha\right\} \wedge T,$$

and then letting $R \to \infty$, we obtain that $Z = B(\cdot, \bar{X})$. Next, we take $\phi = \bar{X} - \varepsilon \tilde{\phi} v$ for $\tilde{\phi} \in L^\infty([0, T] \times \Omega; \, dt \otimes P; \mathbb{R})$, $v \in V$ and $\tau^\phi := \tau_R^X$, $R > 0$. Then we divide by ε and let $\varepsilon \to 0$ to derive by $(H1)$ and the hemicontinuity of ρ that

$$0 \geq E\left[\int_0^T \psi(t)\left(\int_0^{t \wedge \tau^\phi} e^{-\int_0^s (f(r) + \rho(\bar{X}(r)))\, dr} \tilde{\phi}(s)_{V^*}\langle Y(s) - A(s, \bar{X}(s)), v\rangle_V \, ds\right) dt\right]. \tag{5.13}$$

We note that interchanging the limit $\varepsilon \to 0$ with the integrals is indeed justified by (5.2), $(H4')$ and the definition of τ_R^X. By the arbitrariness of ψ and $\tilde{\phi}$, we conclude that $Y = A(\cdot, \bar{X})$ on $[0, \tau_R^X]$. Letting $R \to \infty$ we obtain that $Y = A(\cdot, \bar{X})$.

Hence X is a solution of (5.1).

(2) Uniqueness: Suppose X, Y are solutions of (5.1) with initial conditions X_0, Y_0 respectively, i.e.

$$X(t) = X_0 + \int_0^t A(s, \bar{X}(s))\, ds + \int_0^t B(s, \bar{X}(s))\, dW(s), \ t \in [0, T];$$

$$Y(t) = Y_0 + \int_0^t A(s, \bar{Y}(s))\, ds + \int_0^t B(s, \bar{Y}(s))\, dW(s), \ t \in [0, T].$$

Then by the product rule, Itô's formula from Theorem 4.2.5 and $(H2')$ we have for $t \in [0, T]$

$$e^{-\int_0^t (f(s) + \rho(\bar{Y}(s)))\, ds} \|X(t) - Y(t)\|_H^2 \leq \|X_0 - Y_0\|_H^2$$

$$+ 2 \int_0^t e^{-\int_0^s (f(r) + \rho(\bar{Y}(r)))\, dr} \langle X(s) - Y(s), \big(B(s, \bar{X}(s)) - B(s, \bar{Y}(s))\big)\, dW(s)\rangle_H.$$

By Proposition D.0.1(ii) the real-valued local martingale on the right-hand side is a martingale. So, taking expectation we arrive at

$$E\left[e^{-\int_0^t (f(s) + \rho(\bar{Y}(s)))\, ds} \|X(t) - Y(t)\|_H^2\right] \leq E\|X_0 - Y_0\|_H^2, \ t \in [0, T].$$

If $X_0 = Y_0$, then

$$E\left[e^{-\int_0^t (f(s) + \rho(\bar{Y}(s)))\, ds} \|X(t) - Y(t)\|_H^2\right] = 0, \ t \in [0, T].$$

Since $Y \in \mathcal{M}$, (5.2) implies that

$$\int_0^T (f(s) + \rho(\bar{Y}(s)))\, ds < \infty \ P\text{-}a.s.$$

Consequently, for every $t \in [0, T]$

$$X(t) = Y(t) \ P\text{-}a.s.$$

Therefore, the pathwise uniqueness follows from the path continuity of X, Y in H.

(3) The proof of the Markov property is similar to the proof of Proposition 4.3.5,1.

$\square$

5.1.3 Application to Examples

Below we present some examples where the coefficients are only locally monotone, hence the classical result of monotone operators cannot be applied.

In this section we use the notation D_i to denote the spatial derivative $\frac{\partial}{\partial x_i}$ and $\Lambda \subset \mathbb{R}^d$ is an open bounded domain. For the standard Sobolev space $H_0^{1,p}(\Lambda)$, $p \geq 1$, as before we always use the following Sobolev norm:

$$\|u\|_{1,p} := \left(\int_\Lambda |\nabla u(\xi)|^p \, d\xi \right)^{1/p} .$$

For simplicity we only consider examples where the coefficients are time independent, but one can easily adapt those examples to the time dependent case.

Lemma 5.1.6 *Consider the Gelfand triple*

$$V := H_0^{1,2}(\Lambda) \subset H := L^2(\Lambda) \subset \left(H_0^{1,2}(\Lambda) \right)^* = V^*$$

and the operator

$$A(u) := \Delta u + \langle f(u), \nabla u \rangle,$$

where $f = (f_1, \ldots, f_d) : \mathbb{R} \to \mathbb{R}^d$ is a Lipschitz function and $\langle \ , \ \rangle$ denotes the inner product in $\mathbb{R}^d$. Let $Lip(f)$ denote the corresponding Lipschitz constant.

(0) *If $d \leq 4$, there exists a $C \in]0, \infty[$ such that for all $u, v, w \in V$*

$$\int_\Lambda |u||\nabla w||v| \, d\xi \leq C\|u\|_V\|w\|_V\|v\|_V.$$

In particular $A : V \to V^*$. Furthermore, if $d = 1$ or f is bounded, A satisfies (H4') with $\alpha = 1$ and $\beta = 2, \beta = 0$ respectively.

(1) *If $d = 1$ or if each f_i is bounded and $d = 2$, then there exists a $C \in (0, \infty)$ such that*

$$2_{V^*}\langle A(u) - A(v), u - v\rangle_V \leq -\|u - v\|_V^2 + \left(C + C\|v\|_V^2\right)\|u - v\|_H^2, \ u, v \in V.$$

(2) *If $d = 3$, then there exists a $C \in (0, \infty)$ such, that*

$$2_{V^*}\langle A(u) - A(v), u - v\rangle_V \leq -\|u - v\|_V^2 + \left(C + C\|v\|_V^4\right)\|u - v\|_H^2, \ u, v \in V.$$

(3) *Let Λ be possibly unbounded and let $b_i \in L^d(\Lambda) + L^\infty(\Lambda)$, $1 \leq i \leq d$, $b := (b_1, \ldots, b_d)$ and*

$$A(u) := \Delta u + \langle b, \nabla u\rangle.$$

Then for $d \geq 3$ there exists a $C \in (0, \infty)$ such that

$$2_{V^*}\langle A(u) - A(v), u - v\rangle_V \leq -\|u - v\|_V^2 + C\|u - v\|_H^2, \ u, v \in V,$$

(i.e. in this case (H2) holds).

Proof

(0):   We have for all $u, v \in V$

$$\int_\Lambda |\langle f(u), \nabla u\rangle|\,|v|\ \mathrm{d}\xi \leq \int_\Lambda (|f(0)| + \mathrm{Lip}(f)|u|)\,|\nabla u||v|\ \mathrm{d}\xi.$$

Hence the second part of the assertion follows from the first and Example 4.1.7. To prove the first part, we note that for all $u, v, w \in V$

$$\int_\Lambda |u||\nabla w||v|\ \mathrm{d}\xi \leq \|uv\|_{L^2}\|w\|_V,$$

and by the generalized Hölder inequality the right-hand side is dominated by

(a) $\|u\|_{L^2}\|v\|_{L^\infty}\|w\|_V$,
(b) $\|u\|_{L^4}\|v\|_{L^4}\|w\|_V$,
(c) $\|u\|_{L^d}\|v\|_{L^{\frac{2d}{d-2}}}\|w\|_V$.

In case $d = 1$, $H_0^{1,2}(\Lambda) \subset L^\infty(\Lambda)$ continuously. Hence assertion (0) follows from (a) if $d = 1$.
In case $d = 2$, $H_0^{1,2}(\Lambda) \subset L^p(\Lambda)$ continuously for all $p \in [1, \infty[$. Hence assertion (0) follows from (b) if $d = 2$.
In case $d \leq 3$, $H_0^{1,2}(\Lambda) \subset L^{\frac{2d}{d-2}}(\Lambda)$ continuously, and $L^{\frac{2d}{d-2}}(\Lambda) \subset L^d(\Lambda)$ continuously if $d \leq 4$. Hence assertion (0) follows from (c) if $d = 3$ or 4.

To prove the last part of the assertion we note that this is trivially true if f is bounded. If $d = 1$ and if f is merely Lipschitz continuous it follows immediately from (a).

(1) and (2):   We have

$$_{V^*}\langle A(u) - A(v), u - v\rangle_V$$

$$= -\|u - v\|_V^2 + \sum_{i=1}^{d} \int_\Lambda (f_i(u)D_i u - f_i(v)D_i v)\,(u - v)\; \mathrm{d}\xi.$$

To estimate the second term on the right-hand side, let $F_i : \mathbb{R} \to \mathbb{R}$ be such that $F_i(0) = 0$ and $F_i' = f_i$ and $G_i : \mathbb{R} \to \mathbb{R}$ be such that $G_i(0) = 0$ and $G_i' = F_i$. Then

$$\int_\Lambda (f_i(u)D_i u - f_i(v)D_i v)(u - v)\; \mathrm{d}\xi$$

$$= \int_\Lambda (f_i(u)D_i(u - v) + (f_i(u) - f_i(v))D_i v)\,(u - v)\; \mathrm{d}\xi$$

$$- \int_\Lambda f_i(u - v)D_i(u - v)(u - v)\; \mathrm{d}\xi$$

$$+ \int_\Lambda D_i\,(F_i(u - v))\,(u - v)\; \mathrm{d}\xi,$$

where integrating by parts and using that $u - v \in H_0^{1,2}(\Lambda)$ we see that the last term on the right-hand side is equal to

$$- \int_\Lambda D_i\,(G_i(u - v))\; \mathrm{d}\xi,$$

which in turn after summation from $i = 1$ to d by Gauss's divergence theorem is zero, since $G_i(u - v) = 0$ on $\partial\Lambda$ for all $1 \le i \le d$, because $u, v \in H_0^{1,2}(\Lambda)$. Hence altogether we obtain

$$_{V^*}\langle A(u) - A(v), u - v\rangle_V$$

$$\le -\|u - v\|_V^2 + \int_\Lambda \langle f(u) - f(u - v),\, \nabla(u - v)\rangle(u - v)\; \mathrm{d}\xi$$

$$+ \int_\Lambda \langle f(u) - f(v),\, \nabla v\rangle(u - v)\; \mathrm{d}\xi. \tag{5.14}$$

Now let us first consider the case $d = 1$. Then using that f is Lipschitz and applying the Cauchy–Schwarz and Young inequalities, we estimate the right-hand side of (5.14) by

$$-\|u - v\|_V^2 + \text{Lip}(f)\left(\|u - v\|_V \|v\|_{L^\infty} \|u - v\|_{L^2} + \|v\|_V \|u - v\|_{L^4}^2\right)$$

$$\leq -\frac{3}{4}\|u - v\|_V^2 + C\left(\|v\|_V^2 \|u - v\|_{L^2}^2 + \|v\|_V \|u - v\|_{L^4}^2\right), \tag{5.15}$$

where $C \in (0, \infty)$ is independent of u, v and we used that $H_0^{1,2}(\Lambda) \subset L^\infty(\Lambda)$ continuously, since $d = 1$.

In case $d = 2, 3$ and f is bounded, we similarly obtain that the right-hand side of (5.14) is dominated by

$$-\|u - v\|_V^2 + 2\|f\|_{L^\infty} \|u - v\|_V \|u - v\|_{L^2} + \text{Lip}(f)\|v\|_V \|u - v\|_{L^4}^2$$

$$\leq -\frac{3}{4}\|u - v\|_V^2 + C\left(\|u - v\|_{L^2}^2 + \|v\|_V \|u - v\|_{L^4}^2\right), \tag{5.16}$$

where $C \in (0, \infty)$ is independent of u, v.

For $d = 2$, we have the following well-known interpolation inequality:

$$\|u\|_{L^4}^2 \leq 2\|u\|_{L^2} \|\nabla u\|_{L^2}, \ u \in H_0^{1,2}(\Lambda), \tag{5.17}$$

(see Lemma H.0.1 (i) in Appendix H).

Hence combining this with (5.15), (5.16) and using Young's inequality we deduce that for some $C \in (0, \infty)$

$$_{V^*}\langle A(u) - A(v), u - v\rangle_V \leq -\frac{1}{2}\|u - v\|_V^2 + \left(C + C\|v\|_V^2\right)\|u - v\|_H^2 \text{ for all } u, v \in V,$$

and assertion (1) is proved.

For $d = 3$ we use the following interpolation inequality

$$\|u\|_{L^4}^2 \leq 2\sqrt{2}\|u\|_{L^2}^{\frac{1}{2}} \|\nabla u\|_{L^2}^{\frac{3}{2}}, \ u \in H_0^{1,2}(\Lambda), \tag{5.18}$$

(see Lemma H.0.1 (ii) in Appendix H).

Then assertion (2) can be derived similarly from (5.16) and Young's inequality.

(3): In this case A is linear and the assertion follows from the following lemma.
□

Lemma 5.1.7 *Let* $g \in L^d(\Lambda) + L^\infty(\Lambda)$ *and* $\varepsilon \in]0, \infty[$.
Define

$$\alpha := \alpha(\varepsilon) := \varepsilon^{-1} \inf\left\{\beta \in (0, \infty) \big| \|1_{\{|g|^2 > \beta\}} g\|_{L^d} \leq \frac{\varepsilon}{\lambda}\right\},$$

where λ is the constant of the Sobolev embedding $H_0^{1,2}(\Lambda) \subset L^{\frac{2d}{d-2}}(\Lambda)$, i.e.

$$\|u\|_{L^{\frac{2d}{d-2}}} \leq \lambda \left(\int_\Lambda |\nabla u|^2 \, d\xi \right)^{\frac{1}{2}} \quad \text{for all } u \in H_0^{1,2}(\Lambda).$$

Then $\alpha < \infty$ and for all $u, v \in H_0^{1,2}(\Lambda)$ we have

$$\int_\Lambda |g||u||\nabla v| \, d\xi \leq \left(\varepsilon \int_\Lambda |\nabla u|^2 \, d\xi + \alpha \|u\|_{L^2}^2 \right)^{\frac{1}{2}} \left(\varepsilon \int_\Lambda |\nabla v|^2 \, d\xi + \alpha \|v\|_{L^2}^2 \right)^{\frac{1}{2}}.$$

Proof Since $g \in L^d(\Lambda) + L^\infty(\Lambda)$, we have that $\alpha < \infty$. Furthermore, by the Cauchy–Schwarz inequality

$$\int_\Lambda |g||u||\nabla v| \, d\xi \leq \left(\int_\Lambda |g|^2 u^2 \, d\xi \right)^{\frac{1}{2}} \left(\int_\Lambda |\nabla v|^2 \, d\xi \right)^{\frac{1}{2}}. \tag{5.19}$$

The first factor on the right-hand side can be estimated by

$$\left(\int 1_{\{|g|^2 > \alpha\varepsilon\}} |g|^2 u^2 \, d\xi + \alpha\varepsilon \|u\|_{L^2}^2 \right)^{\frac{1}{2}}$$

$$\leq \left(\|1_{\{|g|^2 > \alpha\varepsilon\}} g\|_{L^d}^2 \|u\|_{L^{\frac{2d}{d-2}}}^2 + \alpha\varepsilon \|u\|_{L^2}^2 \right)^{\frac{1}{2}}$$

$$\leq \left(\varepsilon^2 \int_\Lambda |\nabla u|^2 \, d\xi + \alpha\varepsilon \|u\|_{L^2}^2 \right)^{\frac{1}{2}},$$

where we used Hölder's inequality and the definitions of λ, α respectively. Therefore, the right-hand side of (5.19) is dominated by

$$\varepsilon \left(\int_\Lambda |\nabla u|^2 \, d\xi + \frac{\alpha}{\varepsilon} \|u\|_{L^2}^2 \right)^{\frac{1}{2}} \left(\int_\Lambda |\nabla v|^2 \, d\xi + \frac{\alpha}{\varepsilon} \|v\|_{L^2}^2 \right)^{\frac{1}{2}}$$

which implies the assertion. $\qquad\square$

Example 5.1.8 (Semilinear Stochastic Equations) Let $d \leq 3$ and consider the situation of Lemma 5.1.6, (1), (3) with the operator A as defined there and the semilinear stochastic equation

$$dX(t) = (A(X(t)) + g(X(t))) \, dt + B(X(t)) \, dW(t), \tag{5.20}$$

where W is a cylindrical Wiener process in $L^2(\Lambda)$ defined on a probability space $(\Omega, \mathcal{F}, P)$ with normal filtration $\mathcal{F}_t$, $t \in [0, T]$. Let g, B satisfy the following conditions:

(i) $g : \mathbb{R} \to \mathbb{R}$ is a continuous function with $g(0) = 0$ and such that for some
$C, r, s \in [0, \infty[$

$$|g(x)| \leq C(|x|^r + 1), \; x \in \mathbb{R}; \tag{5.21}$$

$$(g(x) - g(y))(x - y) \leq C(1 + |y|^s)(x - y)^2, \; x, y \in \mathbb{R}. \tag{5.22}$$

(ii) $B : [0, T] \times V \times \Omega \to L_2(L^2(\Lambda))$ is as in Theorem 5.1.3.

Then the following holds:

(1) If $d = 1, r = 3, s = 2$, then for any $p \in [6, \infty[$ and $X_0 \in L^p(\Omega, \mathcal{F}_0, P; H)$,
(5.20) has a unique solution $(X(t))_{t \in [0, T]}$ and this solution satisfies

$$E\left(\sup_{t \in [0,T]} \|X(t)\|_H^p + \int_0^T \|X(t)\|_V^2 \, dt \right) < \infty. \tag{5.23}$$

(2) If $d = 2, r < 3, s = 2$, then for $p \geq \max(2r, 4)$ and any $X_0 \in L^p(\Omega, \mathcal{F}_0, P; H)$,
(5.20) has a unique solution $(X(t))_{t \in [0, T]}$ and this solution satisfies (5.23).
(3) If $d = 3, r = \frac{7}{3}, s = \frac{4}{3}$, then for $p \geq \frac{16}{3}$ and any $X_0 \in L^p(\Omega, \mathcal{F}_0, P; H)$, (5.20)
has a unique solution $(X(t))_{t \in [0, T]}$ and this solution satisfies (5.23).

Proof We define the operator

$$A_g(u) = A(u) + g(u), \; u \in V.$$

Since $d \leq 3$, by Sobolev embedding (cf. the proof of Lemma 5.1.6,(0)) we have
$V \subset L^6(\Lambda)$, hence $g(u) \in L^2(\Lambda) \subset V^*$ for all $u \in V$, and, therefore, $A_g : V \to V^*$. The hemicontinuity $(H1)$ follows easily from the conditions on f, g and by
Lemma 5.1.6, (0) in case (1) and (2), respectively the linearity of A in case (3).

(1):   Let $d = 1$. Then by (5.22) for all $u, v \in V$

$$_{V^*}\langle g(u) - g(v), u - v \rangle_V = \int_\Lambda (g(u) - g(v)) (u - v) \, d\xi$$

$$\leq C \left(1 + \|v\|_\infty^s\right) \|u - v\|_H^2.$$

Since $s = 2$ and $H_0^{1,2}(\Lambda) \subset L^\infty(\Lambda)$ continuously, it follows from
Lemma 5.1.6,(1) that A_g satisfies (H2') with $\rho(v) := C(1 + \|v\|_V^2)$ for some
constant $C \in]0, \infty[$. Likewise $(H3)$ then follows with $\alpha = 2$, since $g(0) = 0$.

Furthermore, for all $u \in V$ by (5.21), because $r = 3$,

$$|_{V^*}\langle g(u), v\rangle_V| \leq C \int_\Lambda (1 + |u|^r)|v| \, d\xi$$

$$\leq C \left(\|v\|_{L^1} + \|v\|_{L^\infty} \|u\|_{L^\infty} \|u\|_{L^2}^2 \right),$$

which by the above embedding is up to a constant dominated by

$$\|v\|_V \left(1 + \|u\|_V \|u\|_H^2 \right).$$

Hence Lemma 5.1.6, (0) implies that $(H4')$ and hence (5.2) holds (with $\alpha = 2$) and $\beta = 4$. So, Theorem 5.1.3 applies, with $p \geq 6$.

(2):   Let $d = 2$. We recall that in this case we assume that f is bounded. We have for all $u, v \in V$ by (5.22)

$$_{V^*}\langle g(u) - g(v), u - v\rangle_V \leq C \int_\Lambda (1 + |v|^s)(u - v)^2 \, d\xi$$

$$\leq C\|u - v\|_H^2 + C\|v\|_{L^{2s}}^s \|u - v\|_{L^4}^2$$

$$\leq C\|u - v\|_H^2 + 2C\|v\|_{L^{2s}}^s \|u - v\|_H \|u - v\|_V,$$

where we used (5.17) in the last step. Applying Young's inequality we obtain that for some constant $\tilde{C} \in]0, \infty[$ the right-hand side of the above inequality is bounded by

$$\frac{1}{4}\|u - v\|_V^2 + C\|u - v\|_H^2 + \tilde{C}\|v\|_{L^{2s}}^{2s} \|u - v\|_H^2.$$

Since $s = 2$, we can use (5.17) again to conclude by Lemma 5.1.6,(1) that A_g satisfies $(H2')$ with

$$\rho(v) := C \left(1 + \|v\|_V^2 \right) \left(1 + \|v\|_H^2 \right)$$

for some constant $C \in]0, \infty[$. And again since $g(0) = 0$, $(H3)$ also holds with $\alpha = 2$. Furthermore, for all $u \in V$ by (5.21) for $\varepsilon \in]0, 1[$

$$|_{V^*}\langle g(u), v\rangle_V| \leq C \int_\Lambda (1 + |u|^r)|v| \, d\xi$$

$$\leq C \left(\|v\|_{L^1} + \|u\|_{L^{r(1+\varepsilon)}}^r \|v\|_{L^{(1+\varepsilon)/\varepsilon}} \right),$$

which (since $d = 2$) up to a constant is dominated by

$$\left(1 + \|u\|_{L^q}^r \right) \|v\|_V$$

with $q := r(1 + \varepsilon)$. Now since $r < 3$ we can choose ε so small that $r < 1 + 2(1 + \varepsilon)^{-1}$, hence

$$p := \frac{2}{2 - (r - 1)(1 + \varepsilon)} \in]1, \infty[.$$

Let $p' := \frac{p}{p-1} = \frac{2}{(r-1)(1+\varepsilon)}$ and $\lambda \in]0, 1[$. Then by interpolation and choosing $\lambda := r^{-1}$ (assuming without loss of generality that $r > 1$) we obtain

$$\|u\|_{L^q}^r \leq \|u\|_{L^{\lambda qp}}^{\lambda r} \|u\|_{L^{(1-\lambda)qp'}}^{(1-\lambda)r}$$

$$\leq \|u\|_{L^{(1+\varepsilon)p}} \|u\|_{L^2}^{r-1},$$

which is up to a constant dominated by $\|u\|_V \|u\|_H^{r-1}$. Hence Lemma 5.1.6, (0) implies that $(H4')$ holds (with $\alpha = 2$) and $\beta = 2(r - 1)$. So, (5.2) holds with $\alpha = 2$, $\beta = 2 \max(r - 1, 1)$ and Theorem 5.1.3 applies with $p \geq \max(2r, 4)$.

(3):   Let $d = 3$. We recall that in this case as in Lemma 5.1.6,(3) we assume that $A = \Delta + \langle b, \nabla \rangle$ with b as given there. We have for all $u, v \in V$ by (5.22) as in the proof of assertion (2) that

$$_{V^*}\langle g(u) - g(v), u - v \rangle_V \leq C\|u - v\|_H^2 + C\|v\|_{L^{2s}}^s \|u - v\|_{L^4}^2$$

$$\leq C\|u - v\|_H^2 + 2\sqrt{2}C\|v\|_{L^{2s}}^s \|u - v\|_{L^2}^{\frac{1}{2}} \|u - v\|_V^{\frac{3}{2}},$$

where we used (5.18) in the last step. So, by Young's inequality for some $\tilde{C} \in]0, \infty[$ the right-hand side is bounded by

$$\frac{1}{4}\|u - v\|_V^2 + C\|u - v\|_H^2 + \tilde{C}\|v\|_{L^{2s}}^{4s} \|u - v\|_H^2.$$

By interpolation for $\lambda := \frac{1}{2s} \in]0, 1[$ and $p := 6$, $p' := \frac{6}{5}$ we have, because $s = \frac{4}{3}$,

$$\|v\|_{L^{2s}}^{4s} \leq \|v\|_{L^{2\lambda sp}}^{4\lambda s} \|v\|_{L^{2(1-\lambda)sp'}}^{4(1-\lambda)s}$$

$$= \|v\|_{L^6}^2 \|v\|_{L^{(2s-1)\frac{6}{5}}}^{2(2s-1)}$$

$$= \|v\|_{L^6}^2 \|v\|_H^{\frac{10}{3}},$$

which is up to a constant dominated by $\|v\|_V^2 \|v\|_H^{\frac{10}{3}}$. Hence Lemma 5.1.7 implies that A_g satisfies $(H2')$ with

$$\rho(v) := C\left(1 + \|v\|_V^2\right)\left(1 + \|v\|_H^{\frac{10}{3}}\right)$$

for some constant $C \in]0, \infty[$. And, since $g(0) = 0$, (H3) also holds with $\alpha = 2$. Furthermore, for all $u \in V$ by (5.21)

$$|_{V^*}\langle g(u), v \rangle_V| \le C \int_\Lambda (1 + |u|^r)|v| \, d\xi$$

$$\le C \left(\|v\|_{L^1} + \|u\|_{L^q}^r \|v\|_{L^{\frac{2d}{d-2}}} \right)$$

with $q := \frac{2dr}{d+2}$, which, since $H_0^{1,2}(\Lambda) \subset L^{\frac{2d}{d-2}}(\Lambda)$ continuously, is up to a constant dominated by

$$\left(1 + \|u\|_{L^q}^r \right) \|v\|_V.$$

For $\lambda := \frac{1}{r} \in]0, 1[$ by interpolation we have

$$\|u\|_{L^q}^r \le \|u\|_{L^{\lambda qp}}^{\lambda r} \|u\|_{L^{(1-\lambda)qp'}}^{(1-\lambda)r}$$

$$\le \|u\|_{L^{\frac{2dp}{(d+2)}}} \|u\|_{L^{\frac{(r-1)qp'}{r}}}^{r-1} .$$

Choosing $p := \frac{d+2}{d-2}$ and $p' = \frac{p}{p-1} = \frac{d+2}{4}$, the right-hand side is equal to

$$\|u\|_{L^{\frac{2d}{d-2}}} \|u\|_{L^{\frac{(r-1)d}{2}}}^{r-1} ,$$

which for $r = 1 + \frac{4}{d} = \frac{7}{3}$, since $d = 3$, is up to a constant dominated by $\|u\|_V \|u\|_H^{\frac{4}{d}} = \|u\|_V \|u\|_H^{\frac{4}{3}}$. Hence Lemma 5.1.7 implies that (H4$'$) holds (with $\alpha = 2$) and $\beta = \frac{8}{3}$. So, (5.2) holds with $\alpha = 2$, $\beta = \frac{10}{3}$ and Theorem 5.1.3 applies with $p \ge \frac{16}{3}$. □

Remark 5.1.9

(i) Equation (5.20), in particular for $g \equiv 0$, is often called the stochastic generalized Burgers equation. In the case where $d = 1$ and $f(r) = r, r \in \mathbb{R}$, (5.20) is called the stochastic Burgers equation.

(ii) One obvious generalization is that one can replace Δ in (5.20) by the p-Laplace operator $\text{div}(|\nabla u|^{p-2}\nabla u)$ or the more general quasi-linear differential operator

$$\sum_{|\alpha| \le m} (-1)^{|\alpha|} D_\alpha A_\alpha(Du),$$

where $Du = (D_\beta u)_{|\beta| \le m}$. Under certain assumptions (cf. [82, Proposition 30.10]) this operator satisfies the monotonicity and coercivity condition. Then, according to Theorem 5.1.3, we can obtain the existence and uniqueness of solutions to this type of quasi-linear SPDE with non-monotone perturbations (e.g. some locally Lipschitz lower order terms).

Example 5.1.10 (Stochastic 2 D Navier–Stokes Equation) Let Λ be a bounded domain in $\mathbb{R}^2$ with smooth boundary. Define

$$V := \left\{ v \in H_0^{1,2}(\Lambda; \mathbb{R}^2) : \nabla \cdot v = 0 \ a.e. \ \text{in} \ \Lambda \right\}, \ \|v\|_V := \left(\int_\Lambda |\nabla v|^2 \, d\xi \right)^{1/2},$$

and let H be the closure of V in $L^2(\Lambda; \mathbb{R}^2)$. The linear operator P_H (Helmholtz–Leray projection) and A (Stokes operator with viscosity constant v) are defined by

$$P_H : L^2(\Lambda; \mathbb{R}^2) \to H \quad \text{orthogonal projection;}$$

$$A : H^{2,2}(\Lambda; \mathbb{R}^2) \cap V \to H, \ Au = v P_H \Delta u,$$

where $H^{2,2}(\Lambda; \mathbb{R}^2)$ is the standard Sobolev space of order 2 in $L^2(\Lambda; \mathbb{R}^2)$ (see e.g. [77, Chap. I, 3.6]) Then the Navier–Stokes equation can be formulated as follows

$$u' = Au + F(u) + f, \ u(0) = u_0 \in H, \tag{5.24}$$

where $f \in L^2([0, T]; V^*)$ denotes some external force and

$$F : V \times V \to V^*, \ F(u, v) := -P_H \left[(u \cdot \nabla) v \right], F(u) := F(u, u),$$

where $u \cdot \nabla = \sum_{i=1}^2 u^i \partial_i$ and $u = (u^1, u^2)$. That $F : V \times V \to V^*$ is indeed well-defined and even continuous follows by Lemma 5.1.6, (0). Using the Gelfand triple

$$V \subset H \equiv H^* \subset V^*,$$

as in Example 4.1.7, one sees that A extends by continuity to a map

$$A : V \to V^*,$$

so that for some $C \in]0, \infty[$, $\|Au\|_{V^*} \leq C\|u\|_V$, $u \in V$. In particular, we have

$$_{V^*}\langle F(u, v), w \rangle_V = -_{V^*}\langle F(u, w), v \rangle_V, \ _{V^*}\langle F(u, v), v \rangle_V = 0, \ u, v, w \in V. \tag{5.25}$$

Now we consider the *stochastic* 2-D Navier–Stokes equation

$$dX(t) = (AX(t) + F(X(t)) + f(t)) \ dt + B(X(t)) \ dW(t), \ X(0) = X_0, \tag{5.26}$$

where W is a cylindrical Wiener process in H on a probability space $(\Omega, \mathcal{F}, P)$ with normal filtration $\mathcal{F}_t$, $t \in [0, T]$.

Claim Suppose that $p \geq 4$, $X_0 \in L^p(\Omega, \mathcal{F}_0, P; H)$ and $B : [0, T] \times V \times \Omega$ is as in Theorem 5.1.3. Then (5.26) has a unique solution $(X(t))_{t \in [0,T]}$ and this solution satisfies

$$E\left(\sup_{t \in [0,T]} \|X(t)\|_H^p + \int_0^T \|X(t)\|_V^2 \, dt \right) < \infty.$$

Proof The hemicontinuity $(H1)$ is obvious since A is a linear and F is a bilinear map.

By (5.25) and (5.17), respectively, it follows that for some constant $C \in]0, \infty[$

$$|_{V^*}\langle F(w), v \rangle_V| \leq C\|w\|_{L^4(\Lambda; \mathbb{R}^2)}^2 \|v\|_V;$$

$$|_{V^*}\langle F(w), v \rangle_V| \leq C\|w\|_V^{3/2}\|w\|_H^{1/2}\|v\|_{L^4(\Lambda; \mathbb{R}^2)}, \quad v, w \in V. \tag{5.27}$$

Then by (5.25) and (5.27) it follows that for some $C, \tilde{C} \in]0, \infty[$

$$\begin{aligned}
_{V^*}\langle F(u) - F(v), u - v \rangle_V &= -_{V^*}\langle F(u, u - v), v \rangle_V + _{V^*}\langle F(v, u - v), v \rangle_V \\
&= -_{V^*}\langle F(u - v), v \rangle_V \\
&\leq C\|u - v\|_V^{3/2}\|u - v\|_H^{1/2}\|v\|_{L^4(\Lambda; \mathbb{R}^2)} \\
&\leq \frac{v}{2}\|u - v\|_V^2 + \frac{\tilde{C}}{v^3}\|v\|_{L^4(\Lambda; \mathbb{R}^2)}^4 \|u - v\|_H^2, \quad u, v \in V.
\end{aligned} \tag{5.28}$$

Hence

$$_{V^*}\langle Au + F(u) - Av - F(v), u - v \rangle_V \leq -\frac{v}{2}\|u - v\|_V^2 + \frac{\tilde{C}}{v^3}\|v\|_{L^4(\Lambda; \mathbb{R}^2)}^4 \|u - v\|_H^2$$

and $(H2')$ holds with $\rho(v) := \frac{\tilde{C}}{v^3}\|v\|_{L^4(\Lambda; \mathbb{R}^2)}^4$ which by (5.17) is up to a constant dominated by $\|u\|_H^2\|u\|_V^2$. By (5.25) it is also easy to show $(H3)$ with $\alpha = 2$. Indeed, for some $C \in]0, \infty[$

$$\begin{aligned}
_{V^*}\langle Av + F(v) + f(t), v \rangle_V &\leq -v\|v\|_V^2 + \|f(t)\|_{V^*}\|v\|_V \\
&\leq -\frac{v}{2}\|v\|_V^2 + C\|f(t)\|_{V^*}^2, \quad v \in V,
\end{aligned}$$

$$\|B(v)\|_{L_2}^2 \leq 2C\|v\|_H^2 + 2\|B(0)\|_{L_2}^2, \quad v \in V.$$

(5.27) and (5.17) imply that $(H4')$ and (5.2) hold with $\beta = 2$.

Therefore, the existence and uniqueness of solutions to (5.26) follow from Theorem 5.1.3 by taking $p \geq 4$. $\qquad \square$

Remark 5.1.11

(1) Let us now consider the 3D Navier–Stokes equation. We note that (5.25) still holds, since it is dimension independent. Hence using (5.18) (instead of (5.17)), as above we obtain for some $C, \tilde{C} \in]0, \infty[$

$$
\begin{aligned}
_{V^*}\langle F(u) - F(v), u - v \rangle_V &= -_{V^*}\langle F(u - v), v \rangle_V \\
&\le C\|u - v\|_V^{7/4}\|u - v\|_H^{1/4}\|v\|_{L^4(\Lambda;\mathbb{R}^3)} \\
&\le \frac{v}{2}\|u - v\|_V^2 + \frac{\tilde{C}}{v^7}\|v\|_{L^4(\Lambda;\mathbb{R}^3)}^8\|u - v\|_H^2, \quad u, v \in V.
\end{aligned}
$$

Hence we have the following local monotonicity (*H2'*):

$$
_{V^*}\langle Au + F(u) - Av - F(v), u - v \rangle_V \le -\frac{v}{2}\|u - v\|_V^2 + \frac{\tilde{C}}{v^7}\|v\|_{L^4(\Lambda;\mathbb{R}^3)}^8\|u - v\|_H^2.
$$

Using an interpolation formula for the norm $\| \ \|_{L^3}$ (see [78, Theorem 2.1]), another form of local monotonicity can be derived similarly:

$$
\begin{aligned}
_{V^*}\langle F(u) - F(v), u - v \rangle_V &= -_{V^*}\langle F(u - v), v \rangle_V \\
&\le C\|u - v\|_V^{3/2}\|u - v\|_H^{1/2}\|v\|_{L^6(\Lambda;\mathbb{R}^3)} \\
&\le \frac{v}{2}\|u - v\|_V^2 + \frac{\tilde{C}}{v^3}\|v\|_{L^6(\Lambda;\mathbb{R}^3)}^4\|u - v\|_H^2, \quad u, v \in V.
\end{aligned}
$$

(2) Clearly, (H3) holds with $\alpha = 2$ by (5.25). However, concerning the growth condition, by (5.18) we have in the 3D case that for some $C, \tilde{C} \in]0, \infty[$

$$
\|F(u)\|_{V^*} \le C\|u\|_{L^4(\Lambda;\mathbb{R}^3)}^2 \le \tilde{C}\|u\|_H^{1/2}\|u\|_V^{3/2}, \quad u \in V.
$$

Unfortunately, this is not enough to verify (*H4'*), since $\alpha = 2$.

(3) By similar arguments as in the case of the 2D Navier–Stokes equation it can be shown that Theorem 5.1.3 applies to incompressible non-Newtonian fluids subject to random forces. For details we refer to [59, Sect. 3.5].

(4) Besides the stochastic 2D Navier–Stokes equation, many other hydrodynamical systems also satisfy the local monotonicity condition (*H2'*) and the coercivity condition (H3). For example, in [17] the authors have studied the well-posedness and large deviation principle for an abstract stochastic semilinear equation which covers a wide class of fluid dynamical models. In fact, the Conditions (C1) and (C2) in [17] imply that the assumptions in Theorem 5.1.3 hold. More precisely, (2.2) in [17] implies (H3) and the local monotonicity (*H2'*) follows from (2.4) (or (2.8)) in [17]. The other assumptions in Theorem 5.1.3 can also be verified easily. Therefore, Theorem 5.1.3 can be applied to show the well-posedness of all stochastic hydrodynamical models in [17] such as the

stochastic magneto-hydrodynamic equations, the stochastic Boussinesq model
for the Bénard convection, the stochastic 2D magnetic Bénard problem, the
stochastic 3D Leray-α model and some stochastic shell models of turbulence.

5.2 Generalized Coercivity

5.2.1 Main Results

Let

$$V \subset H \equiv H^* \subset V^*$$

be a Gelfand triple as in Sect. 4.1 and consider the general nonlinear evolution
equation

$$u'(t) = A(t, u(t)), \ t \in]0, T[, \ u(0) = u_0 \tag{5.29}$$

in H, where $T > 0$ and $A : [0, T] \times V \to V^*$ is $\mathcal{B}([0, T] \times V)/\mathcal{B}(V^*)$-measurable.

In this section we establish the existence, uniqueness and continuous dependence
on initial conditions of solutions to (5.29) (see Theorems 5.2.2 and 5.2.4 below),
replacing the local monotonicity condition $(H2')$ from the previous section by an
even weaker condition (see $(H2'')$ below) and also relaxing the coercivity condition
$(H3)$ (see $(H3')$ below). An analogous result for stochastic PDEs with general
additive noise is also obtained (see Theorem 5.2.6 below). In Sect. 5.2.3 below the
main result will then be applied to establish local/global existence and uniqueness of
solutions for a large class of classical (stochastic) nonlinear evolution equations such
as the stochastic 2D and 3D Navier–Stokes equations, the tamed 3D Navier–Stokes
equation and the Cahn–Hilliard equation. Through our generalized framework we
give new and significantly simpler proofs for all these well known results. Moreover,
the main result is also applied to less well-studied stochastic surface growth PDEs
and stochastic power law fluids to obtain existence and uniqueness results also for
these models (see Sect. 5.2.3 for details). We emphasize that by applying our main
results (see Theorems 5.2.2 and 5.2.6) we obtain both the classical local existence
and uniqueness of *strong* solutions to the stochastic 3D Navier–Stokes equation and
quite recent local existence and uniqueness results for stochastic surface growth
PDE (see [59]). Here the meaning of *strong* solution is in the sense of both PDE and
probability theory.

Let us now formulate the precise conditions on the coefficients in (5.29).

Suppose that there exist constants $\alpha \in]1, \infty[, \ \beta, C \in [0, \infty[, \ \theta \in]0, \infty[$ and
nonnegative functions $f, h \in L^1([0, T]; \mathbb{R})$ such that the following conditions hold
for all $t \in [0, T]$ and $u, v, w \in V$:

(*H*1) (Hemicontinuity) The map $\lambda \mapsto {}_{V^*}\langle A(t, u + \lambda v), w\rangle_V$ is continuous on $\mathbb{R}$.
(*H*2'') (Local monotonicity)

$${}_{V^*}\langle A(t, u) - A(t, v), u - v\rangle_V \leq (f(t) + \eta(u) + \rho(v)) \|u - v\|_H^2,$$

where $\eta, \rho : V \to [0, +\infty[$ are measurable and locally bounded functions
on V.
(*H*3') (Generalized coercivity)

$$2 {}_{V^*}\langle A(t, v), v\rangle_V \leq h(t) g(\|v\|_H^2) - \theta\|v\|_V^\alpha + f(t),$$

where $g : [0, \infty[\to [0, \infty[$ is a non-decreasing continuous function such that
$g(]0, \infty[) \subset]0, \infty[$.
(*H*4') (Growth)

$$\|A(t, v)\|_{V^*} \leq \left(f(t)^{\frac{\alpha-1}{\alpha}} + C\|v\|_V^{\alpha-1}\right)\left(1 + \|v\|_H^\beta\right).$$

Remark 5.2.1 Since η and ρ are locally bounded on V, Remark 5.1.1,(4) holds correspondingly. Hence (*H*1) and (*H*2'') imply that $A(t, \cdot) : V \to V^*$ is demicontinuous for every $t \in [0, T]$.

Define

$$L(t) := \int_0^t h(s) \, ds + G\left(\|u_0\|_H^2 + \int_0^t f(s) \, ds\right), \quad t \in [0, T], \tag{5.30}$$

where f, g, h are as in (*H*3') and $G(x) := \int_{x_0}^x \frac{1}{g(s)} \, ds$, $x \in]0, \infty[$, for some fixed $x_0 \in]0, \infty[$. Note that since $G(x)$ is not defined for $x = 0$, we have to make sure that the argument of G in (5.30) is not zero. But we may always replace f in (*H*3)' by a bigger function, so that $\int_0^t f(s) \, ds > 0$ for all $t \in]0, T]$, which we shall tacitly do below.

Now we can state the first main result of this section, which provides a more general framework to analyze various classes of nonlinear evolution equations.

Theorem 5.2.2 *Suppose that $V \subset H$ is compact and that (H1), (H2''), (H3'), (H4') hold.*

(*i*) *For every $u_0 \in H$ and every $T_0 \in]0, T]$ such that*

$$L(T_0) < \sup_{x \in]0, \infty[} G(x), \tag{5.31}$$

(5.29) has a solution on $[0, T_0]$, i.e.

$$u \in L^\alpha([0, T_0]; V) \cap C([0, T_0]; H),$$

and

$$u(t) = u_0 + \int_0^t A(s, u(s)) \, ds, \ t \in [0, T_0],$$

where the integral is a V^*-valued Bochnel integral.

(ii) If (H3') holds with $g(x) = Cx, x \geq 0$, for some constant C, then one can take $T_0 = T$ in (i).

Remark 5.2.3 If $h \equiv 1$ and $g(x) = C_0|x|^\gamma (\gamma > 1)$, then for $T_0 \in]0, T]$ we have $L(T_0) < \sup_{x \in]0,\infty[} G(x)$ if and only if

$$T_0 < \frac{C_0}{(\gamma - 1) \left(\|u_0\|_H^2 + \int_0^{T_0} f(s) \, ds \right)^{\gamma - 1}},$$

where the right-hand side tends to $+\infty$ if $\gamma \downarrow 1$, proving that (ii) follows from (i) in Theorem 5.2.2.

The next result shows the continuous dependence of solutions to (5.29) on the initial condition u_0.

Theorem 5.2.4 *Suppose that $V \subset H$ is compact and that (H1), (H2''), (H3'), (H4') hold.*

(i) Let $T_0 \in]0, T]$ and let u_i be solutions of (5.29) on $[0, T_0]$ for initial conditions $u_{i,0} \in H$, $i = 1, 2$ respectively. Then there exists a $C \in [0, \infty[$ such that

$$\|u_1(t) - u_2(t)\|_H^2 \leq \|u_{1,0} - u_{2,0}\|_H^2 \exp\left[C \int_0^t \left(f(s) + \eta(u_1(s)) + \rho(u_2(s)) \right) ds \right],$$

$$t \in [0, T_0]. \tag{5.32}$$

In particular, $u_{1,0} = u_{2,0}$ implies $u_1 = u_2$ provided the integral in (5.32) is finite for $t = T_0$, which in turn holds if there exist $C, \gamma \in [0, \infty[$ such that

$$\eta(v) + \rho(v) \leq C(1 + \|v\|_V^\alpha)(1 + \|v\|_H^\gamma), \ v \in V. \tag{5.33}$$

(ii) Suppose (5.33) holds. If $u, u^{(n)}, n \in \mathbb{N}$, are solutions of (5.29) on $[0, T_0]$ such that $u^{(n)}(0) \to u(0)$ in H as $n \to \infty$, then

$$\lim_{n\to\infty} \sup_{t\in[0,T_0]} \|u^{(n)}(t) - u(t)\|_H = 0.$$

Now we formulate the analogous result for SPDEs on Hilbert spaces with additive type noise. Suppose that U is another Hilbert space and $W(t)$ is a U-valued cylindrical Wiener process defined on a probability space $(\Omega, \mathcal{F}, P)$ with normal filtration $\mathcal{F}_t$, $t \in [0, T]$. We consider the following type of stochastic evolution equations on H,

$$dX(t) = [A_1(t, X(t)) + A_2(t, X(t))] \, dt + B(t) \, dW(t), \ t \in [0, T], \ X(0) = X_0,$$
$$(5.34)$$

where $A_1, A_2 : [0, T] \times V \to V^*$ and $B : [0, T] \to L_2(U, H)$ are measurable.

Now we give the definition of a local solution to (5.34).

Definition 5.2.5

(i) For an $(\mathcal{F}_t)$-stopping time $\tau : \Omega \mapsto [0, T]$ a process $(X(t))_{t \in [0, \tau]}$ is called $(\mathcal{F}_t)$-adapted if the process $(X(t))_{t \in [0, T]}$ defined by

$$\widetilde{X}(t) := \begin{cases} X(t) & , t \in [0, \tau] \\ X(\tau) & , t \in \,]\tau, T] \end{cases}$$

is $(\mathcal{F}_t)$-adapted.

(ii) An H-valued $(\mathcal{F}_t)$-adapted process $(X(t))_{t \in [0, \tau]}$ is called a local solution of (5.34) if $X(\cdot, \omega) \in L^1([0, \tau(\omega)]; V) \cap C([0, \tau(\omega)]; H)$ and P-a.s. $\omega \in \Omega$,

$$X(t) = X_0 + \int_0^t [A_1(s, X(s)) + A_2(s, X(s))] \, ds + \int_0^t B(s) \, dW(s), \ t \in [0, \tau(\omega)],$$

where τ is an $(\mathcal{F}_t)$- stopping time satisfying $\tau(\omega) > 0$ for P-a.e. $\omega \in \Omega$ and $X_0 \in L^2(\Omega, \mathcal{F}_0, P; H)$.

(iii) A local solution $(X(t))_{t \in [0, \tau^X]}$ is called unique if for any other local solution $(Y(t))_{t \in [0, \tau^Y]}$ we have

$$P\left(X(t) = Y(t), \ t \in [0, \tau^X \wedge \tau^Y]\}\right) = 1.$$

Theorem 5.2.6 *Suppose that $V \subset H$ is compact, A_1 satisfies (H1), (H2''), (H3'), (H4') with $\eta \equiv 0$, $\beta = 0$, $h = 1$ and $g(x) = Cx$, A_2 satisfies (H1), (H2''), (H3'), (H4') with the same $\alpha \in]1, \infty[$ in (H3'), (H4') as A_1. Furthermore, suppose that $B \in L^2([0, T]; L_2(U, H))$, and that there exist $C, \gamma \in [0, \infty[$ such that*

$$g(x + y) \le C(g(x) + g(y)), \ x, y \in [0, \infty[;$$

$$\eta(u + v) \le C(\eta(u) + \eta(v)), \ u, v \in V;$$

$$\rho(u + v) \le C(\rho(u) + \rho(v)), \ u, v \in V;$$

$$\eta(v) + \rho(v) \le C(1 + \|v\|_V^\alpha)(1 + \|v\|_H^\gamma), \ v \in V.$$

*Then for any $X_0 \in L^2(\Omega, \mathcal{F}_0, P; H)$, there exists a unique local solution $(X(t))_{t\in[0,\tau]}$
to (5.34) satisfying*

$$X(\cdot) \in L^\alpha([0, \tau]; V) \ P\text{-}a.s.$$

*Moreover, if A_2 also satisfies $(H3')$ with $g(x) = Cx$ and if $\alpha\beta \le 2$, then all assertions
above hold for $\tau \equiv T$.*

Remark 5.2.7

(1) The main idea of the proof is to use a shift transformation to reduce (5.34) to a
deterministic evolution equation (with random parameter) which Theorem 5.2.2
can be applied to. More precisely, we consider the process Y which solves the
following stochastic differential equation:

$$dY(t) = A_1(t, Y(t)) \, dt + B(t) \, dW(t), \ t \in [0, T], \ Y(0) = 0. \tag{5.35}$$

Since A_1 satisfies $(H1), (H2''), (H3'), (H4')$ with $\eta \equiv 0, \beta = 0, h \equiv 1$ and
$g(x) = Cx$, the existence and uniqueness of $Y(t)$ follows from Theorem 5.1.3
with $p = 2$. Let $u := X - Y$. Then it is easy to show that u satisfies a
deterministic evolution equation of type (5.29) for each fixed $\omega \in \Omega$.

(2) Unlike in [38], here we do not need to assume the noise to take values in V (i.e.
$B \in L^2([0, T]; L_2(U; V)))$. The reason is that here we use the auxiliary process Y
instead of subtracting the noise part directly as in [38] and that $A_1 \ne 0$ because
it satisfies $(H3')$.

5.2.2 Proofs of the Main Theorems

The proof of Theorem 5.2.2 is split into several lemmas. First, however, we
need some preparations. We shall start with the Galerkin approximation to (5.29).
However, even in the finite dimensional case, existence and uniqueness of solutions
to (5.29) do not immediately follow from the standard results because of the
generalized coercivity condition $(H3')$. We shall prove existence below by using
a classical theorem of Carathéodory for ordinary differential equations. Another
difficulty is that we cannot apply Gronwall's lemma directly for $(H3')$. Instead, we
will use Bihari's inequality, which is a generalized version of Gronwall's lemma.

Lemma 5.2.8 (Bihari's Inequality) *Let $g : [0, \infty[\to [0, \infty[$ be a non-decreasing
continuous function such that $g(]0, \infty[) \subset]0, \infty[$. If $p, h : [0, \infty[\to [0, \infty[$ are
measurable functions with $h \in L^1_{loc}([0, \infty[)$ and $K_0 \in]0, \infty[$ such that*

$$p(t) \le K_0 + \int_0^t h(s) \, g(p(s)) \, ds, \ t \ge 0.$$

Then we have

$$p(t) \leq G^{-1} \left(G(K_0) + \int_0^t h(s) \, ds \right), \ 0 \leq t \leq T_0, \tag{5.36}$$

where $G(x) := \int_{x_0}^x \frac{1}{g(s)} \, ds < \infty$ for all $x \in]0, \infty[$ and some fixed $x_0 \in]0, \infty[$, $G^{-1} :$ $G(]0, \infty[) \to]0, \infty[$ is the inverse function of G and $T_0 \in]0, \infty[$ such that $G(K_0) +$ $\int_0^{T_0} h(s) \, ds$ belongs to the domain of G^{-1}.

Remark 5.2.9 G is continuous and strictly increasing, hence

$$G(]0, \infty[) =] \inf_{x \in]0,\infty[} G(x), \ \sup_{x \in]0,\infty[} G(x)[.$$

In particular, $G(K_0) < \sup_{x \in]0,\infty[} G(x)$ and the interval $[G(K_0), \sup_{x \in]0,\infty[} G(x)[$ is contained in $G(]0, \infty[)$, i.e. in the domain of G^{-1}. Hence (5.36) holds for $t \in [0, T_0]$, where $T_0 \in]0, \infty[$ satisfies

$$\int_0^{T_0} h(s) \, ds + G(K_0) < \sup_{x \in]0,\infty[} G(x).$$

In particular, if $h \equiv 1$ and $g(x) = C_0 x^\gamma$ for some constants $C_0 > 0$ and $\gamma > 1$, then

$$G(x) = \frac{C_0}{\gamma - 1} \left(x_0^{1-\gamma} - x^{1-\gamma} \right); \ G^{-1}(x) = \left| x_0^{1-\gamma} - \frac{\gamma - 1}{C_0} x \right|^{\frac{1}{1-\gamma}} \text{sign} \left(x_0^{1-\gamma} - \frac{\gamma - 1}{C_0} x \right).$$

Hence (5.36) holds on $[0, T_0]$ for any $T_0 \in [0, \frac{C_0}{\gamma-1} K_0^{1-\gamma}[$ (in particular, for any $T_0 \in [0, \infty[$ if $\gamma = 1$).

We first recall the definition of pseudo-monotone operators, which is a very useful generalization of monotone operators and was first introduced by Brézis in [11]. It will be crucial for us below. For abbreviation we use the notation "$\rightharpoonup$" for weak convergence in a Banach space.

Definition 5.2.10 The operator $A : V \to V^*$ is called pseudo-monotone if $u_n \rightharpoonup u$ in V and

$$\liminf_{n \to \infty} {}_{V^*}\langle A(u_n), u_n - u \rangle_V \geq 0$$

implies that for all $v \in V$

$${}_{V^*}\langle A(u), u - v \rangle_V \geq \limsup_{n \to \infty} {}_{V^*}\langle A(u_n), u_n - v \rangle_V.$$

Lemma 5.2.11 *Let A be pseudo-monotone and $u_n \rightharpoonup u$ in V. Then*

$$\limsup_{n \to \infty} {}_{V^*}\langle A(u_n), u_n - u\rangle_V \leq 0.$$

Proof If the conclusion is not true, then we can extract a subsequence such that $u_{n_k} \rightharpoonup u$ in V and

$$\liminf_{k \to \infty} {}_{V^*}\langle A(u_{n_k}), u_{n_k} - u\rangle_V > 0.$$

Since A is pseudo-monotone, this implies that

$$
\begin{aligned}
0 &= {}_{V^*}\langle A(u), u - u\rangle_V \\
&\geq \limsup_{k \to \infty} {}_{V^*}\langle A(u_{n_k}), u_{n_k} - u\rangle_V \\
&\geq \liminf_{k \to \infty} {}_{V^*}\langle A(u_{n_k}), u_{n_k} - u\rangle_V.
\end{aligned}
$$

This contradiction proves the assertion. $\qquad\square$

Remark 5.2.12 In [12] Browder introduced a different definition of pseudo-monotone operators: An operator $A : V \to V^*$ is called pseudo-monotone if $u_n \rightharpoonup u$ in V and

$$\liminf_{n \to \infty} {}_{V^*}\langle A(u_n), u_n - u\rangle_V \geq 0$$

implies

$$A(u_n) \rightharpoonup A(u) \text{ in } V^* \text{ and } \lim_{n \to \infty} {}_{V^*}\langle A(u_n), u_n\rangle_V = {}_{V^*}\langle A(u), u\rangle_V.$$

This definition, however, turns out to be equivalent to Definition 5.2.10. Indeed, obviously the above definition implies that the operator A is pseudo-monotone in the sense of Definition 5.2.10. Conversely, if A is as in Definition 5.2.10 and $u_n \rightharpoonup u$ in V such that

$$\liminf_{n \to \infty} {}_{V^*}\langle A(u_n), u_n - u\rangle_V \geq 0,$$

then by Lemma 5.2.11

$$\lim_{n \to \infty} {}_{V^*}\langle A(u_n), u_n - u\rangle_V = 0.$$

Hence for all $v \in V$

$$\limsup_{n\to\infty} {}_{V^*}\langle A(u_n), u - v\rangle_V$$

$$= \limsup_{n\to\infty} {}_{V^*}\langle A(u_n), u - v\rangle_V + \lim_{n\to\infty} {}_{V^*}\langle A(u_n), u_n - u\rangle_V$$

$$= \limsup_{n\to\infty} {}_{V^*}\langle A(u_n), u_n - v\rangle_V$$

$$\leq {}_{V^*}\langle A(u), u - v\rangle_V.$$

Hence for all $w \in V$

$$\limsup_{n\to\infty} {}_{V^*}\langle A(u_n), w\rangle_V \leq {}_{V^*}\langle A(u), w\rangle_V.$$

Replacing w by $-w$ we deduce that $A(u_n) \rightharpoonup A(u)$ in V^*. Likewise, then we also obtain that

$$\lim_{n\to\infty} {}_{V^*}\langle A(u_n), u_n\rangle_V = \lim_{n\to\infty} {}_{V^*}\langle A(u_n), u_n - u\rangle_V + \lim_{n\to\infty} {}_{V^*}\langle A(u_n), u\rangle_V = {}_{V^*}\langle A(u), u\rangle_V.$$

Lemma 5.2.13 *If the embedding $V \subset H$ is compact, then (H1) and (H2″) imply that $A(t, \cdot)$ is pseudo-monotone for any $t \in [0, T]$.*

Proof We fix $t \in [0, T]$ and denote $A(t, \cdot)$ by $A(\cdot)$.
  Suppose $u_n \rightharpoonup u$ in V and

$$\liminf_{n\to\infty} {}_{V^*}\langle A(u_n), u_n - u\rangle_V \geq 0, \tag{5.37}$$

then for any $v \in V$ we have to show

$${}_{V^*}\langle A(u), u - v\rangle_V \geq \limsup_{n\to\infty} {}_{V^*}\langle A(u_n), u_n - v\rangle_V. \tag{5.38}$$

Let $v \in V$. We set

$$C_0 := \|u\|_V + \|v\|_V + \sup_n \|v_n\|_V;$$

$$C_1 := \sup \{f(t) + \eta(v) + \rho(v) : v \in V, \|v\|_V \leq 2C_0\} \, (< \infty).$$

Since the embedding $V \subset H$ is compact, we have $u_n \to u$ in V^* and

$${}_{V^*}\langle C_1 u, u - v\rangle_V = \lim_{n\to\infty} {}_{V^*}\langle C_1 u_n, u_n - v\rangle_V.$$

Hence for proving (5.38) it is sufficient to show that

$$_{V^*}\langle A_0(u), u - v\rangle_V \geq \limsup_{n\to\infty} {}_{V^*}\langle A_0(u_n), u_n - v\rangle_V,$$

where $A_0 = A - C_1 I$ and I is the identity operator.

Then $(H2'')$ implies that

$$\limsup_{n\to\infty} {}_{V^*}\langle A_0(u_n), u_n - u\rangle_V \leq \limsup_{n\to\infty} {}_{V^*}\langle A_0(u), u_n - u\rangle_V = 0.$$

By (5.37) we obtain

$$\lim_{n\to\infty} {}_{V^*}\langle A_0(u_n), u_n - u\rangle_V = 0. \tag{5.39}$$

Let $z = u + t(v - u)$ with $t \in (0, \frac{1}{2})$, then the local monotonicity $(H2'')$ implies that

$$_{V^*}\langle A_0(u_n) - A_0(z), u_n - z\rangle_V \leq 0,$$

i.e.

$$t_{V^*}\langle A_0(z), u - v\rangle_V - (1 - t)_{V^*}\langle A_0(u_n), u_n - u\rangle_V$$
$$\geq t_{V^*}\langle A_0(u_n), u_n - v\rangle_V - {}_{V^*}\langle A_0(z), u_n - u\rangle_V.$$

By taking lim sup on both sides and using (5.39) we have

$$_{V^*}\langle A_0(z), u - v\rangle_V \geq \limsup_{n\to\infty} {}_{V^*}\langle A_0(u_n), u_n - v\rangle_V.$$

Then letting $t \to 0$, by the hemicontinuity $(H1)$ we obtain

$$_{V^*}\langle A_0(u), u - v\rangle_V \geq \limsup_{n\to\infty} {}_{V^*}\langle A_0(u_n), u_n - v\rangle_V.$$

Therefore, A is pseudo-monotone. □

Remark 5.2.14 As we shall see in Sect. 5.2.3 below, for some concrete operators, it is easier to check the local monotonicity $(H2'')$ instead of the definition of pseudo-monotonicity itself.

Let $e_n \in V, n \in \mathbb{N}$, be an orthonormal basis in H such that $span\{e_n : n \in \mathbb{N}\}$ is dense in V. Let $H_n := span\{e_1, \cdots, e_n\}$ and let $P_n : V^* \to H_n$ be defined by

$$P_n y := \sum_{i=1}^{n} {}_{V^*}\langle y, e_i\rangle_V e_i, \quad y \in V^*.$$

For each finite $n \in \mathbb{N}$ we consider the following evolution equation on H_n:

$$u_n(t) = P_n u_0 + \int_0^t P_n A(s, u_n(s)) \, ds, \ t \in [0, T]. \tag{5.40}$$

From now on, we fix $T_0 \in]0, T]$ as in the assertion of Theorem 5.2.2, i.e. such that $L(T_0) < \sup_{x \in]0, \infty[} G(x)$ with L as defined in (5.30).

Set $X := L^\alpha([0, T_0]; V)$ and $\| \ \|_X := \| \ \|_{L^\alpha([0,T_0];V)}$. Then $X^* = L^{\frac{\alpha}{\alpha-1}}([0, T_0]; V^*)$ with norm $\| \ \|_{X^*} := \| \ \|_{L^{\frac{\alpha}{\alpha-1}}([0,T_0];V^*)}$.

Lemma 5.2.15 *Suppose that $V \subset H$ is compact and that $(H1), (H2''), (H3'), (H4')$ hold. Then for every solution u_n to (5.40) on $[0, T_0]$ we have*

$$\|u_n(t)\|_H^2 + \theta \int_0^t \|u_n(s)\|_V^\alpha \, ds \le G^{-1} \left(G \left(\|u_0\|_H^2 + \int_0^{T_0} f(s) \, ds \right) + \int_0^t h(s) \, ds \right),$$

$$t \in [0, T_0]. \tag{5.41}$$

In particular, there exists a $C_0 \in]0, \infty[$ such that

$$\|u_n\|_X + \sup_{t \in [0,T_0]} \|u_n(t)\|_H + \|A(\cdot, u_n)\|_{X^*} \le C_0 \text{ for all } n \in \mathbb{N}. \tag{5.42}$$

Proof By the Itô formula (4.29), (4.30) (with $Z \equiv 0$) and $(H3')$ we have

$$\|u_n(t)\|_H^2 - \|u_n(0)\|_H^2$$

$$= 2 \int_0^t {}_{V^*}\langle u_n'(s), u_n(s) \rangle_V \, ds$$

$$= 2 \int_0^t {}_{V^*}\langle P_n A(s, u_n(s)), u_n(s) \rangle_V \, ds$$

$$= 2 \int_0^t {}_{V^*}\langle A(s, u_n(s)), u_n(s) \rangle_V \, ds$$

$$\le \int_0^t \left(-\theta \|u_n(s)\|_V^\alpha + h(s) \, g\left(\|u_n(s)\|_H^2\right) + f(s) \right) \, ds. \tag{5.43}$$

Hence we have for $t \in [0, T_0]$,

$$\|u_n(t)\|_H^2 + \theta \int_0^t \|u_n(s)\|_V^\alpha \, ds \le \|u_0\|_H^2 + \int_0^{T_0} f(s) \, ds + \int_0^t h(s) \, g(\|u_n(s)\|_H^2) \, ds.$$

Then by Lemma 5.2.8 and Remark 5.2.9 we know that (5.41) holds.

Therefore, there exists a constant C_2 such that

$$\|u_n\|_X + \sup_{t \in [0,T_0]} \|u_n(t)\|_H \leq C_2, \ n \geq 1.$$

Then by $(H4')$ there exists a constant C_3 such that

$$\|A(\cdot, u_n)\|_{X^*} \leq C_3, \ n \geq 1.$$

Hence the proof is complete. $\square$

Lemma 5.2.16 *Suppose that $V \subset H$ is compact and that $(H1), (H2''), (H3'), (H4')$ hold. Then (5.40) has a solution u_n on $[0, T_0]$.*

Proof For any $t \in [0, T]$, by Remark 5.2.1 we know that $A(t, \cdot)$ is demicontinuous, i.e.

$$u_n \to u \text{ (strongly) in } V \text{ as } n \to \infty$$

implies that

$$A(t, u_n) \rightharpoonup A(t, u) \text{ in } V^* \text{ as } n \to \infty.$$

The demicontinuity implies that $P_n A(t, \cdot) : H_n \to H_n$ is continuous. If $u : [0, T_0] \to H_n$ is a solution of (5.40), then by Lemma 5.2.15 for some $R \in]0, \infty[$

$$\|u(t)\|_H \leq R \text{ for all } t \in [0, T_0]. \tag{5.44}$$

Hence we may replace $P_n A(t, \cdot)$ by $\chi_{n,R} P_n A(t, \cdot)$ with $\chi_{n,R} \in C_0^\infty(H_n)$, which is identically equal to 1 on a ball in H_n, large enough in comparison with R from (5.44). Then obviously (3.1) and (3.4) from Chap. 3 are fulfilled with $\sigma \equiv 0$ and $b := \chi_{n,R} P_n A(t, \cdot)$ by $(H4')$. Also (3.3) follows from $(H2'')$. Hence by Theorem 3.1.1 there exists a $u_n : [0, T] \longrightarrow H_n$ such that u_n is the unique solution to (5.40) on $[0, T_0]$. $\square$

Remark 5.2.17 In the proof of Theorem 5.2.6 below we shall use Lemma 5.2.16 in the case where $b := \chi_{n,R} P_n A(t, \cdot)$ also depends on $\omega \in \Omega$ in a progressively measurable way and T_0 is an $(\mathcal{F}_t)$-stopping time. Then by Theorem 3.1.1 every $u_n : [0, T] \to H_n$ from the proof of Lemma 5.2.16 above will be $(\mathcal{F}_t)$-adapted and so will be $\widetilde{u}_n := u_n(\cdot \wedge T_0)$. This and (5.45) below implies that for u defined in Lemma 5.2.18 below we have that $(u(t))_{t \in [0,T_0]}$ is $(\mathcal{F}_t)$-adapted in the sense of Definition 5.2.5 (i).

Note that X and X^* are reflexive Banach spaces. Hence by Lemma 5.2.15 there exists a subsequence of $u_n, n \in \mathbb{N}$, from Lemma 5.2.16, again denoted by u_n, such that as $n \to \infty$

$$u_n \to \bar{u} \text{ weakly in } X \text{ and weakly}^* \text{ in } L^\infty([0, T_0]; H);$$

$$A(\cdot, u_n) \to w \text{ weakly in } X^*. \tag{5.45}$$

Recall that $u_n(0) = P_n u_0 \to u_0$ in H as $n \to \infty$.

Lemma 5.2.18 *Suppose that $V \subset H$ is compact and that $(H1), (H2''), (H3'), (H4')$ hold. Define*

$$u(t) := u_0 + \int_0^t w(s) \, ds, \ t \in [0, T_0]. \tag{5.46}$$

Then $u \in C([0, T_0]; H)$ and $u = \bar{u}$ dt-a.e.

Proof (Cf. the proof of Theorem 4.2.4) From (5.40) for all $v \in \bigcup_{n \in \mathbb{N}} H_n (\subset V)$, $\varphi \in L^\infty([0, T_0])$ by Fubini's theorem we obtain

$$\int_0^{T_0} {}_{V^*}\langle \bar{u}(t), \varphi(t)v \rangle_V \, dt$$

$$= \lim_{n \to \infty} \int_0^{T_0} {}_{V^*}\langle u_n(t), \varphi(t)v \rangle_V \, dt$$

$$= \lim_{n \to \infty} \left(\int_0^{T_0} {}_{V^*}\langle P_n u_0, \varphi(t)v \rangle_V \, dt + \int_0^{T_0} \int_0^t {}_{V^*}\langle P_n A(s, u_n(s)), \varphi(t)v \rangle_V \, ds \, dt \right)$$

$$= \lim_{n \to \infty} \left({}_{V^*}\langle P_n u_0, v \rangle_V \int_0^{T_0} \varphi(t) \, dt + \int_0^{T_0} {}_{V^*}\left\langle A(s, u_n(s)), \int_s^{T_0} \varphi(t) \, dt \, v \right\rangle_V ds \right)$$

$$= \int_0^{T_0} {}_{V^*}\left\langle u_0 + \int_0^t w(s) \, ds, \varphi(t)v \right\rangle_V \, dt.$$

Therefore, we have $u = \bar{u}$ dt-a.e., and applying Theorem 4.2.5 (with $Z \equiv 0$), we obtain that $u \in C([0, T_0]; H)$. $\qquad\qquad\qquad\qquad\qquad\qquad\qquad\qquad\qquad\qquad \square$

Remark 5.2.19 We would like to emphasize that below we shall always work with this fixed version u of $\bar{u}$ defined in (5.46). Furthermore, since $A(\cdot, u_n) \to w$ in X^*, (5.46) implies that

$$u_n(t) \to u(t) \text{ in } V^* \text{ for all } t \in [0, T_0].$$

Since $\sup_n \|u_n(t)\|_H < \infty$, it follows that also

$$u_n(t) \to u(t) \text{ in } H \text{ for all } t \in [0, T_0]. \tag{5.47}$$

Hence, by (5.45) and the lower sequence semicontinuity of the norm in a Banach space with respect to weak convergence it follows that

$$\sup_{t \in [0,T_0]} \|u(t)\|_H^2 + \theta \int_0^{T_0} \|u(s)\|_V^{\alpha} \, ds \leq G^{-1} \left(G \left(\|u_0\|_H^2 + \int_0^{T_0} f(s) \, ds \right) + \int_0^{T_0} h(s) \, ds \right).$$

(5.48)

The next crucial step in the proof of Theorem 5.2.2 is to verify that $w = A(u)$. In the case of monotone operators, this is the well known Minty's lemma or monotonicity trick (cf. [82, Lemma 30.6] or Claim 3 in the proof of Remark 4.1.1). In the case of locally monotone operators, we use the following integrated version of Minty's lemma which holds due to pseudo-monotonicity.

Lemma 5.2.20 *Suppose that $V \subset H$ is compact and $(H1), (H2''), (H3'), (H4')$ hold, and assume that*

$$\liminf_{n \to \infty} \int_0^{T_0} {}_{V^*}\langle A(t, u_n(t)), u_n(t) - u(t)\rangle_V \, dt \geq 0.$$

(5.49)

Then for every $v \in X$ we have

$$\int_0^{T_0} {}_{V^*}\langle A(t, u(t)), u(t) - v(t)\rangle_V \, dt \geq \limsup_{n \to \infty} \int_0^{T_0} {}_{V^*}\langle A(t, u_n(t)), u_n(t) - v(t)\rangle_V \, dt.$$

(5.50)

In particular, the limit inferior in (5.49) and the limit superior in (5.50) are in fact limits and thus $A(t, u(t)) = w(t)$ for dt-a.e. $t \in [0, T_0]$.

Proof

Claim 1: For all $t \in [0, T_0]$ such that $u(t) \in V$, we have

$$\limsup_{n \to \infty} {}_{V^*}\langle A(t, u_n(t)), u_n(t) - u(t)\rangle_V \leq 0.$$

(5.51)

Indeed, suppose there exist $t_0 \in [0, T_0]$ such that $u(t_0) \in V$ and a subsequence such that

$$\lim_{i \to \infty} {}_{V^*}\langle A(t_0, u_{n_i}(t_0)), u_{n_i}(t_0) - u(t_0)\rangle_V > 0.$$

By $(H3')$ and $(H4')$ there exists a $C_0 \in]0, \infty[$ such that after applying Young's inequality we obtain

$$2_{V^*}\langle A(t_0, u_{n_i}(t_0)), u_{n_i}(t_0) - u(t_0)\rangle_V \leq -\frac{\theta}{2}\|u_{n_i}(t_0)\|_V^{\alpha}$$
$$+ C_0 \left(f(t_0) + h(t_0) \, g(\|u_{n_i}(t_0)\|_H^2) \right)$$
$$+ C_0 \left(1 + \|u_{n_i}(t_0)\|_H^{\alpha\beta} \right) \|u(t_0)\|_V^{\alpha}.$$

(5.52)

Hence by (5.47) we conclude that $u_{n_i}(t_0)$, $i \in \mathbb{N}$, is a bounded sequence in V (w.r.t. $\| \ \|_V$), which again by (5.47) then converges to $u(t_0)$ weakly in V. Since $A(t_0, \cdot)$ is pseudo-monotone, by Lemma 5.2.11 this, however, implies that

$$\limsup_{i \to \infty} \ {}_{V^*}\langle A(t_0, u_{n_i}(t_0)), u_{n_i}(t_0) - u(t_0)\rangle_V \leq 0,$$

which is a contradiction to the definition of the subsequence $u_{n_i}(t_0)$, $i \in \mathbb{N}$. Hence (5.51) holds.

As above, by $(H3')$ and $(H4')$ there exists a $C_0 \in]0, \infty[$ such that for every $v \in X$

$$2_{V^*}\langle A(t, u_n(t)), u_n(t) - v(t)\rangle_V$$

$$\leq -\frac{\theta}{2}\|u_n(t)\|_V^\alpha + C_0\left(f(t) + h(t)\, g(\|u_n(t)\|_H^2)\right)$$

$$+ C_0\left(1 + \|u_n(t)\|_H^{\alpha\beta}\right)\|v(t)\|_V^\alpha \quad \text{for } dt\text{-a.e. } t \in [0, T_0]. \tag{5.53}$$

Hence by Lemma 5.2.15, Fatou's lemma and (5.49) we have

$$0 \leq \liminf_{n \to \infty} \int_0^{T_0} {}_{V^*}\langle A(t, u_n(t)), u_n(t) - u(t)\rangle_V \ dt$$

$$\leq \limsup_{n \to \infty} \int_0^{T_0} {}_{V^*}\langle A(t, u_n(t)), u_n(t) - u(t)\rangle_V \ dt$$

$$\leq \int_0^{T_0} \limsup_{n \to \infty} {}_{V^*}\langle A(t, u_n(t)), u_n(t) - u(t)\rangle_V \ dt \tag{5.54}$$

which by (5.51) is negative.

Hence

$$\lim_{n \to \infty} \int_0^{T_0} {}_{V^*}\langle A(t, u_n(t)), u_n(t) - u(t)\rangle_V \ dt = 0, \tag{5.55}$$

and, therefore, for every $v \in X$

$$\lim_{n \to \infty} \int_0^{T_0} {}_{V^*}\langle A(t, u_n(t)), u_n(t) - v(t)\rangle_V \ dt = \int_0^{T_0} {}_{V^*}\langle w(t), u(t) - v(t)\rangle_V \ dt. \tag{5.56}$$

Claim 2:   There exists a subsequence u_{n_i}, $i \in \mathbb{N}$, such that

$$\lim_{i \to \infty} {}_{V^*}\langle A(t, u_{n_i}(t)), u_{n_i}(t) - u(t)\rangle_V = 0 \quad \text{for } dt\text{-a.e. } t \in [0, T_0]. \tag{5.57}$$

Define $g_n(t) := 1_{\{u \in V\}}(t) \,_{V^*}\langle A(t, u_n(t)), u_n(t) - u(t)\rangle_V$, $t \in [0, T]$. Then by (5.55), (5.51) respectively

$$\lim_{n \to \infty} \int_0^{T_0} g_n(t) \, dt = 0, \quad \limsup_{n \to \infty} g_n(t) \le 0 \quad \text{for } dt\text{-a.e. } t \in [0, T_0].$$

The latter clearly implies that $\lim_{n \to \infty} g_n^+(t) = 0$ for dt-a.e. $t \in [0, T_0]$, where $g_n^+(t) := max\{g_n(t), 0\}$.

Furthermore, (5.53) (with $v := u$) and (5.42) imply that for some $C \in]0, \infty[$ and all $n \in \mathbb{N}$

$$g_n(t) \le C \left(f(t) + h(t) + \|u(t)\|_V^\alpha \right) \quad \text{for } dt\text{-a.e. } t \in [0, T_0].$$

Hence by Lebesgue's dominated convergence theorem we have

$$\lim_{n \to \infty} \int_0^{T_0} g_n^+(t) \, dt = 0.$$

Note that $|g_n(t)| = 2g_n^+(t) - g_n(t)$. Hence we have

$$\lim_{n \to \infty} \int_0^{T_0} |g_n(t)| \, dt = 0.$$

Therefore, we can take a subsequence $g_{n_i}(t)$, $i \in \mathbb{N}$, such that

$$\lim_{i \to \infty} g_{n_i}(t) = 0 \quad \text{for } dt\text{-a.e. } t \in [0, T_0],$$

i.e. (5.57) holds.

Let $t \in \{s \in [0, T_0] : u(s) \in V\}$ such that the convergence in (5.57) holds. Then by (5.53) (with $v := u$), $\{u_{n_i}(t) : i \in \mathbb{N}\}$ is bounded in V, hence by (5.47) $u_{n_i}(t) \rightharpoonup u(t)$ in V. Since A is pseudo-monotone, (5.57) implies that for all $v \in X$

$$_{V^*}\langle A(t, u(t)), u(t) - v(t)\rangle_V \ge \limsup_{i \to \infty} {}_{V^*}\langle A(t, u_{n_i}(t)), u_{n_i}(t) - v(t)\rangle_V.$$

By (5.53) and Fatou's lemma it follows for every $v \in X$ that

$$\int_0^{T_0} {}_{V^*}\langle A(t, u(t)), u(t) - v(t)\rangle_V \, dt$$

$$\ge \int_0^{T_0} \limsup_{i \to \infty} {}_{V^*}\langle A(t, u_{n_i}(t)), u_{n_i}(t) - v(t)\rangle_V \, dt$$

$$\ge \limsup_{i \to \infty} \int_0^{T_0} {}_{V^*}\langle A(t, u_{n_i}(t)), u_{n_i}(t) - v(t)\rangle_V \, dt$$

$$= \lim_{n \to \infty} \int_0^{T_0} {}_{V^*}\langle A(t, u_n(t)), u_n(t) - v(t) \rangle_V \, dt$$

$$= \int_0^{T_0} {}_{V^*}\langle w(t), u(t) - v(t) \rangle_V \, dt, \tag{5.58}$$

where we used (5.56) in the last two steps.

Hence (5.50) is proved. But since $v \in X$ was arbitrary, (5.58) implies that $A(\cdot, u) = w$ as elements in X^*, which completes the proof. □

Now we can give the complete proof of Theorem 5.2.2.

Proof of Theorem 5.2.2 Since (ii) was already proved in Remark 5.2.3 it remains to prove (i). Using the Itô formula (4.29) (with $Z \equiv 0$) we conclude from Lemmas 5.2.16 and 5.2.18 that

$$\|u_n(T_0)\|_H^2 - \|u_n(0)\|_H^2 = 2 \int_0^{T_0} {}_{V^*}\langle A(t, u_n(t)), u_n(t) \rangle_V \, dt$$

and

$$\|u(T_0)\|_H^2 - \|u_0\|_H^2 = 2 \int_0^{T_0} {}_{V^*}\langle w(t), u(t) \rangle_V \, dt,$$

respectively. Since by (5.47), $u_n(T_0) \rightharpoonup u(T_0)$ in H, by the lower semicontinuity of $\| \ \|_H$ we have

$$\liminf_{n \to \infty} \|u_n(T_0)\|_H^2 \geq \|u(T_0)\|_H^2.$$

Hence we have by (5.45)

$$\liminf_{n \to \infty} \int_0^{T_0} {}_{V^*}\langle A(t, u_n(t)), u_n(t) \rangle_V \, dt$$

$$\geq \frac{1}{2} \left(\|u(T_0)\|_H^2 - \|u(0)\|_H^2 \right)$$

$$= \int_0^{T_0} {}_{V^*}\langle w(t), u(t) \rangle_V \, dt$$

$$= \lim_{n \to \infty} \int_0^{T_0} {}_{V^*}\langle A(t, u_n(t)), u(t) \rangle_V \, dt.$$

By Lemma 5.2.20 it now follows that u is a solution to (5.29). □

Proof of Theorem 5.2.4

(i) Let

$$a(s) := f(s) + \eta(u_1(s)) + \rho(u_2(s)), \; s \in [0, T_0].$$

Since, otherwise, (5.32) is trivially true, we may assume that $a \in L^1([0, T_0])$. Then again by (4.29) with $Z \equiv 0$ and the product rule we have for $t \in [0, T_0]$

$$e^{-\int_0^t a(s)\,\mathrm{d}s} \|u_1(t) - u_2(t)\|_H^2$$

$$= \|u_{1,0} - u_{2,0}\|_H^2 + 2\int_0^t e^{-\int_0^s a(r)\,\mathrm{d}r}{}_{V^*}\langle A(s, u_1(s)) - A(s, u_2(s)), u_1(s) - u_2(s)\rangle_V \, \mathrm{d}s$$

$$- \int_0^t e^{-\int_0^s a(r)\,\mathrm{d}r} a(s)\|u_1(s) - u_2(s)\|_H^2 \, \mathrm{d}s,$$

which by ($H2''$) is dominated by $\|u_{1,0} - u_{2,0}\|_H^2$.

Hence (5.32) follows. The remaining parts of the assertion then follow immediately.

(ii) The assertion is an immediate consequence of (i), (5.48) and (5.33). □

Proof of Theorem 5.2.6 We first consider the process Y which solves the following SPDE:

$$dY(t) = A_1(t, Y(t)) \, dt + B(t) \, dW(t), \; t \in [0, T], \; Y(0) = 0.$$

By Theorem 5.1.3 we know that there exists a unique solution Y to the above equation in the sense of Definition 5.1.2 and it satisfies

$$Y \in L^\alpha([0, T]; V) \cap C([0, T]; H), \; P\text{-a.s.}$$

Fix $\omega \in \Omega$ and consider the transformed equation:

$$u(t) = u_0 + \int_0^t \tilde{A}(s, u(s)) \, \mathrm{d}s, \; t \in [0, T], \tag{5.59}$$

where $\tau = \tau(\omega) \in \,]0, T]$ will be determined later, $u_0 := X_0(\omega)$ and

$$\tilde{A}(t, v) := A_1(t, v + \bar{Y}(t, \omega)) - A_1(t, \bar{Y}(t, \omega)) + A_2(t, v + \bar{Y}(t, \omega)), \; v \in V, \; t \in 0, T].$$

Here $\bar{Y}$ is a fixed V-valued progressively measurable $\mathrm{d}t \otimes P$-version of Y (see Definition 5.1.2). Clearly, if $u(= u(\omega))$ satisfies (5.59) up to some time $\tau(= \tau(\omega))$, then $X := u + Y$ is a solution to (5.34) on $[0, \tau]$ in the sense of Definition 5.2.5.

To obtain the existence (and uniqueness) of a solution to (5.59) we only need to show that $\tilde{A}$ satisfies all the assumptions of Theorem 5.2.2 for some properly chosen $\tau = \tau(\omega)$ replacing T_0.

Clearly, $\tilde{A}$ is $\mathcal{B}([0, T] \times V)/\mathcal{B}(V^*)$-measurable. Furthermore, $\tilde{A}$ is hemicontinuous since $(H1)$ holds for both A_1 and A_2.

For $u, v \in V$ and $t \in [0, T]$ by our assumptions on η, ρ (writing $Y(t)$ instead of $\bar{Y}(t, \omega)$ for simplicity) we have

$$
\begin{aligned}
&{}_{V^*}\langle \tilde{A}(t, u) - \tilde{A}(t, v), u - v \rangle_V \\
={}&{}_{V^*}\langle A_1(t, u + Y(t)) - A_1(t, v + Y(t)), u - v \rangle_V \\
&+ {}_{V^*}\langle A_2(t, u + Y(t)) - A_2(t, v + Y(t)), u - v \rangle_V \\
\leq{}& (f(t) + \rho(v + Y(t))) \|u - v\|_H^2 \\
&+ (f(t) + \eta(v + Y(t)) + \rho(v + Y(t))) \|u - v\|_H^2 \\
\leq{}& C \left[f(t) + \eta(Y(t)) + \rho(Y(t)) + \eta(v) + \rho(v) \right] \|u - v\|_H^2,
\end{aligned}
$$

i.e. $(H2'')$ holds for $\tilde{A}$ with

$$
\tilde{f}(t) := C \left[f(t) + \eta(Y(t)) + \rho(Y(t)) \right] \in L^1([0, T]).
$$

Since A_2 satisfies $(H3')$ and $(H4')$, by Young's inequality we have for $v \in V, t \in [0, T]$

$$
\begin{aligned}
2 {}_{V^*}\langle A_2(t, v + Y(t)), v \rangle_V ={}& 2 {}_{V^*}\langle A_2(t, v + Y(t)), v + Y(t) - Y(t) \rangle_V \\
\leq{}& -\theta \|v + Y(t)\|_V^\alpha + h(t) \, g(\|v + Y(t)\|_H^2) + f(t) - 2 {}_{V^*}\langle A_2(t, v + Y(t)), Y(t) \rangle_V \\
\leq{}& -\theta \|v + Y(t)\|_V^\alpha + h(t) \, g(\|v + Y(t)\|_H^2) + f(t) \\
&+ C \left(f(t)^{\frac{\alpha-1}{\alpha}} + \|v + Y(t)\|_V^{\alpha-1} \right) \left(1 + \|v + Y(t)\|_H^\beta \right) \|Y(t)\|_V \\
\leq{}& -\frac{\theta}{2} \|v + Y(t)\|_V^\alpha + h(t) \, g(\|v + Y(t)\|_H^2) + (1 + \frac{\theta}{2}) f(t) \\
&+ C \|Y(t)\|_V^\alpha \left(1 + \|v + Y(t)\|_H^{\alpha\beta} \right) \\
\leq{}& -\frac{\theta}{2} \left(2^{1-\alpha} \|v\|_V^\alpha - \|Y(t)\|_V^\alpha \right) + h(t) \, g(2\|v\|_H^2 + 2\|Y(t)\|_H^2) \\
&+ (1 + \frac{\theta}{2}) f(t) + C \|Y(t)\|_V^\alpha \left(1 + \|v\|_H^{\alpha\beta} + \|Y(t)\|_H^{\alpha\beta} \right) \\
\leq{}& -2^{-\alpha} \theta \|v\|_V^\alpha + C h(t) \left(g(\|v\|_H^2) + g(\|Y(t)\|_H^2) \right) + C \|Y(t)\|_V^\alpha \|v\|_H^{\alpha\beta} \\
&+ (1 + \frac{\theta}{2}) f(t) + C \|Y(t)\|_V^\alpha \left(1 + \|Y(t)\|_H^{\alpha\beta} \right), \quad v \in V,
\end{aligned}
$$

where C is some constant changing from line to line (but independent of t and ω).

Similarly, we have $v \in V, t \in [0, T]$

$$2_{V^*}\langle A_1(t, v + Y(t)) - A_1(t, Y(t)), v\rangle_V$$

$$= 2_{V^*}\langle A_1(t, v + Y(t)), v + Y(t) - Y(t)\rangle_V - 2_{V^*}\langle A_1(t, Y(t)), v\rangle_V$$

$$\leq -\theta\|v + Y(t)\|_V^\alpha + C\|v + Y(t)\|_H^2 + f(t)$$

$$+ 2\|Y(t)\|_V \left(f(t)^{\frac{\alpha-1}{\alpha}} + C\|v + Y(t)\|_V^{\alpha-1}\right) + 2\|v\|_V\|A_1(t, Y(t))\|_{V^*}$$

$$\leq -\frac{\theta}{2}\|v + Y(t)\|_V^\alpha + C\|v + Y(t)\|_H^2 + \left(1 + \frac{\theta}{2}\right)f(t)$$

$$+ C\|Y(t)\|_V^\alpha + 2\|v\|_V\|A_1(t, Y(t))\|_{V^*}$$

$$\leq -\frac{\theta}{2}\left(2^{1-\alpha}\|v\|_V^\alpha - \|Y(t)\|_V^\alpha\right) + C\left(\|v\|_H^2 + \|Y(t)\|_H^2\right)$$

$$+ C(f(t) + \|Y(t)\|_V^\alpha) + \|v\|_V\left(f(t)^{\frac{\alpha-1}{\alpha}} + C\|Y(t)\|_V^{\alpha-1}\right)$$

$$\leq -2^{-\alpha-1}\theta\|v\|_V^\alpha + C\|v\|_H^2 + C\left(f(t) + \|Y(t)\|_V^\alpha + \|Y(t)\|_H^2\right), \quad v \in V.$$

Since $Y \in L^\alpha([0, T]; V) \cap C([0, T]; H)$, we conclude that $\tilde{A}$ satisfies $(H3')$ with

$$\tilde{f}(t, \omega) := C\Big(f(t) + \|Y(t, \omega)\|_V^\alpha + \|Y(t, \omega)\|_H^2 + \|Y(t, \omega)\|_V^\alpha\|Y(t, \omega)\|_H^{\alpha\beta}$$

$$+ h(t) g(\|Y(t, \omega)\|_H^2)\Big), \quad t \in [0, T],$$

$$\tilde{g}(x) := g(x) + x^{\alpha\beta/2} + x, \quad x \in [0, \infty[,$$

$$\tilde{h}(t, \omega) := C\left(h(t) + \|Y(t, \omega)\|_V^\alpha + 1\right), \quad t \in [0, T],$$

replacing f, g, h.

The growth condition $(H4')$ also holds for $\tilde{A}$, since for some $C \in]0, \infty[$ independent of ω and t (which may change from line to line) we have for all $v \in V$, $t \in [0, T]$

$$\|\tilde{A}(t, v)\|_{V^*} \leq \|A_1(t, v + Y(t))\|_{V^*} + \|A_1(t, Y(t))\|_{V^*} + \|A_2(t, v + Y(t))\|_{V^*}$$

$$\leq C\left(f(t)^{\frac{\alpha-1}{\alpha}} + \|v + Y(t)\|_V^{\alpha-1}\right)\left(1 + \|v + Y(t)\|_H^\beta\right)$$

$$+ f(t)^{\frac{\alpha-1}{\alpha}} + C\|Y(t)\|_V^{\alpha-1}$$

$$\leq \left(Cf(t)^{\frac{\alpha-1}{\alpha}} + C\|Y(t)\|_V^{\alpha-1} + C\|v\|_V^{\alpha-1}\right)\left(1 + \|Y(t)\|_H^\beta + \|v\|_H^\beta\right)$$

$$\leq C\left(1 + \sup_{t \in [0,T]} \|Y(t)\|_H^\beta\right)\left(\tilde{f}(t)^{\frac{\alpha-1}{\alpha}} + C\|v\|_V^{\alpha-1}\right)\left(1 + \|v\|_H^\beta\right).$$

Now for $\varepsilon \in]0,\ \sup_{x\in]0,\infty[} \tilde{G}(x) - \tilde{G}(\|u_0\|_H^2)[$ we define

$$\tau_\varepsilon(\omega) := \inf\left\{t \in [0, T] : \tilde{L}(t, \omega) \geq \sup_{x\in]0,\infty[} \tilde{G}(x) - \varepsilon\right\} \wedge T,$$

where $\tilde{L}(\cdot, \omega)$ is as defined in (5.30) with $\tilde{f}(\cdot, \omega), \tilde{g}, \tilde{h}(\cdot, \omega), \tilde{G}$ replacing f, g, h, G
· respectively and $\tilde{G}(x) := \int_{x_0}^{x} \frac{1}{\tilde{g}(s)} \, ds$, $x \in]0, \infty[$, with $x_0 \in]0, \infty[$ fixed.

Since $\tilde{f}$ and $\tilde{h}$ are $(\mathcal{F}_t)$-adapted processes, the continuous real-valued process
$\tilde{L}(t)$, $t \in [0, T]$, is also $(\mathcal{F}_t)$-adapted. Hence τ_ε is an $(\mathcal{F}_t)$-stopping time.

Clearly,

$$\tilde{L}(\tau_\varepsilon) \leq \sup_{x\in]0,\infty[} \tilde{G}(x) - \varepsilon \text{ on } \{\tau_\varepsilon > 0\}.$$

But by the choice of ε, obviously $\{\tau_\varepsilon > 0\} = \Omega$. So,

$$\tilde{L}(\tau_\varepsilon) < \sup_{x\in]0,\infty[} \tilde{G}(x),$$

i.e. (5.31) holds for $\tilde{L}(\cdot, \omega)$ with $\tau_\varepsilon(\omega)$ replacing T_0.

Therefore, according to Theorems 5.2.2, (5.59) has a solution u on $[0, \tau_\varepsilon(\omega)]$ for
$P\text{-}a.e.\ \omega \in \Omega$.

Define

$$X(t) := u(t) + Y(t),\ t \in [0, \tau_\varepsilon].$$

By Remark 5.2.17 $(u(t))_{t\in[0,\tau_\varepsilon]}$ and hence $(X(t))_{t\in[0,\tau_\varepsilon]}$ is $(\mathcal{F}_t)$-adapted. Furthermore,
obviously $(X(t))_{t\in[0,\tau_\varepsilon]}$ is a local solution to (5.34).

To prove uniqueness let $(X_i(t))_{t\in[0,\tau^{(i)}]}$, $i = 1, 2$, be two local solutions to (5.34).
Define

$$u_i(t) := X_i(t) - Y(t),\ t \in [0, \tau^{(i)}].$$

Then obviously each u_i solves (5.59) on $[0, \tau^{(i)}]$. Hence it follows by Theorem 5.2.4
(i) that $u_1 = u_2$ on $[0, \tau^{(1)} \wedge \tau^{(2)}]$, i. e. $X_1 = X_2$ on $[0, \tau^{(1)} \wedge \tau^{(2)}]$. The last part of
the assertion follows by the definition of $\tilde{g}$ above, since it implies that (5.59) has a
(global) solution on $[0, T]$.

Now the proof is complete.   □

Remark 5.2.21 It follows from the above proof that, of course, we have a unique
solution to (5.34) on $[0, \sup_{\varepsilon>0} \tau_\varepsilon[$.

5.2.3 Application to Examples

In this section we will first apply our general results to some classical examples (see the first three examples below), but which have not been covered by the more restricted framework in Sect. 5.1 and Chap. 4. Subsequently, in the last example of this section we apply our results to an example, which, at least in such generality, for the first time was solved in [57, 59] in the deterministic and stochastic case respectively. Recall that in this section we use C to denote a generic constant which may change from line to line.

3D Navier–Stokes Equation

First we want to apply Theorems 5.2.2 and 5.2.4 to the 3D Navier–Stokes equation, which is a classical model to describe the time evolution of an incompressible fluid, given as follows:

$$\frac{\partial}{\partial t}u(t) = \nu\Delta u(t) - (u(t) \cdot \nabla)u(t) + \nabla p(t) + f(t),$$

$$\mathrm{div}(u) = 0, \; u|_{\partial\Lambda} = 0, \; u(0) = u_0, \tag{5.60}$$

where Λ is a bounded open domain in $\mathbb{R}^3$ with smooth boundary, $u(t, \xi) = (u^1(t, \xi), u^2(t, \xi), u^3(t, \xi))$, $\xi \in \Lambda$, represents the velocity field of the fluid, ν is the viscosity constant, the pressure $p(t, \xi)$ is an unknown scalar function and f is a (known) external force field acting on the fluid. In the pioneering work [55] Leray proved the existence of a weak solution for the 3D Navier–Stokes equation in the whole space. However, up to now, the uniqueness and regularity of weak solutions are still open problems (cf. e.g. [79]).

Let $C_{0,\sigma}^\infty(\Lambda; \mathbb{R}^3)$ denote the set of all divergence free smooth vector fields from Λ to $\mathbb{R}^3$ with compact support equipped with the following norms respectively:

$$\|u\|_{H^{1,2}} := \left(\int_\Lambda |\nabla u|^2 \, d\xi\right)^{1/2}; \; \|u\|_{H^{2,2}} := \left(\int_\Lambda |\Delta u|^2 \, d\xi\right)^{1/2}.$$

For $p \geq 1$, let $L^p := L^p(\Lambda; \mathbb{R}^3)$ be the vector valued L^p-space with the usual norm $\| \; \|_{L^p}$. We note that by Poincaré's inequality there exists a $C \in \,]0, \infty[$ such that for all $u \in C_{0,\sigma}^\infty(\Lambda; \mathbb{R}^3)$

$$\|u\|_{H^{1,2}}^2 = \sum_{i,j=1}^3 \int_\Lambda \left(\partial_i u^j\right)^2 \, d\xi = -\sum_{i,j=1}^3 \int_\Lambda u^j \partial_i^2 u^j \, d\xi$$

$$\leq \left(\sum_{j=1}^{3} \int_{\Lambda} \left(u^j \right)^2 \, d\xi \right)^{\frac{1}{2}} \|u\|_{H^{2,2}}$$

$$\leq C \left(\sum_{j=1}^{3} \int_{\Lambda} |\nabla u^j|^2 \, d\xi \right)^{\frac{1}{2}} \|u\|_{H^{2,2}} = C \|u\|_{H^{1,2}} \|u\|_{H^{2,2}},$$

hence

$$\|u\|_{H^{1,2}} \leq C \|u\|_{H^{2,2}}.$$

Now we define

$$L_\sigma^2 := \text{completion of } C_{0,\sigma}^\infty(\Lambda; \mathbb{R}^3) \ w.r.t. \ \| \ \|_{L^2};$$

$$H := \text{completion of } C_{0,\sigma}^\infty(\Lambda; \mathbb{R}^3) \ w.r.t. \ \| \ \|_{H^{1,2}};$$

$$V := \text{completion of } C_{0,\sigma}^\infty(\Lambda; \mathbb{R}^3) \ w.r.t. \ \| \ \|_{H^{2,2}}.$$

Here the norm $\| \ \|_{H^{2,2}}$ (resp. $\| \ \|_{H^{1,2}}$) restricted to V (resp. H) will be also denoted by $\| \ \|_V$ (resp. $\| \ \|_H$).

In the literature it is standard to use the Gelfand triple $H \subset L_\sigma^2 \subset H^*$ to analyze the Navier–Stokes equation and it works very well in the 2D case even with general stochastic perturbations (cf. Example 5.1.10). However, as pointed out in Remark 5.1.11(2), the growth condition $(H4')$ fails to hold on this triple for the 3D Navier–Stokes equation.

Motivated by some recent works on the (stochastic) tamed 3D Navier–Stokes equation (see the next section and the references mentioned there), we will use the new Gelfand triple $V \subset H \subset V^*$ defined above to verify the growth condition $(H4')$. One further ingredient is to use the following inequality in the 3D case (see e.g. [45, Theorem 2.1]):

$$\sup_{\Lambda} |u|^2 \leq C \|\Delta u\|_{L^2} \|\nabla u\|_{L^2}, \ u \in V. \tag{5.61}$$

Before we proceed, we note that for $v \in V$, $\varphi \in C_{0,\sigma}^\infty(\Lambda; \mathbb{R}^3)$

$$_{V^*}\langle v, \varphi \rangle_V = \langle v, \varphi \rangle_H = \langle v, -\Delta\varphi \rangle_{L^2},$$

hence

$$|_{V^*}\langle v, \varphi \rangle_V| \leq \|v\|_{L_\sigma^2} \|\varphi\|_V,$$

so the linear map $v \mapsto {}_{V^*}\langle v, \cdot \rangle_V$ from V into V^* extends by continuity to a continuous linear map from L^2_σ to V^*, which we also denote by f. We have for all $f \in L^2_\sigma$

$$_{V^*}\langle f, \varphi \rangle_V = \langle f, -\Delta\varphi \rangle_{L^2_\sigma} \text{ for all } \varphi \in C^\infty_{0,\sigma}(\Lambda; \mathbb{R}^3) \tag{5.62}$$

and

$$\|f\|_{V^*} \leq \|f\|_{L^2_\sigma}. \tag{5.63}$$

Let P_H be the orthogonal (Helmholtz–Leray) projection from $L^2(\Lambda; \mathbb{R}^3)$ to L^2_σ. Then by means of the divergence free Hilbert spaces V, H and the orthogonal projection P_H, the classical 3D Navier–Stokes equation (5.60) can be reformulated in the following abstract form:

$$u' = Au + F(u) + \tilde{f}, \ u(0) = u_0 \in H, \tag{5.64}$$

where

$$F(u) := F(u, u),$$

$$\tilde{f} : [0, T] \to L^2_\sigma, \ \tilde{f} := P_H f$$

and $A : V \to V^*$, $F : V \times V \to V$ are defined as follows:
For $u, v \in C^\infty_{0,\sigma}(\Lambda; \mathbb{R}^3)$ set

$$Au := \nu P_H \Delta u \ \text{ and } \ F(u, v) := P_H[(u \cdot \nabla)v].$$

We note that for $u \in C^\infty_{0,\sigma}(\Lambda; \mathbb{R}^3)$ also $\Delta u \in C^\infty_{0,\sigma}(\Lambda; \mathbb{R}^3)$, so $Au = \nu\Delta u$. Then

$$|_{V^*}\langle Au, v \rangle_V| = |\langle Au, v \rangle_H| = |\langle Au, (-\Delta)v \rangle_{L^2}| \leq \|u\|_V \|v\|_V, \ u, v \in C^\infty_{0,\sigma}(\Lambda).$$

Hence, A can be extended by continuity to a linear map from V to V^* such that

$$\|Au\|_{V^*} \leq \|u\|_V, \ u \in V.$$

Furthermore, for all $u, v, \varphi \in C^\infty_{0,\sigma}(\Lambda; \mathbb{R}^3)$ by (5.63)

$$\|F(u, v)\|_{V^*} \leq \|F(u, v)\|_{L^2_\sigma}.$$

But for some $C \in]0, \infty[$ (independent of u, v)

$$\int_\Lambda |F(u, v)|^2 \, d\xi \leq \int_\Lambda |(u \cdot \nabla)v|^2 \, d\xi \leq C\|u\|^2_{L^\infty} \|v\|^2_V,$$

which by (5.61) is up to a constant dominated by $\|u\|_V \|u\|_H \|v\|_V^2$. So, altogether there exist $C_1, C_2, C_3 \in]0, \infty[$ such that

$$
\begin{aligned}
\|F(u,v)\|_{V^*} &\leq C_1 \|u\|_{L^\infty} \|v\|_H \leq C_2 \|u\|_V^{1/2} \|u\|_H^{1/2} \|v\|_H \\
&\leq C_3 \|u\|_V \|v\|_H \text{ for all } u, v \in C_{0,\sigma}^\infty(\Lambda; \mathbb{R}^3).
\end{aligned}
\tag{5.65}
$$

Hence, since $F : C_{0,\sigma}^\infty(\Lambda; \mathbb{R}^3) \times C_{0,\sigma}^\infty(\Lambda; \mathbb{R}^3) \to V^*$ is bilinear, F has a unique continuous bilinear extensions from $V \times V \to V^*$ such that (5.65) holds for all $u, v \in V$.

Remark 5.2.22 We note in contrast to (5.25) under the current Gelfand triple

$$
{}_{V^*}\langle F(u,v), v \rangle_V = \langle F(u,v), (-\Delta)v \rangle_{L^2}, \ u, v, w \in V,
$$

which might not be equal to 0 in general.

For simplicity we only apply Theorems 5.2.2 and 5.2.4 to the deterministic 3D Navier–Stokes equation and give a simple proof for this well known result. But we can also add an additive noise to (5.64) and obtain the corresponding result in the stochastic case by applying Theorem 5.2.6. We refer to [77, 79] for historical remarks and references on the classical local existence and uniqueness results for the 3D Navier–Stokes equation.

Example 5.2.23 (3D Navier–Stokes Equation) If $\tilde{f} \in L^2(0, T; L_\sigma^2)$ and $u_0 \in H$, then there exists a constant $T_0 \in (0, T]$ such that (5.64) has a unique strong solution $u \in L^2([0, T_0]; V) \cap C([0, T_0]; H)$.

Proof For simplicity, let us assume that $\nu = 1$.

The hemicontinuity (H1) is obvious since A is linear and F is bilinear. Below we prove (H2''), (H3'), (H4') for $u, v \in C_{0,\sigma}^\infty(\Lambda; \mathbb{R}^3)$ which by the continuity of $A : V \to V^*$ and $F : V \times V \to V^*$ is sufficient. Furthermore, C will denote a constant that may change from line to line. We have

$$
\begin{aligned}
{}_{V^*}\langle Au - Av, u - v \rangle_V &= \langle A(u - v), (-\Delta)(u - v) \rangle_{L^2} \\
&= -\|u - v\|_V^2.
\end{aligned}
$$

By (5.65) and Young's inequality we have

$$
\begin{aligned}
&{}_{V^*}\langle F(u) - F(v), u - v \rangle_V \\
&= {}_{V^*}\langle F(u, u - v), u - v \rangle_V + {}_{V^*}\langle F(u - v, v), u - v \rangle_V \\
&\leq C\|u - v\|_V \cdot \left(\|u\|_{L^\infty} \|u - v\|_H + \|u - v\|_V^{1/2} \|u - v\|_H^{1/2} \|v\|_H \right) \\
&\leq \frac{1}{2} \|u - v\|_V^2 + C \left(\|u\|_{L^\infty}^2 + \|v\|_H^4 \right) \|u - v\|_H^2.
\end{aligned}
\tag{5.66}
$$

Combining these two estimates we obtain

$$_{V^*}\langle Au + F(u) - Av - F(v), u - v\rangle_V$$

$$\leq -\frac{1}{2}\|u - v\|_V^2 + C\left(\|u\|_{L^\infty}^2 + \|v\|_H^4\right)\|u - v\|_H^2. \tag{5.67}$$

Hence $(H2'')$ follows with $\eta(u) := C\|u\|_{L^\infty}^2$ and $\rho(v) := C\|v\|_H^4$. In particular, taking $u = 0$ we obtain that

$$_{V^*}\langle Av + F(v), v\rangle_V \leq -\frac{1}{2}\|v\|_V^2 + C\|v\|_H^6.$$

Therefore, $(H3')$ holds with $g(x) = Cx^3$ and $h \equiv 1$, because

$$_{V^*}\langle Av + F(v) + \tilde{f}, v\rangle_V \leq -\frac{1}{2}\|v\|_V^2 + C\|v\|_H^6 + \|\tilde{f}\|_{V^*}\|v\|_V$$

$$\leq -\frac{1}{4}\|v\|_V^2 + C\|v\|_H^6 + C\|\tilde{f}\|_{L^2}^2, \quad v \in V.$$

Moreover, by (5.63) and (5.65)

$$\|A(v) + F(v) + \tilde{f}\|_{V^*} \leq \|v\|_V + C\|v\|_V\|v\|_H + \|\tilde{f}\|_{L^2}$$

$$\leq \left(\|\tilde{f}\|_{L^2} + C\|v\|_V\right)(1 + \|v\|_H), \quad v \in V, \tag{5.68}$$

i.e. $(H4')$ holds with $\beta = 1$.

Finally, we note that $\rho + \eta$ satisfies (5.33) by (5.61) and both ρ and η are subadditive up to a constant. Therefore, the local existence and uniqueness of solutions to (5.64) on $[0, T_0]$ with T_0 satisfying (5.31) follows from Theorems 5.2.2 and 5.2.4. □

Remark 5.2.24

(1) By Remark 5.2.3 we can take any $T_0 \in (0, T]$ such that

$$T_0 < \frac{C}{(\|u_0\|_H^2 + \int_0^{T_0}(1 + \|\tilde{f}(t)\|_{L^2}^2)\, dt)^2},$$

for a suitable $C \in]0, \infty[$.

(2) Note that the solution here is a strong solution in the sense of PDEs. It is obvious that we can also allow $\tilde{f}$ in (5.64) to depend on the unknown solution u provided $\tilde{f}$ satisfies some local monotonicity condition.

A Tamed 3D Navier–Stokes Equation

In the case of the 3D Navier–Stokes equation we have seen that the generalized coercivity condition $(H3')$ holds with $g(x) = Cx^3$, hence we only get local existence and uniqueness of solutions. In this section we consider a tamed version of the 3D Navier–Stokes equation, which was proposed in [73, 74, 83]. The main feature of this tamed equation is that if there is a bounded strong solution to the classical 3D Navier–Stokes equation (5.60), then this smooth solution must also satisfy the following tamed equation (for N large enough):

$$\partial_t u(t) = \nu \Delta u(t) - (u(t) \cdot \nabla)u(t) + \nabla p(t) - g_N \left(\|u(t)\|_{L^\infty}^2 \right) u(t) + f(t),$$

$$\mathrm{div}(u) = 0, \ u|_{\partial \Lambda} = 0, \ u(0) = u_0, \tag{5.69}$$

where the taming function $g_N : \mathbb{R}_+ \to \mathbb{R}_+$ is smooth and satisfies for some $N > 0$,

$$\begin{cases} g_N(r) = 0, \text{ if } r \leq N, \\ g_N(r) = (r - N)/\nu, \text{ if } r \geq N + 1, \\ 0 \leq g_N'(r) \leq C, \ r \geq 0. \end{cases}$$

We note that our "taming term" $g_N\left(\|u(t)\|_{L^\infty}^2 \right)u(t)$ is slightly different and, in fact, simpler than the one used in [73, 74, 83].

We will use the same Gelfand triple as for the 3D Navier–Stokes equation, i.e.,

$$H := \text{completion of } C_{0,\sigma}^\infty(\Lambda, \mathbb{R}^3) \ w.r.t. \ \| \ \|_{H^{1,2}};$$

$$V := \text{completion of } C_{0,\sigma}^\infty(\Lambda, \mathbb{R}^3) \ w.r.t. \ \| \ \|_{H^{2,2}}.$$

Example 5.2.25 (Tamed 3D Navier–Stokes Equation) For $f \in L^2(0, T; L_\sigma^2)$ and $u_0 \in H$, (5.69) has a unique strong solution $u \in L^2([0, T]; V) \cap C([0, T]; H)$.

Proof We assume $\nu = 1$ for simplicity.

Similarly as in (5.64), using the Gelfand triple

$$V \subset H \subset V^*$$

(5.69) can be rewritten in the following abstract form (a priori as an equation in V^*):

$$u(t) = \int_0^t \left[Au(s) + F(u)(s) - g_N(\|u(s)\|_{L^\infty}^2)u(s) + \tilde{f}(s) \right] \ ds, \ t \in [0, T_0], \ u_0 \in H.$$

Again $(H1)$ is easy to check since we already know that $A + F + \tilde{f}$ is hemicontinuous and because g_N is Lipschitz continuous. By (5.61) and Young's inequality we also have for all $u, v \in C_{0,\sigma}^{\infty}(\Lambda; \mathbb{R}^3)$

$$- {}_{V^*}\langle g_N(\|u\|_{L^\infty}^2)u - g_N(\|v\|_{L^\infty}^2)v, u - v\rangle_V$$

$$= - \langle g_N(\|u\|_{L^\infty}^2)u - g_N(\|v\|_{L^\infty}^2)v, (-\Delta)(u - v)\rangle_{L^2}$$

$$= - \langle g_N(\|u\|_{L^\infty}^2)\nabla u - g_N(\|v\|_{L^\infty}^2)\nabla v, \nabla(u - v)\rangle_{L^2}$$

$$= - g_N(\|u\|_{L^\infty}^2)\|u - v\|_H^2 - \langle\left(g_N(\|u\|_{L^\infty}^2) - g_N(\|v\|_{L^\infty}^2)\right)\nabla v, \nabla(u - v)\rangle_{L^2}$$

$$\leq |g_N(\|u\|_{L^\infty}^2) - g_N(\|v\|_{L^\infty}^2)| \cdot \|v\|_H \cdot \|u - v\|_H$$

$$\leq C\|u - v\|_{L^\infty}(\|u\|_{L^\infty} + \|v\|_{L^\infty})\|v\|_H\|u - v\|_H$$

$$\leq C\|u - v\|_{L^\infty}\left(\|u\|_{L^\infty}^2 + \|v\|_H^2 + \|v\|_{L^\infty}\|v\|_H\right)\|u - v\|_H$$

$$\leq C\|u - v\|_V\left(\|u\|_{L^\infty}^2 + \|v\|_H^2 + \|v\|_{L^\infty}\|v\|_H\right)\|u - v\|_H$$

$$\leq \frac{1}{4}\|u - v\|_V^2 + C\left(\|u\|_{L^\infty}^4 + \|v\|_H^4 + \|v\|_{L^\infty}^2\|v\|_H^2\right)\|u - v\|_H^2,$$

where we have used that $g_N \geq 0$ and $0 \leq g_N' \leq C$ in the above estimates.

Hence by (5.67) we have the following estimate for all $u, v \in C_{0,\sigma}^{\infty}(\Lambda; \mathbb{R}^3)$

$$_{V^*}\langle Au + F(u) - g_N(\|u\|_{L^\infty}^2)u - Av - F(v) + g_N(\|v\|_{L^\infty}^2)v, \ u - v\rangle_V$$

$$\leq -\frac{1}{4}\|u - v\|_V^2 + C\left(1 + \|u\|_{L^\infty}^4 + \|v\|_H^4 + \|v\|_{L^\infty}^2\|v\|_H^2\right)\|u - v\|_H^2$$

$$\leq -\frac{1}{4}\|u - v\|_V^2 + C\left(1 + \|u\|_{L^\infty}^4 + \|v\|_H^4 + \|v\|_{L^\infty}^4\right)\|u - v\|_H^2,$$

i.e. $(H2'')$ holds with $\eta(u) = C\|u\|_{L^\infty}^4$ and $\rho(v) = C\left(\|v\|_H^4 + \|v\|_{L^\infty}^4\right)$.

For all $v \in C_{0,\sigma}(\Lambda; \mathbb{R}^d)$ we have

$$_{V^*}\langle Av, v\rangle_V = \langle P_H\Delta v, (-\Delta)v\rangle_{L^2} = -\|v\|_V^2$$

and by (5.65)

$$_{V^*}\langle F(v), v\rangle_V \leq \|F(v)\|_{V^*}\|v\|_V \leq \frac{1}{4}\|v\|_V^2 + \|v\|_{L^\infty}^2\|v\|_H^2.$$

Furthermore,

$$-{}_{V^*}\langle g_N(\|v\|_{L^\infty}^2)v, v\rangle_V = -g_N(\|v\|_{L^\infty}^2)\langle v, (-\Delta)v\rangle_{L^2}$$

$$= -g_N(\|v\|_{L^\infty}^2) \cdot \int_\Lambda |\nabla v|^2 \, dx$$

$$\leq -\|v\|_{L^\infty}^2\|v\|_H^2 + (N + 1)\|v\|_H^2,$$

where we used that $g_N(r) \geq r - (N + 1)$ in the last inequality.

Adding these three inequalities we obtain

$$_{V^*}\langle Av + F(v) - g_N(\|v\|_{L^\infty}^2)v + \tilde{f}, v \rangle_V$$

$$\leq -\frac{3}{4}\|v\|_V^2 + (N+1)\|v\|_H^2 + \|\tilde{f}\|_{L^2}\|v\|_V$$

$$\leq -\frac{1}{2}\|v\|_V^2 + (N+1)\|v\|_H^2 + C\|\tilde{f}\|_{L^2}^2, \quad v \in V,$$

where we used Young's inequality in the last step. This implies $(H3')$ with $g(x) = Nx$ and $h \equiv 1$.

By (5.61) it easily follows that

$$\|g_N(\|v\|_{L^\infty}^2)v\|_{V^*} \leq \|g_N(\|v\|_{L^\infty}^2)v\|_{L^2} \leq C\|v\|_{L^\infty}^2\|v\|_{L^2} \leq C\|v\|_V\|v\|_H^2, \quad v \in V.$$

Then, combining this with (5.68), we obtain that $(H4')$ holds with $\alpha = 2$, $\beta = 2$.

Equation (5.61) implies that $\|u\|_{L^\infty}^4 \leq C\|u\|_H^2\|u\|_V^2$, hence $\rho + \eta$ satisfies (5.33). Clearly, both ρ and η are subadditive up to a constant. Hence the global existence and uniqueness of solutions to (5.69) follows from Theorems 5.2.2 and 5.2.4. □

Remark 5.2.26 As for the classical 3D Navier–Stokes equation we can also add an additive noise and apply Theorem 5.2.6 to get a unique local solution for its stochastic version. However, since as shown above, $\alpha = 2$ and $\beta = 2$ in this case, the last part of Theorem 5.2.6 does not apply to get global solutions by our techniques. It is, however, true that a unique global solution exists in the stochastic case. This can be proved by different methods (see [72, 73]).

The Cahn–Hilliard Equation

The Cahn–Hilliard equation is a classical model to describe phase separation in a binary alloy and some other media, we refer to [62] for a survey on this model (see also [21, 30] for the stochastic case). Let Λ be a bounded open domain in $\mathbb{R}^d$ with $d \leq 3$ with smooth boundary. The Cahn–Hilliard equation has the following form:

$$\partial_t u = -\Delta^2 u + \Delta\varphi(u), \; u(0) = u_0,$$

$$\nabla u \cdot n = \nabla(\Delta u) \cdot n = 0 \text{ on } \partial\Lambda, \tag{5.70}$$

where Δ is the Laplace operator and ∇ is the gradient, n is the outward unit normal vector on the boundary $\partial\Lambda$ and the nonlinear term φ is some function which will be specified below.

Let $C^4(\bar{\Lambda})$ denote the set of all real valued functions on Λ, which are restrictions of functions in $C^4(\mathbb{R}^d)$. Define

$$V_0 := \{u \in C^4(\bar{\Lambda}) : \nabla u \cdot n = \nabla(\Delta u) \cdot n = 0 \text{ on } \partial\Lambda\}.$$

Now we consider the following Gelfand triple

$$V \subset H := L^2(\Lambda) \subset V^*,$$

where

$$V := \text{closure of } V_0 \text{ in } H^{2,2}(\Lambda)$$

and

$$\|u\|_{H^{2,2}} := \left(\int_\Lambda |\Delta u|^2 \, dx \right)^{1/2}.$$

For the reader's convenience, we recall the following Gagliardo–Nirenberg interpolation inequality (cf. [78, Theorems 2.25 and 2.21]) for bounded open $\Lambda \subset \mathbb{R}^d$, $d \geq 1$, with smooth boundary.

If $m, j \in \mathbb{Z}$ such that $m > j \geq 0$ and $q \in [1, \infty]$ such that

$$\frac{1}{q} = \frac{1}{2} + \frac{j}{d} - \frac{ma}{d}, \quad \frac{j}{m} \leq a \leq 1,$$

with $a \neq 1$, if $m - j - \frac{d}{2} = 0$, then there exists a $C \in]0, \infty[$ such that

$$\|u\|_{H^{j,q}} \leq C\|u\|_{H^{m,2}}^a \|u\|_{L^2}^{1-a}, \quad u \in H^{m,2}(\Lambda). \tag{5.71}$$

Here, for nonnegative integers j, q, the space $H^{j,q}(\Lambda)$ denotes the classical Sobolev space of order j in $L^q(\Lambda)$.

For $j := 0, q := 0, m := 2$ and hence $a := \frac{d}{4}$, this implies, since $d \leq 3$ the classical Sobolev embedding

$$H^{2,2}(\Lambda) \subset L^\infty(\Lambda) \quad \text{continuously.} \tag{5.72}$$

Then we get the following existence and uniqueness result for (5.70).

Example 5.2.27 Suppose that $\varphi \in C^1(\mathbb{R})$ and that there exist $C \in]0, \infty[$, $p \in \left[2, \frac{d+4}{d}\right]$ such that

$$\varphi'(x) \geq -C, \ |\varphi(x)| \leq C(1 + |x|^p), \ x \in \mathbb{R};$$

$$|\varphi(x) - \varphi(y)| \leq C(1 + |x|^{p-1} + |y|^{p-1})|x - y|, \ x, y \in \mathbb{R}.$$

Then for every $u_0 \in L^2(\Lambda)$, there exists a unique global solution to (5.70).

Proof Let

$$A(u) := A_1 u + A_2 u, \quad u \in V_0,$$

where

$$A_1(u) := -\Delta^2 u \text{ and } A_2(u) := \Delta\varphi(u), \ u \in V_0.$$

Note that for $u \in V_0$ by the boundary conditions imposed on the elements of V_0 we have for $v \in V_0$

$$|_{V^*}\langle A_1(u), v \rangle_V| = |\langle -\Delta u, \Delta v \rangle_{L^2}|$$
$$\leq \|v\|_V \|u\|_V.$$

Hence A_1 extends to a linear continuous map from V to V^* such that

$$\|A_1(u)\|_{V^*} \leq \|u\|_V \text{ for all } u \in V_0. \tag{5.73}$$

Furthermore, again by the boundary conditions on elements in V_0, by the local Lipschitz property of φ and by (5.71) we obtain for all $u_1, u_2 \in V_0$

$$_{V^*}\langle A_2(u_1) - A_2(u_2), v \rangle_V = \langle \varphi(u_1) - \varphi(u_2), \Delta v \rangle_{L^2}$$
$$\leq \|v\|_V C\|(1 + |u_1|^{p-1} + |u_2|^{p-1}) |u_1 - u_2|\|_{L^2}$$
$$\leq \|v\|_V C(1 + \|u_1\|_{L^\infty}^{p-1} + \|u_2\|_{L^\infty}^{p-1}) \|u_1 - u_2\|_{L^2}$$
$$\leq \|v\|_V \|u_1 - u_2\|_V C(1 + \|u_1\|_V^{p-1} + \|u_2\|_V^{p-1}). \tag{5.74}$$

Therefore, by uniform continuity A_2 extends to a continuous map from V to V^*. In particular, (H1) holds and (H2''), (H3') and (H4') only have to be checked for all $u, v \in V_0$.

For all $u, v \in V_0$, we have

$$-_{V^*}\langle \Delta^2 u - \Delta^2 v, u - v \rangle_V = -\|u - v\|_V^2.$$

In addition, as in (5.73) and using Young's inequality we get

$$_{V^*}\langle \Delta\varphi(u) - \Delta\varphi(v), u - v \rangle_V$$
$$\leq \|u - v\|_V \|\varphi(u) - \varphi(v)\|_{L^2}$$
$$\leq \frac{1}{2}\|u - v\|_V^2 + C\left(1 + \|u\|_{L^\infty}^{2p-2} + \|v\|_{L^\infty}^{2p-2}\right) \|u - v\|_H^2, \ u, v \in V_0.$$

Hence $(H2'')$ holds with $\eta(v) = \rho(v) = C\|v\|_{L^\infty}^{2p-2}$.

Similarly, since φ' is lower bounded we have for all $v \in V_0$,

$$_{V*}\langle\Delta\varphi(v), v\rangle_V = -\int_\Lambda \varphi'(v)|\nabla v|^2 \, d\xi \leq C\|v\|_{H^{1,2}}^2$$

$$= -C\int_\Lambda \Delta v v \, d\xi \leq \frac{1}{2}\|v\|_V^2 + C\|v\|_H^2.$$

It follows that $(H3')$ holds with $\alpha = 2$, $h \equiv 1$ and $g(x) = Cx$.

By (5.71) with $j := 0, q := 2p, m := 2$ and thus $a := \frac{(p-1)d}{4p}$ it follows that for all $v \in V_0$

$$\|\Delta\varphi(v)\|_{V*} \leq \|\varphi(v)\|_H$$

$$\leq C\left(1 + \|v\|_{L^{2p}}^p\right)$$

$$\leq C\left(1 + \|v\|_V^{ap}\|v\|_H^{(1-a)p}\right)$$

$$= C\left(1 + \|v\|_V^{ap}\|v\|_H^{1-ap}\|v\|_H^{p-1}\right).$$

Since $p \leq \frac{4}{d} + 1$ (i.e. $ap = \frac{(p-1)d}{4} \leq 1$) and $\|v\|_H \leq C\|v\|_V$, this implies that

$$\|\Delta\varphi(v)\|_{V*} \leq C(1 + \|v\|_V)\left(1 + \|v\|_H^{p-1}\right) \text{ for all } v \in V_0.$$

Hence by (5.73) $(H4')$ follows with $\beta = p - 1$.

Furthermore, (5.71) with $j = 0, q = \infty, m = 2$, and thus $a = \frac{d}{4}$ implies that for all $v \in V$

$$\eta(v)+\rho(v) = C\|v\|_{L^\infty}^{2p-2} \leq C\|v\|_V^{2a(p-1)}\|v\|_H^{2(1-a(p-1))} \leq C\|v\|_V^2\|v\|_H^{2p-4},$$

because $a(p-1) = \frac{d}{4}(p-1) \leq 1$ and $\|v\|_H \leq C\|v\|_V$. Hence (5.33) also holds and both η and ρ are subadditive up to a constant.

Therefore, the assertion follows from Theorems 5.2.2 and 5.2.4.   □

Remark 5.2.28 Again we can apply Theorem 5.2.6 to obtain local existence and uniqueness of solutions for (5.70) perturbed by additive noise. These solutions are global if $p \leq 2$, since, as shown above $\alpha = 2$, $\beta = p - 1$ in this case. So, the last part of Theorem 5.2.6 applies, because then $\alpha\beta \leq 2$.

Surface Growth PDE with Noise

We consider a model which appears in the theory of growth of surfaces, which describes an amorphous material deposited on an initially flat surface in high vacuum (cf. [7] and the references therein). Taking into account random noises the equation is formulated on the interval $\Lambda :=]0, L[$ as follows:

$$dX(t) = \left[-\partial_\xi^4 X(t) - \partial_\xi^2 X(t) + \partial_\xi^2 (\partial_\xi X(t))^2 \right] dt + B(t) \, dW(t),$$

$$X(t)|_{\partial \Lambda} = 0, \ X(0) = x_0, \ t \in [0, T], \tag{5.75}$$

where $\partial_\xi, \partial_\xi^2, \partial_\xi^4$ denote the first, second and fourth spatial derivatives in $\xi \in]0, L[$ respectively.

Recall that $W(t)$, $t \in [0, T]$, is a U-valued cylindrical Wiener process defined on a probability space $(\Omega, \mathcal{F}, P)$ with normal filtration $\mathcal{F}_t$, $t \in [0, T]$. We shall use the following Gelfand triple

$$V := H_0^{4,2}(]0, L[) \subset H := H_0^{2,2}(]0, L[) \subset V^*,$$

where as usual for nonnegative integers j, q the space $H_0^{j,q}(]0, L[)$ denotes the closure of $C_0^\infty(]0, L[)$ in $H^{j,q}(]0, L[)$. We obtain the following local existence and uniqueness of strong solutions for (5.75).

Example 5.2.29 Suppose that $B \in L^2([0, T]; L_2(U; H))$. For any $X_0 \in L^2(\Omega, \mathcal{F}_0, P; H)$, there exists a unique local solution $(X(t))_{t \in [0, \tau]}$ to (5.75) satisfying

$$X(\cdot) \in L^2([0, \tau]; V) \cap C([0, \tau]; H), P\text{-}a.s.$$

Proof It is sufficient to verify $(H1), (H2''), (H3'), (H4')$ for (5.75). Then the assertion follows from Theorem 5.2.6. For $u \in C_0^\infty(]0, L[)$ we have for all $v \in C_0^4(]0, L[)$

$$\left| {}_{V^*}\langle -(\partial_\xi^4 u + \partial_\xi^2 u), v \rangle_V \right| = \left| \langle \partial_\xi^6 u + \partial_\xi^4 u, \partial_\xi^2 v \rangle_{L^2} \right|$$

$$= \left| \langle \partial_\xi^4 u + \partial_\xi^2 u, \partial_\xi^4 v \rangle_{L^2} \right|$$

$$\leq (\|u\|_V + \|u\|_H)\|v\|_V,$$

hence $-(\partial_\xi^4 + \partial_\xi^2)$ extends to a continuous linear map from V to V^* such that

$$\| -(\partial_\xi^4 + \partial_\xi^2)u\|_{V^*} \leq \|u\|_V + \|u\|_H \text{ for all } u \in V.$$

Furthermore, recalling that

$$H_0^{1,2}(]0, L[) \subset L^\infty(]0, L[)$$

continuously for $u_1, u_2 \in C_0^\infty(]0, L[)$ by the Leibnitz rule we have for all $v \in C_0^\infty(]0, L[)$

$$
{}_{V^*}\langle \partial_\xi^2(\partial_x u_1)^2 - \partial_\xi^2(\partial_\xi u_2)^2, v \rangle_V
$$

$$
= \langle \partial_\xi^2(\partial_\xi u_1)^2 - \partial_\xi^2(\partial_\xi u_2)^2, \partial_\xi^4 v \rangle_{L^2}
$$

$$
\leq \|v\|_V \|\partial_\xi^2(\partial_\xi u_1)^2 - \partial_\xi^2(\partial_\xi u_2)^2\|_{L^2}
$$

$$
\leq 2\|v\|_V \left[\|(\partial_\xi^2 u_1)^2 - (\partial_\xi^2 u_2)^2\|_{L^2} + \|\partial_\xi u_1 \partial_\xi^3 u_1 - \partial_\xi u_2 \partial_\xi^3 u_2\|_{L^2} \right]
$$

$$
\leq 2\|v\|_V \left[(\|\partial_\xi^2 u_1\|_{L^\infty} + \|\partial_\xi^2 u_2\|_{L^\infty}) \|u_1 - u_2\|_H \right.
$$

$$
\left. + \|\partial_\xi u_1\|_{L^\infty} \|\partial_\xi^3 u_1 - \partial_\xi^3 u_2\|_{L^2} + \|\partial_\xi^3 u_2\|_{L^2} \|\partial_\xi u_1 - \partial_\xi u_2\|_{L^\infty} \right]
$$

$$
\leq C\|v\|_V \left[(\|\partial_\xi^3 u_1\|_{L^2} + \|\partial_\xi^3 u_2\|_{L^2}) \|u_1 - u_2\|_H + \|u_1\|_H \|\partial_\xi^3(u_1 - u_2)\|_{L^2} \right]
$$

$$
= C\|v\|_V \left[(\|\partial_\xi^2 u_1\|_{H^{1,2}} + \|\partial_\xi^2 u_2\|_{H^{1,2}}) \|u_1 - u_2\|_H + \|u_1\|_H \|\partial_\xi^2(u_1 - u_2)\|_{H^{1,2}} \right]
$$

$$
\leq C\|v\|_V \left[(\|\partial_\xi^2 u_1\|_{H^{2,2}}^{1/2} \|u_1\|_H^{1/2} + \|\partial_\xi^2 u_2\|_{H^{2,2}}^{1/2} \|u_2\|_H^{1/2}) \|u_1 - u_2\|_H \right.
$$

$$
\left. + \|u_1\|_H \|\partial_\xi^2(u_1 - u_2)\|_{H^{2,2}}^{1/2} \|u_1 - u_2\|_H^{1/2} \right]
$$

$$
= C\|v\|_H \left[(\|u_1\|_V^{1/2} \|u_1\|_H^{1/2} + \|u_2\|_V^{1/2} \|u_2\|_H^{1/2}) \|u_1 - u_2\|_H \right.
$$

$$
\left. + \|u_1\|_H \|u_1 - u_2\|_V^{1/2} \|u_1 - u_2\|_H^{1/2} \right], \tag{5.76}
$$

where we used (5.71) in the second to last step and $C \in]0, \infty[$, independent of u_1, u_2, v, but possibly changing from line to line. This implies local uniform continuity and hence extendability of $\partial_\xi^2(\partial_\xi \cdot)^2$ to a continuous map from V to V^*.

We define the maps $A_1, A_2 : V \to V^*$ by

$$
A_1(u) := -\frac{1}{2}(\partial_\xi^4 u + \partial_\xi^2 u), \quad A_2(u) := -\frac{1}{2}(\partial_\xi^4 u + \partial_\xi^2 u) + \partial_\xi^2(\partial_\xi u)^2, \quad u \in V,
$$

which by continuity satisfy (H1). It is also easy to check that A_1 satisfies all assumptions in Theorem 5.2.6. Furthermore, it suffices to check that A_2 satisfies (H2''), (H3') and (H4') for $u, v \in C_0^\infty(]0, L[)$

By (5.76) and Young's inequality there exists a $C \in]0, \infty[$ such that for all $u, v \in C_0^\infty(]0, L[)$

$$
{}_{V^*}\langle \partial_\xi^2(\partial_\xi u)^2 - \partial_\xi^2(\partial_\xi v)^2, u - v \rangle_V
$$

$$
\leq \frac{1}{4}\|u - v\|_V^2 + C(\|u\|_V \|u\|_H + \|v\|_V \|v\|_H + \|u\|_H^4) \|u - v\|_H^2. \tag{5.77}
$$

Furthermore, for all $u, v \in C_0^\infty(]0, L[)$

$$\frac{1}{2} {}_{V^*}\langle -\partial_\xi^4 u - \partial_\xi^2 u + \partial_\xi^4 v + \partial_\xi^2 v, u - v \rangle_V$$

$$\leq -\frac{1}{2}\|u - v\|_V^2 + \frac{1}{2}\|u - v\|_V \|u - v\|_H$$

$$\leq -\frac{3}{8}\|u - v\|_V^2 + 8\|u - v\|_H^2. \tag{5.78}$$

(5.77) and (5.78) imply that $(H2'')$ holds for A_2 with

$$\eta(u) := C\left(\|u\|_V^2 + \|u\|_H^4\right), \quad \rho(v) := C\|v\|_V^2.$$

Taking $u = 0$ in (5.77), (5.78) and applying Young's inequality we obtain that for some $C \in]0, \infty[$

$$2_{V^*}\langle A_2 v, v \rangle_V \leq -\frac{1}{10}\|v\|_V^2 + C\|v\|_H^6 + C \text{ for all } u \in C_0^\infty(]0, L[),$$

which implies that $(H3')$ holds for A_2 with $\alpha = 2$, $g(x) = Cx^3$ and $h \equiv 1$. Furthermore, (5.76) implies that $(H4')$ holds for A_2 with $\alpha = 2$ and $\beta = 1$.

Now the proof is complete.   □

Remark 5.2.30

(1) It is known in the literature that the surface growth model has some similar features of difficulty as the 3D Navier–Stokes equation. The uniqueness of analytically weak solutions for this model is still an open problem in both the deterministic and stochastic cases.

(2) The solution obtained here for the stochastic surface growth model is a strong solution both in the analytic and probabilistic sense. For the space-time white noise case, i.e. $B \equiv I$, the existence of a martingale solution was obtained in [7] for this model, and the existence of a Markov selection and ergodicity properties were also proved there.

(3) In [8, 9] the authors established local existence and uniqueness of solutions for the surface growth model with more general initial conditions in the critical Hilbert space $H^{1/2}$ or some Besov space (the largest possible critical space where weak solutions make sense). They used fixed point arguments and a technique introduced by H. Koch and D. Tataru for the Navier–Stokes equation (cf. [9] and the references therein).

Chapter 6
Mild Solutions

This chapter contains a concise introduction to the "semigroup (or mild solution) approach". One difference to the variational approach is that we do not use a Gelfand triple, but just our Hilbert space H. The main idea is to use the linear part (if there is one) of the drift as a "smoothing device".

6.1 Prerequisites for This Chapter

As said before, this course is mainly concentrated on the "variational approach" and this chapter is meant to be merely complementary, presenting another important approach to stochastic partial differential equations. Therefore, and since these prerequisites are only used in this chapter, in contrast to the other parts of this monograph, in this section we do not include proofs, but refer instead to [26].

6.1.1 The Itô Formula

We assume that

- $\Phi \in \mathcal{N}_W(0, T; H)$
- $\varphi \colon \Omega_T \to H$ is a predictable and P-a.s. Bochner integrable process on $[0, T]$
- $X(0) \colon \Omega \to H$ is $\mathcal{F}_0$-measurable

© Springer International Publishing Switzerland 2015
W. Liu, M. Röckner, *Stochastic Partial Differential Equations: An Introduction*,
Universitext, DOI 10.1007/978-3-319-22354-4_6

- $F: [0, T] \times H \to \mathbb{R}$ is twice Fréchet differentiable with derivatives

$$\frac{\partial F}{\partial t} := D_1 F: [0, T] \times H \to \mathbb{R}$$

$$DF := D_2 F: [0, T] \times H \to L(H, \mathbb{R}) \cong H$$

$$D^2 F := D_2^2 F: [0, T] \times H \to L(H),$$

which are uniformly continuous on bounded subsets of $[0, T] \times H$.

Under theses assumptions the process

$$X(t) := X(0) + \int_0^t \varphi(s) \, ds + \int_0^t \Phi(s) \, dW(s), \qquad t \in [0, T],$$

is well defined and we get the following result.

Theorem 6.1.1 (Itô Formula) *There exists a P-null set $N \in \mathcal{F}$ such that the following formula is fulfilled on N^c for all $t \in [0, T]$:*

$$F(t, X(t)) = F(0, X(0)) + \int_0^t \langle DF(s, X(s)), \Phi(s) \, dW(s) \rangle$$

$$+ \int_0^t \frac{\partial F}{\partial t}(s, X(s)) + \langle DF(s, X(s)), \varphi(s) \rangle$$

$$+ \frac{1}{2} \, \mathrm{tr} \left[D^2 F(s, X(s))(\Phi(s) Q^{\frac{1}{2}})(\Phi(s) Q^{\frac{1}{2}})^* \right] \, ds.$$

Proof [26, Theorem 4.17, p. 105]. Note that by Proposition B.0.10 the term involving the trace is indeed finite and is equal to

$$\frac{1}{2} \, \mathrm{tr} \left[(\Phi(s) Q^{\frac{1}{2}})^* D^2 F(s, X(s))(\Phi(s) Q^{\frac{1}{2}}) \right].$$

$\square$

6.1.2 A Burkholder–Davis–Gundy Type Inequality

Theorem 6.1.2 (Burkholder–Davis–Gundy Type Inequality) *Let $p \geq 2$ and $\Phi \in \mathcal{N}_W(0, T; H)$. Then*

$$\left(E \left(\sup_{t \in [0,T]} \left\| \int_0^t \Phi(s) \, dW(s) \right\|^p \right) \right)^{\frac{1}{p}} \leq p \left(\frac{p}{2(p-1)} \right)^{\frac{1}{2}} \left(\int_0^T \left(E \left(\| \Phi(s) \|_{L_2^0}^p \right) \right)^{\frac{2}{p}} \, ds \right)^{\frac{1}{2}}.$$

Proof [26, Lemma 7.2, p. 182].   □

Remark 6.1.3 If $\Phi \in \mathcal{N}_W^2(0, T)$ we get that $\int_0^t \Phi(s) \, dW(s)$, $t \in [0, T]$, is a martingale and therefore

$$\sup_{t \in [0,T]} E\left(\left\|\int_0^t \Phi(s) \, dW(s)\right\|^2\right) = E\left(\left\|\int_0^T \Phi(s) \, dW(s)\right\|^2\right).$$

6.1.3 Stochastic Fubini Theorem

We assume that

1. $(E, \mathcal{E}, \mu)$ is a measure space where μ is finite.
2. $\Phi: \Omega_T \times E \to L_2^0$, $(t, \omega, x) \mapsto \Phi(t, \omega, x)$ is $\mathcal{P}_T \otimes \mathcal{E}/\mathcal{B}(L_2^0)$-measurable, thus in particular $\Phi(\cdot, \cdot, x)$ is a predictable L_2^0-valued process for all $x \in E$.

Theorem 6.1.4 (Stochastic Fubini Theorem) *Assume 1., 2. and that*

$$\int_E \|\Phi(\cdot, \cdot, x)\|_t \, \mu(dx) = \int_E \left(E\left(\int_0^T \|\Phi(t, \cdot, x)\|_{L_2^0}^2 \, dt\right)\right)^{\frac{1}{2}} \mu(dx) < \infty.$$

Then

$$\int_E \left[\int_0^T \Phi(t, x) \, dW(t)\right] \mu(dx) = \int_0^T \left[\int_E \Phi(t, x) \, \mu(dx)\right] dW(t) \qquad P\text{-a.s.}$$

Proof [26, Theorem 4.18, p. 109]   □

6.2 Existence, Uniqueness and Continuity with Respect to the Initial Data

As in previous chapters let $(U, \| \, \|_U)$ and $(H, \| \, \|)$ be separable Hilbert spaces. We take $Q = I$ and fix a cylindrical Q-Wiener process $W(t)$, $t \geq 0$, in U on a probability space $(\Omega, \mathcal{F}, P)$ with a normal filtration $\mathcal{F}_t$, $t \geq 0$. Moreover, we fix $T > 0$ and consider the following type of stochastic differential equations in H

$$\begin{cases} dX(t) = [AX(t) + F(X(t))] \, dt + B(X(t)) \, dW(t), & t \in [0, T] \\ X(0) = \xi, \end{cases} \qquad (6.1)$$

where

- $A : D(A) \rightarrow H$ is the infinitesimal generator of a C_0-semigroup $S(t), t \geqslant 0$, of linear operators on H,
- $F : H \rightarrow H$ is $\mathcal{B}(H)/\mathcal{B}(H)$-measurable,
- $B : H \rightarrow L(U, H)$,
- ξ is an H-valued, $\mathcal{F}_0$-measurable random variable.

To motivate the definition of a mild solution below we first note that only in very special cases can one find a solution to (6.1) such that $X \in D(A)$ d$t \otimes P$-a.s. Therefore, one reformulates the equation on the basis of the following heuristics:

Consider the integral form of (6.1) and apply the (not-defined!) operator e^{-tA} for $t \in [0, T]$ to this equation. Applying Itô's product rule (again heuristically), we find

$$e^{-At}X(t) = \xi + \int_0^t e^{-As}(AX(s)) + F(X(s)) \, ds + \int_0^t e^{-As}B(X(s)) \, dW(s)$$

$$- \int_0^t e^{-As}AX(s) \, ds$$

$$\Rightarrow X(t) = S(t)\xi + \int_0^t S(t-s)F(X(s)) \, ds + \int_0^t S(t-s)B(X(s)) \, dW(s).$$

Definition 6.2.1 (Mild Solution) An H-valued predictable process $X(t), t \in [0, T]$, is called a *mild solution* of problem (6.1) if

$$X(t) = S(t)\xi + \int_0^t S(t-s)F(X(s)) \, ds$$

$$+ \int_0^t S(t-s)B(X(s)) \, dW(s) \quad P\text{-a.s.} \qquad (6.2)$$

for each $t \in [0, T]$ (i.e. the P-zero set, where (6.2) does not hold, may depend on t). In particular, the appearing integrals have to be well defined.

To prove the existence of a mild solution on $[0, T]$ we make the following usual assumptions (see [27, Hypothesis 5.1, p. 65]).

Hypothesis M.0

- $A : D(A) \rightarrow H$ is the infinitesimal generator of a C_0-semigroup $S(t), t \geqslant 0$, on H.
- $F : H \rightarrow H$ is Lipschitz continuous, i.e. that there exists a constant $C > 0$ such that

$$\|F(x) - F(y)\| \leqslant C\|x - y\| \quad \text{for all } x, y \in H.$$

- $B : H \rightarrow L(U, H)$ is strongly continuous, i.e. the mapping

$$x \mapsto B(x)u$$

is continuous from H to H for each $u \in U$.

- For all $t \in]0, T]$ and $x \in H$ we have that

$$S(t)B(x) \in L_2(U, H).$$

- There is a square integrable mapping $K : [0, T] \to [0, \infty]$ such that

$$\|S(t)(B(x) - B(y))\|_{L_2} \leqslant K(t)\|x - y\|$$

and

$$\|S(t)B(x)\|_{L_2} \leqslant K(t)(1 + \|x\|)$$

for all $t \in]0, T]$ and $x, y \in H$.

Remark 6.2.2

(i) $M_T := \sup_{t \in [0,T]} \|S(t)\|_{L(H)} < \infty$.

(ii) For the last assumption it is even enough to verify that there exists an $\varepsilon \in]0, T]$ such that the inequalities hold for $0 < t \leqslant \varepsilon$ and

$$\int_0^\varepsilon K^2(s) \ ds < \infty.$$

(iii) The Lipschitz constant of F in Hypothesis M.0 can be chosen in such a way that we also have

$$\|F(x)\| \leqslant C(1 + \|x\|) \quad \text{for all } x \in H.$$

Proof

(i) By the semigroup property it's easy to show there exist constants $\omega \geqslant 0$ and $M \geqslant 1$ such that

$$\|S(t)\|_{L(H)} \leqslant Me^{\omega t} \quad \text{for all } t \geqslant 0.$$

(ii) Let $\tilde{K} : [0, \varepsilon] \to [0, \infty]$ be square integrable such that

$$\|S(t)(B(x) - B(y))\|_{L_2} \leqslant \tilde{K}(t)\|x - y\|$$

and

$$\|S(t)B(x)\|_{L_2} \leqslant \tilde{K}(t)(1 + \|x\|)$$

for all $t \in [0, \varepsilon]$ and $x, y \in H$. Then we choose $N \in \mathbb{N}$ such that $\frac{T}{N} \leqslant \varepsilon$ and set

$$K(t) := M_T \tilde{K}(\frac{t}{N}) \quad \text{for } t \in [0, T],$$

where $M_T = \sup_{t \in [0,T]} \|S(t)\|_{L(H)}$. Then it is clear that $K : [0, T] \to [0, \infty[$ is square integrable and for all $x, y \in H$, $t \in]0, T]$ we get by the semigroup property that

$$\|S(t)(B(x) - B(y))\|_{L_2} = \|S(\frac{Nt-t}{N})S(\frac{t}{N})(B(x) - B(y))\|_{L_2}$$

$$\leq M_T \|S(\frac{t}{N})(B(x) - B(y))\|_{L_2}$$

$$\leq M_T \tilde{K}(\frac{t}{N}) \|x - y\| = K(t)\|x - y\|$$

and

$$\|S(t)B(x)\|_{L_2} \leq M_T \|S(\frac{t}{N})B(x)\|_{L_2}$$

$$\leq M_T \tilde{K}(\frac{t}{N})(1 + \|x\|) = K(t)(1 + \|x\|).$$

(iii) For all $x \in H$ we have that

$$\|F(x)\| \leq \|F(x) - F(0)\| + \|F(0)\|$$

$$\leq C\|x\| + \|F(0)\|$$

$$\leq (C \vee \|F(0)\|)(1 + \|x\|)$$

and, of course, we still have that

$$\|F(x) - F(y)\| \leq (C \vee \|F(0)\|)\|x - y\| \quad \text{for all } x, y \in H.$$

□

Now we introduce the spaces in which we want to find the mild solution of the above problem:

For each $T > 0$ and $p \geq 2$ we define $\mathcal{H}^p(T, H)$ to be the space of all H-valued predictable processes Y such that

$$\|Y\|_{\mathcal{H}^p} := \sup_{t \in [0,T]} \left(E(\|Y(t)\|^p)\right)^{\frac{1}{p}} < \infty.$$

Then $(\mathcal{H}^p(T, H), \|\ \|_{\mathcal{H}^p})$ is a Banach space (after going over to the usual equivalence classes of processes).

For technical reasons we also consider the norms $\|\ \|_{p,\lambda,T}$, $\lambda \geq 0$, on $\mathcal{H}^p(T, H)$ given by

$$\|Y\|_{p,\lambda,T} := \sup_{t \in [0,T]} e^{-\lambda t} \left(E(\|Y(t)\|^p)\right)^{\frac{1}{p}}.$$

Then $\| \ \|_{\mathcal{H}^p} = \| \ \|_{p,0,T}$ and for all $\lambda > 0$, $Y \in \mathcal{H}^p(T,H)$ we get that

$$\|Y\|_{p,\lambda,T} \leqslant \|Y\|_{\mathcal{H}^p} \leqslant e^{\lambda T} \|Y\|_{p,\lambda,T},$$

which means that all norms $\| \ \|_{p,\lambda,T}$, $\lambda \geqslant 0$, are equivalent. For simplicity we introduce the following notations

$$\mathcal{H}^p(T,H) := (\mathcal{H}^p(T,H), \| \ \|_{\mathcal{H}^p})$$

and

$$\mathcal{H}^{p,\lambda}(T,H) := (\mathcal{H}^p(T,H), \| \ \|_{p,\lambda,T}).$$

Theorem 6.2.3 *Under Hypothesis M.0 there exists a unique mild solution $X(\xi) \in \mathcal{H}^p(T,H)$ of problem (6.1) with initial condition*

$$\xi \in L^p(\Omega, \mathcal{F}_0, P; H) =: L_0^p.$$

In addition we even obtain that the mapping

$$X : L_0^p \to \mathcal{H}^p(T,H)$$

$$\xi \mapsto X(\xi)$$

is Lipschitz continuous with Lipschitz constant $L_{T,p}$.

Remark 6.2.4

1. The above result can be found in [27, Theorem 5.3.1, p. 66]. The proof is based on the abstract implicit function Theorem F.0.1. In particular, one has to verify that there is a predictable version of

$$\int_0^t S(t-s)B(Y(s)) \ dW(s), \quad t \in [0,T],$$

for all $Y \in \mathcal{H}^p(T,H)$. In [27, Proposition 6.2, p. 153] this is solved in the case where $B(Y) \in \mathcal{N}_W$. We, however, do not assume that B itself takes values in $L_2(U,H)$.

2. It follows from the Lipschitz continuity of X that there exists a constant $C_{T,p}$ independent of $\xi \in L_0^p$ such that

$$\|X(\xi)\|_{\mathcal{H}^p} \leqslant C_{T,p}(1 + \|\xi\|_{L^p}).$$

Before giving a proof of the theorem we need the following lemmas.

Lemma 6.2.5 *If* $Y : \Omega_T \rightarrow H$ *is* $\mathcal{P}_T/\mathcal{B}(H)$-*measurable and Hypothesis M.0 is fulfilled, then the mapping*

$$\tilde{Y} : \quad \Omega_T \rightarrow H$$

$$(s, \omega) \mapsto 1_{[0,t[}(s)S(t-s)Y(s, \omega)$$

is also $\mathcal{P}_T/\mathcal{B}(H)$-*measurable for each fixed* $t \in [0, T]$.

Proof

Step 1: We prove the assertion for simple processes Y given by

$$Y = \sum_{k=1}^{n} x_k 1_{A_k},$$

where $n \in \mathbb{N}$, $x_k \in H$, $1 \leqslant k \leqslant n$, and $A_k \in \mathcal{P}_T$, $1 \leqslant k \leqslant n$, is a disjoint covering of Ω_T. Then we get that

$$\tilde{Y} : \quad \Omega_T \rightarrow H$$

$$(s, \omega) \mapsto 1_{[0,t[}(s)S(t-s)Y(s, \omega) = 1_{[0,t[}(s) \sum_{k=1}^{n} S(t-s)x_k 1_{A_k}(s, \omega)$$

is $\mathcal{P}_T$-measurable since for each $B \in \mathcal{B}(H)$

$$\tilde{Y}^{-1}(B) = \bigcup_{k=1}^{n} (\underbrace{\{s \in [0, T] | 1_{[0,t[}(s)S(t-s)x_k \in B\}}_{\in \mathcal{B}([0,T])} \times \Omega) \cap A_k,$$

$$\underbrace{\phantom{\bigcup_{k=1}^{n} (\{s \in [0, T] | 1_{[0,t[}(s)S(t-s)x_k \in B\} \times \Omega) \cap A_k}}_{\in \mathcal{P}_T}$$

because of the strong continuity of the semigroup.

Step 2: We prove the assertion for an arbitrary predictable process Y.

If $Y : \Omega_T \rightarrow H$ is $\mathcal{P}_T$-measurable, there exists a sequence Y_n, $n \in \mathbb{N}$, of simple predictable processes such that $Y_n(s, \omega) \longrightarrow Y(s, \omega)$ as $n \rightarrow \infty$ for all $(s, \omega) \in [0, T] \times H$ (see Lemma A.1.4). Since $S(t) \in L(H)$ for all $t \in [0, T]$ we obtain that

$$\tilde{Y}(s, \omega) := 1_{[0,t[}(s)S(t-s)Y(s, \omega) = \lim_{n \rightarrow \infty} \underbrace{1_{[0,t[}(s)S(t-s)Y_n(s, \omega)}_{=: \tilde{Y}_n(s, \omega)}$$

By Step 1 $\tilde{Y}_n$, $n \in \mathbb{N}$, are predictable and, therefore, Proposition A.1.3 implies that $\tilde{Y}$ is also predictable. $\square$

Lemma 6.2.6 *If Y is a predictable H-valued process and Hypothesis M.0 is fulfilled then the mapping*

$$(s, \omega) \mapsto 1_{[0,t[}(s)S(t-s)B(Y(s,\omega))$$

is $\mathcal{P}_T/\mathcal{B}(L_2(U,H))$-measurable.

Proof Let f_k, $k \in \mathbb{N}$, be an orthonormal basis of H and e_k, $k \in \mathbb{N}$, an orthonormal basis of U. Then $f_k \otimes e_j = f_k \langle e_j, \cdot \rangle_U$, $k,j \in \mathbb{N}$, is an orthonormal basis of $L_2(U,H)$ (see Proposition B.0.7). Because of the strong continuity of B we obtain that

$$(s, \omega) \mapsto B(Y(s,\omega))e_j$$

is predictable for all $j \in \mathbb{N}$. Hence the previous lemma implies that

$$(s, \omega) \mapsto \langle f_k \otimes e_j, 1_{[0,t[}(s)S(t-s)B(Y(s,\omega))\rangle_{L_2}$$
$$= \langle f_k, 1_{[0,t[}(s)S(t-s)B(Y(s,\omega))e_j\rangle$$

is predictable for all $j, k \in \mathbb{N}$. This is enough to conclude that

$$(s, \omega) \mapsto 1_{[0,t[}(s)S(t-s)B(Y(s,\omega))$$

is predictable. □

Lemma 6.2.7 *If a mapping $g : \Omega_T \to H$ is $\mathcal{P}_T/\mathcal{B}(H)$-measurable then the mapping*

$$\tilde{Y} : \Omega_T \to H$$
$$(s, \omega) \mapsto 1_{[0,t[}(s)g(s,\omega)$$

is $\mathcal{B}([0,T]) \otimes \mathcal{F}_t/\mathcal{B}(H)$-measurable for each $t \in [0,T]$.

Proof We have to show that $([0, t[\times \Omega) \cap \mathcal{P}_T \subset \mathcal{B}([0,T]) \otimes \mathcal{F}_t$. Let $t \in [0,T]$. If we set

$$\mathcal{A} := \{A \in \mathcal{P}_T \mid A \cap ([0, t[\times \Omega) \in \mathcal{B}([0,T]) \otimes \mathcal{F}_t\}$$

it is clear that $\mathcal{A}$ is a σ-field which contains the predictable rectangles $]s, u] \times F_s$, $F_s \in \mathcal{F}_s$, $0 \leqslant s \leqslant u \leqslant T$ and $\{0\} \times F_0$, $F_0 \in \mathcal{F}_0$. Therefore $\mathcal{A} = \mathcal{P}_T$. □

Lemma 6.2.8 *If a process Φ is adapted to $\mathcal{F}_t$, $t \in [0,T]$, and stochastically continuous with values in a Banach space E, then there exists a predictable version of Φ.*

Proof [26, Proposition 3.6 (ii), p. 76] □

Lemma 6.2.9 *Let Φ be a predictable H-valued process which is P-a.s. Bochner integrable. Then the process given by*

$$\int_0^t S(t-s)\Phi(s,\omega) \ ds, \quad \omega \in \Omega, \quad t \in [0,T],$$

is P-a.s. continuous and adapted to $\mathcal{F}_t$, $t \in [0,T]$. This especially implies that it is predictable (more precisely, has a predictable version).

Proof By Lemma 6.2.5 the process $1_{[0,t[}(s)S(t-s)\Phi(s)$, $s \in [0,T]$, is predictable and in addition $\|1_{[0,t[}(s)S(t-s)\Phi(s)\| \leqslant M_T\|\Phi(s)\|$, $s \in [0,T]$.

Hence the integrals $\int_0^t S(t-s)\Phi(s) \ ds, t \in [0,T]$, are well defined.

First we want to prove the continuity. To this end let $0 \leqslant s \leqslant t \leqslant T$. Then we get that

$$\left\| \int_0^s S(s-u)\Phi(u) \ du - \int_0^t S(t-u)\Phi(u) \ du \right\|$$

$$\leqslant \left\| \int_0^s [S(s-u) - S(t-u)]\Phi(u) \ du \right\| + \left\| \int_s^t S(t-u)\Phi(u) \ du \right\|$$

$$\leqslant \int_0^s \|[S(s-u) - S(t-u)]\Phi(u)\| \ du + \int_s^t \|S(t-u)\Phi(u)\| \ du,$$

where the first summand converges to zero as $s \uparrow t$ or $t \downarrow s$ because by Lebesgue's dominated convergence theorem:

$$\|1_{[0,s[}(u)[S(s-u) - S(t-u)]\Phi(u)\| \to 0 \quad \text{as } s \uparrow t \text{ or } t \downarrow s$$

for all $u \in [0,T]$ because of the strong continuity of the semigroup $S(u)$, $u \in [0,T]$. Moreover

$$\|1_{[0,s[}(u)[S(s-u) - S(t-u)]\Phi(u)\|$$

$$\leqslant 1_{[0,s[}(u)(\|S(s-u)\|_{L(H)} + \|S(t-u)\|_{L(H)})\|\Phi(u)\|$$

$$\leqslant 2M_T\|\Phi(u)\|,$$

where $\|\Phi\| \in L^1([0,T],dx) := L^1([0,T],\mathcal{B}([0,T]), dx; \mathbb{R})$ P-a.s.

Concerning the second summand we get the same result since

$$\int_s^t \|S(t-u)\Phi(u)\| \ du$$

$$\leqslant \int_s^t M_T\|\Phi(u)\| \ du \longrightarrow 0 \quad \text{as } s \uparrow t \text{ or } t \downarrow s$$

P-a.s. by Lebesgue's dominated convergence theorem.

In order to prove that the process given by the integrals is adapted, we fix $t \in [0, T]$. By Lemma 6.2.7 the mapping

$$(s, \omega) \mapsto 1_{[0,t]}(s)S(t - s)\Phi(s, \omega)$$

is $\mathcal{B}([0, T]) \otimes \mathcal{F}_t$-measurable. Hence, by Proposition A.2.2, we get for each $x \in H$ that the mapping

$$
\begin{aligned}
\omega \mapsto & \Big\langle \int_0^t S(t - s)\Phi(s, \omega) \ ds, x \Big\rangle \\
= & \int_0^t \langle S(t - s)\Phi(s, \omega), x \rangle \ ds \\
= & \int_0^T \langle 1_{[0,t]}(s)S(t - s)\Phi(s, \omega), x \rangle \ ds
\end{aligned}
$$

is $\mathcal{F}_t$-measurable by the real Fubini theorem and, therefore, the integral itself is $\mathcal{F}_t$-measurable. $\qquad\square$

Lemma 6.2.10 *Let $(x_{n,m})_{m \in \mathbb{N}}$, $n \in \mathbb{N}$, be sequences of real numbers such that for each $n \in \mathbb{N}$ there exists an $x_n \in \mathbb{R}$ with*

$$x_{n,m} \longrightarrow x_n \quad as \ m \to \infty.$$

If there exists a further sequence y_n, $n \in \mathbb{N}$, such that $|x_{n,m}| \leqslant y_n$ for all $m \in \mathbb{N}$ and $\sum_{n \in \mathbb{N}} y_n < \infty$ then

$$\sum_{n \in \mathbb{N}} x_{n,m} \longrightarrow \sum_{n \in \mathbb{N}} x_n \quad as \ m \to \infty.$$

Proof The assertion is a simple consequence of Lebesgue's dominated convergence theorem with respect to the measure $\mu := \sum_{n \in \mathbb{N}} \delta_n$ where δ_n is the Dirac measure in n. $\qquad\square$

Proof of Theorem 6.2.3

Idea: Let $p \geqslant 2$. For $t \in [0, T]$, $\xi \in L_0^p$ and $Y \in \mathcal{H}^p(T, H)$ we define

$$\mathcal{F}(\xi, Y)(t) := S(t)\xi + \int_0^t S(t - s)F(Y(s)) \ ds + \int_0^t S(t - s)B(Y(s)) \ dW(s)$$

and prove that

$$\mathcal{F} : L_0^p \times \mathcal{H}^p(T, H) \to \mathcal{H}^p(T, H).$$

Since $X(\xi) \in \mathcal{H}^p(T, H)$ is a mild solution of problem (6.1) if and only if $\mathcal{F}(\xi, X(\xi)) = X(\xi)$ we have to look for an implicit function $X : L_0^p \to \mathcal{H}^p(T, H)$ such that the previous equation holds for arbitrary $\xi \in L_0^p$. To this end we show that there exists a $\lambda = \lambda(p) \geqslant 0$ such that

$$\mathcal{F} : L_0^p \times \mathcal{H}^{p, \lambda, T}(T, H) \to \mathcal{H}^{p, \lambda, T}(T, H)$$

is a contraction with respect to the second variable, i.e. that there exists an $\alpha(p) < 1$ such that for all $\xi \in L_0^p$ and $Y, \tilde{Y} \in \mathcal{H}^p(T, H)$

$$\|\mathcal{F}(\xi, Y) - \mathcal{F}(\xi, \tilde{Y})\|_{p, \lambda, T} \leqslant \alpha(p) \|Y - \tilde{Y}\|_{p, \lambda, T}.$$

Setting $G := \mathcal{F}$, $\Lambda := L_0^p$ and $E := \mathcal{H}^p(T, H)$ we are hence in the situation described at the beginning of Appendix F. Therefore, it is clear that the implicit function $X = \varphi$ exists and that it is unique.

To get the Lipschitz continuity of the mapping $X : L_0^p \to \mathcal{H}^p(T, H)$ we verify that the condition of Theorem F.0.1 (ii) is fulfilled. Because of the equivalence of the norms $\| \ \|_{p, \lambda, T}$ and $\| \ \|_{\mathcal{H}^p}$ that means that it is enough to show the existence of a constant $L_{T, p} > 0$ such that

$$\|\mathcal{F}(\xi, Y) - \mathcal{F}(\tilde{\xi}, Y)\|_{\mathcal{H}^p} \leqslant L_{T, p} \|\xi - \tilde{\xi}\|_{L^p}$$

for all $\xi, \tilde{\xi} \in L_0^p$ and $Y \in \mathcal{H}^p(T, H)$.

Step 1:   We prove that the mapping $\mathcal{F}$ is well defined.
Let $\xi \in L_0^p$ and $Y \in \mathcal{H}^p(T, H)$.

1. The Bochner integral $\int_0^t S(t - s) F(Y(s)) \ ds$, $t \in [0, T]$, is well defined by Lemma 6.2.9:

    (i) Because of the continuity of $F : H \to H$ it is clear that $F(Y(t))$, $t \in [0, T]$, is predictable.

    (ii) In addition the process $F(Y(t))$, $t \in [0, T]$, is P-a.s. Bochner integrable since

    $$E\left(\int_0^t \|F(Y(s))\| \ ds\right) \leqslant \int_0^t E(C(1 + \|Y(s)\|)) \ ds \leqslant CT(1 + \|Y\|_{\mathcal{H}^p}) < \infty.$$

2. The stochastic integrals $\int_0^t S(t - s) B(Y(s)) \ dW(s)$, $t \in [0, T]$, are well defined since the processes $1_{[0, t[}(s) S(t - s) B(Y(s))$, $s \in [0, T]$, are in $\mathcal{N}_W^2(0, T)$ for all $t \in [0, T]$:

    (i) The mapping

    $$(s, \omega) \mapsto 1_{[0, t[}(s) S(t - s) B(Y(s, \omega))$$

is $\mathcal{P}_T/\mathcal{B}(L_2(U,H))$-measurable by Lemma 6.2.6.

(ii) For the norm we obtain that

$$\|1_{[0,t[}S(t-\cdot)B(Y)\|_T^2 = E\Big(\int_0^t \|S(t-s)B(Y(s))\|_{L_2}^2 \, ds\Big)$$

$$= \int_0^t E(\|S(t-s)B(Y(s))\|_{L_2}^2) \, ds$$

$$\leqslant \int_0^t K^2(t-s)E\big((1+\|Y(s)\|)^2\big) \, ds$$

$$\leqslant 2\int_0^t K^2(t-s)E(1+\|Y(s)\|^2) \, ds$$

$$\leqslant 2\,(1+\sup_{s\in[0,T]} E(\|Y(s)\|^2))\int_0^t K^2(t-s) \, ds$$

$$\leqslant 2\,(1+\|Y\|_{\mathcal{H}^p}^2)\int_0^t K^2(s) \, ds < \infty.$$

Step 2: We prove that $\mathcal{F}(\xi,Y) \in \mathcal{H}^p(T,H)$ for all $\xi \in L_0^p$ and $Y \in \mathcal{H}^p(T,H)$. Let $\xi \in L_0^p$ and $Y \in \mathcal{H}^p(T,H)$.

1. The first summand $S(t)\xi$, $t \in [0,T]$, is an element of $\mathcal{H}^p(T,H)$:

(i) The mapping

$$(s,\omega) \mapsto S(t)\xi(\omega)$$

is predictable since for fixed $\omega \in \Omega$

$$t \mapsto S(t)\xi(\omega)$$

is a continuous mapping from $[0,T]$ to H and for fixed $t \in [0,T]$

$$\omega \mapsto S(t)\xi(\omega)$$

is not only $\mathcal{F}_t$- but even $\mathcal{F}_0$-measurable.

(ii) For the norm we obtain that

$$\|S(\cdot)\xi\|_{\mathcal{H}^p} = \sup_{t\in[0,T]} (E(\|S(t)\xi\|^p))^{\frac{1}{p}} \leqslant M_T\|\xi\|_{L^p} < \infty.$$

2. There is a version of the second summand $\int_0^t S(t-s)F(Y(s)) \, ds$, $t \in [0,T]$, which is an element of $\mathcal{H}^p(T,H)$:

(i) First we notice that the mapping

$$(s, \omega) \mapsto \int_0^t S(t - s)F(Y(s, \omega)) \ ds$$

has a predictable version because the assumptions of Lemma 6.2.9 are fulfilled.

(ii) Concerning the norm we prove that

$$\left\| \int_0^{\cdot} S(\cdot - s)F(Y(s)) \ ds \right\|_{\mathcal{H}^p} \leq CTM_T(1 + \|Y\|_{\mathcal{H}^p}).$$

To verify the assertion we take t from $[0, T]$ and show that the L^p-norm of

$$\left\| \int_0^t S(t - s)F(Y(s)) \ ds \right\|$$

can be estimated independently of $t \in [0, T]$:

$$\left\| \int_0^t S(t - s)F(Y(s)) \ ds \right\|^p \leq C^p T^{p-1} M_T^p \int_0^t (1 + \|Y(s)\|)^p \ ds \quad P\text{-a.s.}$$

Taking the expectation we get that

$$\left(E(\left\| \int_0^t S(t - s)F(Y(s)) \ ds \right\|^p) \right)^{\frac{1}{p}}$$

$$\leq CT^{\frac{p-1}{p}} M_T \left(E(\int_0^t (1 + \|Y(s)\|)^p \ ds) \right)^{\frac{1}{p}}$$

$$\leq CT^{\frac{p-1}{p}} M_T \left[\left(E(\int_0^T 1 \ ds) \right)^{\frac{1}{p}} + \left(\int_0^T E(\|Y(s)\|^p) \ ds \right)^{\frac{1}{p}} \right]$$

$$\leq CTM_T(1 + \|Y\|_{\mathcal{H}^p}) < \infty$$

and the claimed inequality follows.

3. There is a version of $\int_0^t S(t - s)B(Y(s)) \ dW(s)$, $t \in [0, T]$, which is in $\mathcal{H}^p(T, H)$: (i) First we show that there is a predictable version of the process. To do so we proceed in several steps.

Claim 1: If $\alpha > 1$ the process $\int_0^{\frac{t}{\alpha}} S(t - s)B(Y(s)) \ dW(s)$, $t \in [0, T]$, has a predictable version.

To prove this we first use the semigroup property and get that

$$\int_0^{\frac{t}{\alpha}} S(t-s)B(Y(s)) \ dW(s)$$

$$= \int_0^{\frac{t}{\alpha}} S(t-\alpha s)S((\alpha-1)s)B(Y(s)) \ dW(s), \quad t \in [0,T].$$

We set

$$\Phi^\alpha(s) := (s)1_{]0,T]}(s)S((\alpha-1)s)B(Y(s)).$$

Then it is clear that $\Phi^\alpha(t)$, $t \in [0,T]$, is an element of $\mathcal{N}_W^2(0,T)$:
The fact that there is a predictable version of

$$(s,\omega) \mapsto 1_{]0,T]}(s)S((\alpha-1)s)B(Y(s,\omega))$$

can be proved in the same way as Lemma 6.2.6 and, of course, by Hypothesis
M.0

$$E\Big(\int_0^T \|S((\alpha-1)s)B(Y(s))\|_{L_2}^2 \ ds\Big)$$

$$\leqslant \int_0^T E\big(K^2((\alpha-1)s)(1+\|Y(s)\|)^2\big) \ ds$$

$$\leqslant 2\int_0^T K^2((\alpha-1)s)(1+\sup_{s\in[0,T]} E(\|Y(s)\|^2)) \ ds$$

$$= 2(1+\|Y\|_{\mathcal{H}^2(T,H)}^2)\int_0^{(\alpha-1)T} \frac{1}{\alpha-1}K^2(s) \ ds < \infty.$$

Therefore, we now have to prove that the process

$$\int_0^{\frac{t}{\alpha}} S(t-\alpha s)\tilde{\Phi}(s) \ dW(s), \quad t \in [0,T],$$

has a predictable version for each $\alpha > 1$ and $\tilde{\Phi} \in \mathcal{N}_W^2(0,T)$.

(a) We first consider the case where $\tilde{\Phi}(t)$, $t \in [0,T]$, is a simple process of the
form

$$\tilde{\Phi} = \sum_{k=1}^m L_k 1_{A_k}$$

where $m \in \mathbb{N}$, $L_k \in L_2(U, H)$ and $A_k \in \mathcal{P}_T$, $1 \leqslant k \leqslant m$. To get the required measurability we check the conditions of Lemma 6.2.8.
At first it is clear that

$$\int_0^{\frac{L}{\alpha}} S(t - \alpha s)\tilde{\Phi}(s) \; dW(s)$$

is $\mathcal{F}_{\frac{L}{\alpha}}$- and therefore also $\mathcal{F}_t$-measurable for each $t \in [0, T]$ since the process

$$1_{[0, \frac{L}{\alpha}]}(s)S(t - \alpha s)\tilde{\Phi}(s), \quad s \in [0, T],$$

lies in $\mathcal{N}_W^2(0, T)$ (see the proof of Lemma 6.2.6) and therefore the process

$$\int_0^u 1_{[0, \frac{L}{\alpha}]}(s)S(t - \alpha s)\tilde{\Phi}(s) \; dW(s), \quad u \in [0, T],$$

is an H-valued martingale with respect to $\mathcal{F}_u$, $u \in [0, T]$. Now we show that

$$t \mapsto \int_0^{\frac{L}{\alpha}} S(t - \alpha s)\tilde{\Phi}(s) \; dW(s)$$

is continuous in mean square and, therefore, stochastically continuous.
To this end we take arbitrary $0 \leqslant t < u \leqslant T$ and get that

$$\left(E(\| \int_0^{\frac{u}{\alpha}} S(u - \alpha s)\tilde{\Phi}(s) \; dW(s) - \int_0^{\frac{L}{\alpha}} S(t - \alpha s)\tilde{\Phi}(s) \; dW(s)\|^2)\right)^{\frac{1}{2}}$$

$$\leqslant \left(E(\| \int_0^{\frac{L}{\alpha}} [S(u - \alpha s) - S(t - \alpha s)]\tilde{\Phi}(s) \; dW(s)\|^2)\right)^{\frac{1}{2}}$$

$$+ \left(E(\| \int_0^{\frac{u}{\alpha}} 1_{[\frac{L}{\alpha}, \frac{u}{\alpha}]}(s)S(u - \alpha s)\tilde{\Phi}(s) \; dW(s)\|^2)\right)^{\frac{1}{2}}$$

$$= \left(E(\int_0^{\frac{L}{\alpha}} \|[S(u - \alpha s) - S(t - \alpha s)]\tilde{\Phi}(s)\|_{L_2}^2 \; ds)\right)^{\frac{1}{2}}$$

$$+ \left(E(\int_{\frac{L}{\alpha}}^{\frac{u}{\alpha}} \|S(u - \alpha s)\tilde{\Phi}(s)\|_{L_2}^2 \; ds)\right)^{\frac{1}{2}}$$

$$\leqslant \sum_{k=1}^m \left(E(\int_0^{\frac{L}{\alpha}} 1_{A_k}(s, \cdot)\|[S(u - \alpha s) - S(t - \alpha s)]L_k\|_{L_2}^2 \; ds)\right)^{\frac{1}{2}}$$

$$+ \sum_{k=1}^m \left(E(\int_{\frac{L}{\alpha}}^{\frac{u}{\alpha}} 1_{A_k}(s, \cdot)\|S(u - \alpha s)L_k\|_{L_2}^2 \; ds)\right)^{\frac{1}{2}}$$

$$\leq \sum_{k=1}^{m} \left(\int_{0}^{\frac{L}{\alpha}} \|[S(u - \alpha s) - S(t - \alpha s)]L_k\|_{L_2}^2 \, ds \right)^{\frac{1}{2}}$$

$$+ \sum_{k=1}^{m} \left(\int_{\frac{L}{\alpha}}^{\frac{u}{\alpha}} \|S(u - \alpha s)L_k\|_{L_2}^2 \, ds \right)^{\frac{1}{2}}, \tag{6.3}$$

where the first summand in the right-hand side of (6.3) converges to zero as $t \uparrow u$ or $u \downarrow t$ for the following reason:
Let e_n, $n \in \mathbb{N}$, be an orthonormal basis of U. Then we get for each $s \in [0, T]$ and $1 \leq k \leq m$ that

$$1_{[0, \frac{L}{\alpha}[}(s) \|[S(u - \alpha s) - S(t - \alpha s)]L_k\|_{L_2}^2$$

$$= \sum_{n \in \mathbb{N}} 1_{[0, \frac{L}{\alpha}[}(s) \|[S(u - \alpha s) - S(t - \alpha s)]L_k e_n\|^2,$$

where

$$1_{[0, \frac{L}{\alpha}[}(s) \|[S(u - \alpha s) - S(t - \alpha s)]L_k e_n\|^2 \longrightarrow 0 \quad \text{as } t \uparrow u \text{ or } u \downarrow t$$

and at the same time

$$1_{[0, \frac{L}{\alpha}[}(s) \|[S(u - \alpha s) - S(t - \alpha s)]L_k e_n\|^2 \leq 4M_T^2 \|L_k e_n\|^2$$

for all $n \in \mathbb{N}$, $1 \leq k \leq m$. By Lemma 6.2.10 this result implies the pointwise convergence

$$1_{[0, \frac{L}{\alpha}[}(s) \|[S(u - \alpha s) - S(t - \alpha s)]L_k\|_{L_2}^2 \longrightarrow 0 \quad \text{as } t \uparrow u \text{ or } u \downarrow t.$$

Since there is the following upper bound

$$1_{[0, \frac{L}{\alpha}[}(s) \|[S(u - \alpha s) - S(t - \alpha s)]L_k\|_{L_2}^2 \leq 4M_T^2 \|L_k\|_{L_2}^2 \in L^1([0, T], dx)$$

for all $s \in [0, T]$, $0 \leq t < u \leq T$, we get the required convergence of the integrals $\int_{0}^{\frac{L}{\alpha}} \|[S(u - \alpha s) - S(t - \alpha s)]L_k\|_{L_2}^2 \, ds$, $1 \leq k \leq m$, by Lebesgue's dominated convergence theorem.
The second summand in the right side of (6.3) of the above equation also converges to zero since for each $1 \leq k \leq m$

$$\int_{\frac{L}{\alpha}}^{\frac{u}{\alpha}} \|S(u - \alpha s)L_k\|_{L_2}^2 \, ds \leq \frac{u - t}{\alpha} M_T^2 \|L_k\|_{L_2}^2 \longrightarrow 0 \quad \text{as } u \downarrow t \text{ or } t \uparrow u.$$

Hence Lemma 6.2.8 implies that there is a predictable version of

$$\int_0^{\frac{t}{\alpha}} S(t - \alpha s) \tilde{\Phi}(s) \ dW(s), \quad t \in [0, T],$$

if $\tilde{\Phi}$ is elementary.

(b) Now we generalize this result to arbitrary $\tilde{\Phi} \in \mathcal{N}_W^2$:

If $\tilde{\Phi}$ is an arbitrary process in $\mathcal{N}_W^2(0, T)$ there exists a sequence $\tilde{\Phi}_n$, $n \in \mathbb{N}$, of simple processes of the form we considered in (a) such that

$$E \left(\int_0^T \| \tilde{\Phi}(s) - \tilde{\Phi}_n(s) \|_{L_2}^2 \ ds \right) \xrightarrow[n \to \infty]{} 0$$

(see Lemma A.1.4). Hence let $\Psi_n(t)$, $t \in [0, T]$, $n \in \mathbb{N}$, be a predictable version of $\int_0^{\frac{t}{\alpha}} S(t - \alpha s) \tilde{\Phi}_n(s) \ dW(s)$, $t \in [0, T]$, $n \in \mathbb{N}$, which exists by step (a). To get the predictability of $\int_0^{\frac{t}{\alpha}} S(t - \alpha s) \Phi(s) \ dW(s)$ we prove that there is a subsequence n_k, $k \in \mathbb{N}$, such that

$$\Psi_{n_k}(t) \xrightarrow[k \to \infty]{} \int_0^{\frac{t}{\alpha}} S(t - \alpha s) \tilde{\Phi}(s) \ dW(s) \quad P\text{-a.s. for all } t \in [0, T].$$

To this end we take $c > 0$, $t \in [0, T]$ and obtain that

$$P \left(\left\| \int_0^{\frac{t}{\alpha}} S(t - \alpha s) \tilde{\Phi}(s) \ dW(s) - \Psi_n(t) \right\| > c \right)$$

$$\leqslant \frac{1}{c^2} E \left(\left\| \int_0^{\frac{t}{\alpha}} S(t - \alpha s) [\tilde{\Phi}(s) - \tilde{\Phi}_n(s)] \ dW(s) \right\|^2 \right)$$

$$= \frac{1}{c^2} E \left(\int_0^{\frac{t}{\alpha}} \| S(t - \alpha s) [\tilde{\Phi}(s) - \tilde{\Phi}_n(s)] \|_{L_2}^2 \ ds \right)$$

$$\leqslant \frac{M_T^2}{c^2} E \left(\int_0^{\frac{t}{\alpha}} \| \tilde{\Phi}(s) - \tilde{\Phi}_n(s) \|_{L_2}^2 \ ds \right) \leqslant \frac{M_T^2}{c^2} E \left(\int_0^T \| \tilde{\Phi}(s) - \tilde{\Phi}_n(s) \|_{L_2}^2 \ ds \right).$$

As this upper bound is independent of $t \in [0, T]$ this implies that

$$\sup_{t \in [0,T]} P \left(\left\| \int_0^{\frac{t}{\alpha}} S(t - \alpha s) \tilde{\Phi}(s) \ dW(s) - \Psi_n(t) \right\| > c \right)$$

$$\leqslant \frac{M_T^2}{c^2} E \left(\int_0^T \| \tilde{\Phi}(s) - \tilde{\Phi}_n(s) \|_{L_2}^2 \ ds \right) \xrightarrow[n \to \infty]{} 0.$$

Therefore, there is a sequence n_k, $k \in \mathbb{N}$, such that

$$P(\| \int_0^{\frac{L}{\alpha}} S(t - \alpha s) \tilde{\Phi}(s) \; dW(s) - \Psi_{n_k} \| > 2^{-k}) \leq 2^{-k}$$

for all $t \in [0, T]$, $k \in \mathbb{N}$. By the Borel–Cantelli lemma it follows that

$$\Psi_{n_k}(t) \longrightarrow \int_0^{\frac{L}{\alpha}} S(t - \alpha s) \tilde{\Phi}(s) \; dW(s) \quad P\text{-a.s. as } k \to \infty$$

for all $t \in [0, T]$. If we set now

$$A := \{(t, \omega) \in \Omega_T \mid (\Psi_{n_k}(t, \omega))_{k \in \mathbb{N}} \text{ is convergent in } H\},$$

then $A \in \mathcal{P}_T$ and the process defined by

$$\Psi(t, \omega) := \begin{cases} \lim_{k \to \infty} \Psi_{n_k}(t, \omega) & \text{if } (t, \omega) \in A \\ 0 & \text{otherwise} \end{cases}$$

is a predictable version of $\int_0^{\frac{L}{\alpha}} S(t - \alpha s) \tilde{\Phi}(s) \; dW(s)$, $t \in [0, T]$.
Taking $\tilde{\Phi} = \Phi^\alpha$ we hence obtain that

$$\int_0^{\frac{L}{\alpha}} S(t - \alpha s) \Phi^\alpha(s) \; dW(s) = \int_0^{\frac{L}{\alpha}} S(t - s) B(Y(s)) \; dW(s), \quad t \in [0, T],$$

has a predictable version. By this result we can prove the assertion we are interested in.

Claim 2: The process $\int_0^t S(t - s) B(Y(s)) \; dW(s)$, $t \in [0, T]$, has a predictable version.

Let $(\alpha_n)_{n \in \mathbb{N}}$ be a sequence of real numbers such that $\alpha_n \downarrow 1$ as $n \to \infty$. By Claim 1 there is a predictable version $\Psi_{\alpha_n}(t)$, $t \in [0, T]$, of $\int_0^{\frac{L}{\alpha_n}} S(t - s) B(Y(s)) \; dW(s)$, $t \in [0, T]$, $n \in \mathbb{N}$. If we define

$$B := \{(t, \omega) \in \Omega_T \mid (\Psi_{\alpha_n}(t, \omega))_{n \in \mathbb{N}} \text{ is convergent}\},$$

it is clear that $B \in \mathcal{P}_T$ and the process given by

$$\Psi(t, \omega) := \begin{cases} \lim_{n \to \infty} \Psi_{\alpha_n}(t, \omega) & \text{if } (t, \omega) \in B \\ 0 & \text{otherwise} \end{cases}$$

is predictable. Besides we get that for each $t \in [0, T]$

$$\Psi(t) = \int_0^t S(t-s)B(Y(s)) \ dW(s) \quad P\text{-a.s.,}$$

since P-a.s.

$$\Psi_{\alpha_n}(t) = \int_0^{\frac{t}{\alpha_n}} S(t-s)B(Y(s)) \ dW(s) \xrightarrow[n\to\infty]{} \int_0^t S(t-s)B(Y(s)) \ dW(s),$$

because of the continuity of the stochastic integrals

$$\int_0^u 1_{[0,t[}(s)S(t-s)B(Y(s)) \ dW(s), \quad u \in [0, T].$$

Therefore the predictable version is found.
(ii) Concerning the norm we get that

$$\left\| \int_0^{\cdot} S(\cdot - s)B(Y(s)) \ dW(s) \right\|_{\mathcal{H}^p}$$

$$\leq (\frac{p}{2}(p-1))^{\frac{1}{2}} \left(\int_0^T K^2(s) \ ds \right)^{\frac{1}{2}} (1 + \|Y\|_{\mathcal{H}^p}),$$

since we obtain by Theorem 6.1.2 (Burkholder–Davis–Gundy type inequality) that

$$\left(E(\| \int_0^t S(t-s)B(Y(s)) \ dW(s) \|^p) \right)^{\frac{1}{p}}$$

$$\leq (\frac{p}{2}(p-1))^{\frac{1}{2}} \left(\int_0^t \left(E(\|S(t-s)B(Y(s))\|_{L_2}^p) \right)^{\frac{2}{p}} \ ds \right)^{\frac{1}{2}}$$

$$\leq (\frac{p}{2}(p-1))^{\frac{1}{2}} \left(\int_0^t K^2(t-s) \left(E((1 + \|Y(s)\|)^p) \right)^{\frac{2}{p}} \ ds \right)^{\frac{1}{2}}$$

$$\leq (\frac{p}{2}(p-1))^{\frac{1}{2}} \left(\int_0^t K^2(t-s)(1 + \|Y(s)\|_{L^p})^2 \ ds \right)^{\frac{1}{2}}$$

$$\leq (\frac{p}{2}(p-1))^{\frac{1}{2}} \left(\int_0^T K^2(s) \ ds \right)^{\frac{1}{2}} (1 + \|Y\|_{\mathcal{H}^p}).$$

Therefore, we have finally proved that

$$\mathcal{F} : L_0^p \times \mathcal{H}^p(T, H) \to \mathcal{H}^p(T, H).$$

Step 3: For each $p \geqslant 2$ there is a $\lambda(p) =: \lambda$ such that

$$\mathcal{F}(\xi, \cdot) : \mathcal{H}^{p,\lambda}(T, H) \to \mathcal{H}^{p,\lambda}(T, H)$$

is a contraction for all $\xi \in L_0^p$ where the contraction constant $\alpha(\lambda) < 1$ does not depend on ξ:
Let $Y, \tilde{Y} \in \mathcal{H}^p(T, H)$, $\xi \in L_0^p$ and $t \in [0, T]$. Then we get that

$$\|[\mathcal{F}(\xi, Y) - \mathcal{F}(\xi, \tilde{Y})](t)\|_{L^p}$$

$$\leqslant \left\| \int_0^t S(t - s)[F(Y(s)) - F(\tilde{Y}(s))] \ ds \right\|_{L^p}$$

$$+ \left\| \int_0^t S(t - s)[B(Y(s)) - B(\tilde{Y}(s))] \ dW(s) \right\|_{L^p},$$

where the first summand can be estimated in the following way:

$$\left\| \int_0^t S(t - s)[F(Y(s)) - F(\tilde{Y}(s))] \ ds \right\|^p \leqslant M_T^p C^p T^{p-1} \int_0^t \|Y(s) - \tilde{Y}(s)\|^p \ ds \ \text{P-a.s.},$$

which implies that

$$\left(E(\left\| \int_0^t S(t - s)[F(Y(s)) - F(\tilde{Y}(s))] \ ds \right\|^p) \right)^{\frac{1}{p}}$$

$$\leqslant M_T C T^{\frac{p-1}{p}} \left(\int_0^t E(\|Y(s) - \tilde{Y}(s)\|^p) \ ds \right)^{\frac{1}{p}}$$

$$= M_T C T^{\frac{p-1}{p}} \left(\int_0^t e^{\lambda p s} \underbrace{e^{-\lambda p s} E(\|Y(s) - \tilde{Y}(s)\|^p)}_{\leqslant \|Y - \tilde{Y}\|_{p,\lambda,T}^p} \ ds \right)^{\frac{1}{p}}$$

$$\leqslant M_T C T^{\frac{p-1}{p}} \left(\int_0^t e^{\lambda p s} \ ds \right)^{\frac{1}{p}} \|Y - \tilde{Y}\|_{p,\lambda,T}$$

$$\leqslant M_T C T^{\frac{p-1}{p}} e^{\lambda t} (\frac{1}{\lambda p})^{\frac{1}{p}} \|Y - \tilde{Y}\|_{p,\lambda,T}.$$

Dividing by $e^{\lambda t}$ yields

$$\left\| \int_0^{\cdot} S(\cdot - s)[F(Y(s)) - F(\tilde{Y}(s))] \ ds \right\|_{p,\lambda,T} \leqslant \underbrace{M_T C T^{\frac{p-1}{p}} (\frac{1}{\lambda p})^{\frac{1}{p}}}_{\longrightarrow 0 \ \text{as} \ \lambda \to \infty} \|Y - \tilde{Y}\|_{p,\lambda,T}.$$

By Theorem 6.1.2 we get the following estimate for the second summand:

$$\left(E(\| \int_0^t S(t-s)[B(Y(s)) - B(\tilde{Y}(s))] \ dW(s)\|^p) \right)^{\frac{1}{p}}$$

$$\leqslant (\tfrac{p}{2}(p-1))^{\frac{1}{2}} \left(\int_0^t \left(E(\|S(t-s)[B(Y(s)) - B(\tilde{Y}(s))]\|_{L_2}^p) \right)^{\frac{2}{p}} \ ds \right)^{\frac{1}{2}}$$

$$\leqslant (\tfrac{p}{2}(p-1))^{\frac{1}{2}} \left(\int_0^t K^2(t-s)\|Y(s) - \tilde{Y}(s)\|_{L^p}^2 \ ds \right)^{\frac{1}{2}}$$

$$= (\tfrac{p}{2}(p-1))^{\frac{1}{2}} \left(\int_0^t K^2(t-s)e^{2\lambda s} \underbrace{e^{-2\lambda s}\|Y(s) - \tilde{Y}(s)\|_{L^p}^2}_{\leqslant \|Y - \tilde{Y}\|_{p,\lambda,T}^2} \ ds \right)^{\frac{1}{2}}$$

$$\leqslant (\tfrac{p}{2}(p-1))^{\frac{1}{2}} \left(\int_0^t K^2(t-s)e^{2\lambda s} \ ds \right)^{\frac{1}{2}} \|Y - \tilde{Y}\|_{p,\lambda,T}$$

$$\leqslant (\tfrac{p}{2}(p-1))^{\frac{1}{2}} e^{\lambda t} \left(\int_0^T K^2(s)e^{-2\lambda s} \ ds \right)^{\frac{1}{2}} \|Y - \tilde{Y}\|_{p,\lambda,T}.$$

As for the first summand this implies that

$$\| \int_0^{\cdot} S(\cdot - s)[B(Y(s)) - B(\tilde{Y}(s))] \ dW(s)\|_{p,\lambda,T}$$

$$\leqslant \underbrace{(\tfrac{p}{2}(p-1))^{\frac{1}{2}} \left(\int_0^T K^2(s)e^{-2\lambda s} \ ds \right)^{\frac{1}{2}} \|Y - \tilde{Y}\|_{p,\lambda,T}.}_{\longrightarrow 0 \quad \text{as } \lambda \to \infty}$$

Therefore, we have finally proved that there is a $\lambda = \lambda(p)$ such that there exists an $\alpha(\lambda) < 1$ with

$$\|\mathcal{F}(\xi, Y) - \mathcal{F}(\xi, \tilde{Y})\|_{p,\lambda,T} \leqslant \alpha(\lambda)\|Y - \tilde{Y}\|_{p,\lambda,T}$$

for all $\xi \in L_0^p$, $Y, \tilde{Y} \in \mathcal{H}^{p,\lambda}(T,H)$. Thus the existence of a unique implicit function

$$X : L_0^p \to \mathcal{H}^p(T,H)$$

$$\xi \mapsto X(\xi) = \mathcal{F}(\xi, X(\xi))$$

is verified.

Step 4:   We prove the Lipschitz continuity of $X : L_0^p \to \mathcal{H}^p(T,H)$.

By Theorem F.0.1 (ii) and the equivalence of the norms $\| \ \|_{\mathcal{H}^p}$ and $\| \ \|_{p,\lambda,T}$ we only have to check that the mappings

$$\mathcal{F}(\cdot, Y) : L_0^p \to \mathcal{H}^p(T,H)$$

are Lipschitz continuous for all $Y \in \mathcal{H}^p(T, H)$ where the Lipschitz constant does not depend on Y.

But this assertion holds as for all $\xi, \zeta \in L_0^p$ and $Y \in \mathcal{H}^p(T, H)$

$$\|\mathcal{F}(\xi, Y) - \mathcal{F}(\zeta, Y)\|_{\mathcal{H}^p} = \|S(\cdot)(\xi - \zeta)\|_{\mathcal{H}^p} \leqslant M_T \|\xi - \zeta\|_{L^p}.$$

$\square$

6.3 Smoothing Property of the Semigroup: Pathwise Continuity of the Mild Solution

Let $X(\xi)$ be the mild solution of problem (6.1) with initial condition $\xi \in L_0^p$. The aim of this section is to prove that the mapping $t \mapsto X(\xi)(t)$ has a continuous version. Because of Lemma 6.2.9 we already know that the process of the Bochner integrals

$$\int_0^t S(t - s)F(X(\xi)(s)) \ ds, \quad t \in [0, T],$$

is P-a.s. continuous. Hence it only remains to show that the process

$$\int_0^t S(t - s)B(X(\xi)(s)) \ dW(s), \quad t \in [0, T],$$

has a continuous version. To this end we use the method which is presented in [27, Theorem 5.2.6, p. 59; Proposition A.1.1, p. 307]. In contrast to [27] we do not demand that the semigroup is analytic and therefore we only get continuity instead of Hölder continuity as in [27].

First, we have to introduce the general concept of the stochastic convolution.

Definition 6.3.1 (Stochastic Convolution) If $\Phi(t)$, $t \in [0, T]$, is a $L(U, H)$-valued predictable process such that the stochastic integrals

$$W_A^\Phi(t) := \int_0^t S(t - s)\Phi(s) \ dW(s), \ t \in [0, T],$$

are well defined, then the process $W_A^\Phi(t)$, $t \in [0, T]$, is called stochastic convolution.

The following result (see [27, Theorem 5.2.5, p. 58]) is a corollary of the stochastic Fubini Theorem (i.e. Theorem 6.1.4 above).

Theorem 6.3.2 (Factorization Formula) *Let* $\alpha \in]0, 1[$ *and* Φ *be an* $L(U, H)$-*valued predictable process. Assume that*

1. $S(t - s)\Phi(s)$ *is* $L_2(U, H)$-*valued for all* $s \in [0, t[$, $t \in [0, T]$.
2. $\displaystyle\int_0^s (s - u)^{-2\alpha} E\left(\|S(s - u)\Phi(u)\|_{L_2}^2\right) \ du < \infty$ *for all* $s \in [0, T]$.

3. $\int_0^t (t-s)^{\alpha-1} \left[\int_0^s (s-u)^{-2\alpha} E\left(\|S(s-u)\Phi(u)\|_{L_2}^2\right) \, du \right]^{\frac{1}{2}} \, ds < \infty$ for all $t \in [0, T]$.

Then we have the following representation of the stochastic convolution.

$$\int_0^t S(t-s)\Phi(s) \, dW(s) = \frac{\sin \alpha \pi}{\pi} \int_0^t (t-s)^{\alpha-1} S(t-s) Y_\alpha^\Phi(s) \, ds \qquad (6.4)$$

P-a.s. for all $t \in [0, T]$, where $Y_\alpha^\Phi(s)$, $s \in [0, T]$, is an $\mathcal{B}([0, T]) \otimes \mathcal{F}_T$-measurable version of

$$\int_0^s (s-u)^{-\alpha} S(s-u)\Phi(u) \, dW(u), \ s \in [0, T]. \qquad (6.5)$$

Proof First we check that $\int_0^s (s-u)^{-\alpha} S(s-u)\Phi(u) \, dW(u)$, $s \in [0, T]$, is well defined and has an $\mathcal{B}([0, T]) \otimes \mathcal{F}_T$-measurable version $Y_\alpha^\Phi(s)$, $s \in [0, T]$. But this is true since first we have that the mapping

$$\varphi : (u, \omega, s) \mapsto 1_{[0,s[}(u)(s-u)^{-\alpha} S(s-u)\Phi(u), \ u \in [0, T],$$

is $\mathcal{P}_T \otimes \mathcal{B}([0, T])/\mathcal{B}(L_2)$-measurable by assumption 1. (The proof can be done in a similar way as the proof of Lemma 6.2.5 and the proof of Lemma 6.2.6). Secondly, by assumption 2., we obtain that

$$E\left(\int_0^T \|1_{[0,s[}(u)(s-u)^{-\alpha} S(s-u)\Phi(u)\|_{L_2}^2 \, du\right)$$

$$= \int_0^s (s-u)^{-2\alpha} E\left(\|S(s-u)\Phi(u)\|_{L_2}^2\right) \, du < \infty,$$

by assumption 2.

Therefore the mapping $\varphi : \Omega_T \times [0, T] \to L_2(U, H)$ fulfills the conditions of Theorem 6.1.4 and thus the process in (6.5) is well defined and has a product measurable version Y_α^Φ (see the proof of [26, Theorem 4.18, p. 109]). In addition, for fixed $t \in [0, T]$ the mapping φ_t given by

$$\varphi_t : \Omega_T \times [0, T] \to L_2(U, H)$$

$$(u, \omega, s) \mapsto 1_{[0,t[}(s)(t-s)^{\alpha-1} 1_{[0,s[}(u)(s-u)^{-\alpha} S(t-u)\Phi(u, \omega)$$

also fulfills the conditions of Theorem 6.1.4 for the following reasons:

Now fix $t \in [0, T]$. We have that the mapping

$$\varphi_t : \Omega_T \times [0, T] \to L_2(U, H)$$

is $\mathcal{P}_T \otimes \mathcal{B}([0,T])/\mathcal{B}(L_2)$-measurable. Moreover, we get by assumption 3 that

$$\int_0^T \left(E\left(\int_0^T 1_{[0,t[}(s) 1_{[0,s[}(u)(t-s)^{2(\alpha-1)}(s-u)^{-2\alpha} \right. \right.$$
$$\left. \left. \|S(t-u)\Phi(u)\|_{L_2}^2 \, du \right) \right)^{\frac{1}{2}} \, ds$$

$$= \int_0^T 1_{[0,t[}(s)(t-s)^{\alpha-1} \left(\int_0^T 1_{[0,s[}(u)(s-u)^{-2\alpha} \right.$$
$$\left. E\left(\|S(t-u)\Phi(u)\|_{L_2}^2 \right) du \right)^{\frac{1}{2}} \, ds$$

(by Fubini's theorem)

$$\leq M_T \int_0^T 1_{[0,t[}(s)(t-s)^{\alpha-1} \left(\int_0^T 1_{[0,s[}(u)(s-u)^{-2\alpha} \right.$$
$$\left. E\left(\|S(s-u)\Phi(u)\|_{L_2}^2 \right) du \right)^{\frac{1}{2}} \, ds < \infty$$

(by the semigroup property).

Therefore, there exists a product measurable version of

$$\int_0^s 1_{[0,t[}(s)(t-s)^{\alpha-1}(s-u)^{-\alpha}S(t-u)\Phi(u) \, dW(u), \quad s \in [0,T].$$

Since by Lemma 2.4.1 for each $s \in [0,T]$

$$\int_0^s 1_{[0,t[}(s)(t-s)^{\alpha-1}(s-u)^{-\alpha}S(t-u)\Phi(u) \, dW(u)$$
$$= 1_{[0,t[}(s)(t-s)^{\alpha-1}S(t-s)Y_\alpha^\Phi(s) \quad P\text{-a.s.},$$

for this version again by Fubini's theorem we get that

$$E\left(\| \int_0^t (t-s)^{\alpha-1}S(t-s)Y_\alpha^\Phi(s) \, ds \right.$$
$$\left. - \int_0^T \int_0^s 1_{[0,t[}(s)(t-s)^{\alpha-1}(s-u)^{-\alpha}S(t-u)\Phi(u) \, dW(u) \, ds\| \right)$$

$$\leq \int_0^T E\left(\|1_{[0,t[}(s)(t-s)^{\alpha-1}S(t-s)Y_\alpha^\Phi(s) \right.$$
$$\left. - \int_0^s 1_{[0,t[}(s)(t-s)^{\alpha-1}(s-u)^{-\alpha}S(t-u)\Phi(u) \, dW(u)\| \right) \, ds$$

$$= 0.$$

Furthermore, we can use Theorem 6.1.4 to exchange the integration and thus we finally obtain that

$$\int_0^t (t-s)^{\alpha-1} S(t-s) Y_\alpha^\Phi(s) \ ds$$

$$= \int_0^t \int_0^s (t-s)^{\alpha-1} (s-u)^{-\alpha} S(t-u) \Phi(u) \ dW(u) \ ds$$

$$= \int_0^t \int_u^t (t-s)^{\alpha-1} (s-u)^{-\alpha} S(t-u) \Phi(u) \ ds \ dW(u)$$

$$= \int_0^t \left(\int_u^t (t-s)^{\alpha-1} (s-u)^{-\alpha} \ ds \right) S(t-u) \Phi(u) \ dW(u)$$

$$= \frac{\pi}{\sin \alpha \pi} \int_0^t S(t-u) \Phi(u) \ dW(u) \quad P\text{-a.s.}$$

since $\displaystyle\int_u^t (t-s)^{\alpha-1} (s-u)^{-\alpha} \ ds = \frac{\pi}{\sin \alpha \pi}$ (keyword: Euler's beta function). $\square$

Using this representation of the stochastic convolution we are now able to prove the desired pathwise continuity.

Let $p \geq 2$ and $\alpha \in]\frac{1}{p}, 1[$. For $\varphi \in L^p([0, T]; H) := L^p([0, T], \mathcal{B}([0, T]), dt; H)$ we define

$$R_\alpha \varphi(t) := \int_0^t (t-s)^{\alpha-1} S(t-s) \varphi(s) \ ds, \ t \in [0, T].$$

Then $R_\alpha \varphi$ is well defined since

$$\int_0^t \|(t-s)^{\alpha-1} S(t-s) \varphi(s)\| \ ds \leqslant \left(\int_0^t s^{(\alpha-1)\frac{p}{p-1}} \ ds \right)^{\frac{p-1}{p}} M_T \|\varphi\|_{L^p} < \infty,$$

since $\alpha > \frac{1}{p}$ and therefore $(\alpha - 1)\frac{p}{p-1} > -1$.

Proposition 6.3.3 *Let $\alpha \in]0, 1[$ and $p > \frac{1}{\alpha}$. Then*

$$R_\alpha : L^p([0, T]; H) \to C([0, T]; H).$$

Remark 6.3.4 If one assumes that the semigroup $S(t), t \in [0, T]$, is analytic one even gets that $R_\alpha \varphi$ is Hölder continuous for all $\varphi \in L^p([0, T]; H)$ (see [27, Proposition A.1.1, p. 307]).

Proof of 6.3.3 Let $\varphi \in L^p([0,T];H)$, $t \in [0,T]$, and let t_n, $n \in \mathbb{N}$, be a sequence in $[0,T]$ such that $t_n \underset{n\to\infty}{\longrightarrow} t$. Then

$$\|R_\alpha \varphi(t_n) - R_\alpha \varphi(t)\|$$

$$= \left\| \int_0^{t_n} (t_n - s)^{\alpha-1} S(t_n - s)\varphi(s) \, ds - \int_0^t (t - s)^{\alpha-1} S(t - s)\varphi(s) \, ds \right\|$$

$$\leq \int_0^T \left\| 1_{[0,t_n[}(s)(t_n - s)^{\alpha-1} S(t_n - s)\varphi(s) - 1_{[0,t[}(s)(t - s)^{\alpha-1} S(t - s)\varphi(s) \right\| \, ds.$$

Concerning the inner term we obtain that

$$\left\| 1_{[0,t_n[}(s)(t_n - s)^{\alpha-1} S(t_n - s)\varphi(s) - 1_{[0,t[}(s)(t - s)^{\alpha-1} S(t - s)\varphi(s) \right\| \underset{n\to\infty}{\longrightarrow} 0$$

for ds-a.e. $s \in [0,T]$. Moreover, the family

$$\left(\left\| 1_{[0,t_n[}(t_n - \cdot)^{\alpha-1} S(t_n - \cdot)\varphi(\cdot) - 1_{[0,t[}(t - \cdot)^{\alpha-1} S(t - \cdot)\varphi(\cdot) \right\| \right)_{n\in\mathbb{N}}$$

is uniformly integrable :
For $t \in [0,T]$ we set

$$F_t(s) := 1_{[0,t[}(s)(t - s)^{\alpha-1} \|S(t - s)\varphi(s)\|, \quad s \in [0,T].$$

Since $(\alpha - 1)\frac{p}{p-1} > -1$, there exists an $\varepsilon > 0$ such that

$$(\alpha - 1)(1 + \varepsilon)\frac{p}{p - 1 - \varepsilon} > -1 \text{ and } p > 1 + \varepsilon.$$

Then

$$\int_0^T F_t^{1+\varepsilon}(s) \, ds \leq \left(\int_0^t (t - s)^{(\alpha-1)(1+\varepsilon)\frac{p}{p-1-\varepsilon}} \, ds \right)^{\frac{p-1-\varepsilon}{p}} M_T^{1+\varepsilon} \|\varphi\|_{L^p}^{1+\varepsilon}$$

$$\leq \left(\int_0^T s^{(\alpha-1)(1+\varepsilon)\frac{p}{p-1-\varepsilon}} \, ds \right)^{\frac{p-1-\varepsilon}{p}} M_T^{1+\varepsilon} \|\varphi\|_{L^p}^{1+\varepsilon}$$

$$< \infty.$$

Therefore, $\sup_{t\in[0,T]} \int_0^T F_t^{1+\varepsilon}(s) \, ds < \infty$ and hence F_t, $t \in [0,T]$, is uniformly integrable. Since

$$\left\| 1_{[0,t_n[}(s)(t_n - s)^{\alpha-1} S(t_n - s)\varphi(s) - 1_{[0,t[}(s)(t - s)^{\alpha-1} S(t - s)\varphi(s) \right\|$$

$$\leq F_{t_n}(s) + F_t(s)$$

for all $s \in [0, T]$ the assertion follows.   □

Thus we have found a tool to check whether the process

$$\int_0^t S(t - s)B(X(\xi)(s))\ \mathrm{d}W(s), \quad t \in [0, T],$$

has a P-a.s. continuous version.

Proposition 6.3.5 *Assume that the mappings A, F and B satisfy Hypothesis M.0, and let $p \geq 2$. If there exists an $\alpha \in]\frac{1}{p}, \infty[$ such that*

$$\int_0^T s^{-2\alpha} K^2(s)\ \mathrm{d}s < \infty,$$

then the mild solution $X(\xi)$ of problem (6.1) has a continuous version for all initial conditions $\xi \in L_0^p$.

Proof Without loss of generality we may assume that $\alpha \in]0, 1[$. $S(\cdot)\xi$ is P-a.s. continuous because of the strong continuity of the semigroup.

In Step 2, 2. (i) of the proof of Theorem 6.2.3 we have already shown that the process of the Bochner integrals

$$\int_0^t S(t - s)F(X(\xi)(s))\ \mathrm{d}s, \quad t \in [0, T],$$

has P-a.s. continuous trajectories.

Thus, in fact, it only remains to prove that the process

$$\int_0^t S(t - s)B(X(\xi)(s))\ \mathrm{d}W(s), \quad t \in [0, T],$$

has a continuous version:

Since

$$\int_0^t (t - s)^{\alpha - 1} \Big[\int_0^s (s - u)^{-2\alpha} E\big(\|S(s - u)B(X(\xi)(u))\|_{L_2}^2 \big)\ \mathrm{d}u \Big]^{\frac{1}{2}}\ \mathrm{d}s$$

$$\leq \int_0^t (t - s)^{\alpha - 1} \Big[\int_0^s (s - u)^{-2\alpha} K^2(s - u) E\big((1 + \|X(\xi)(u)\|)^2 \big)\ \mathrm{d}u \Big]^{\frac{1}{2}}\ \mathrm{d}s$$

$$\leq (1 + \|X(\xi)\|_{\mathcal{H}^2}) \int_0^t (t - s)^{\alpha - 1} \Big[\int_0^s (s - u)^{-2\alpha} K^2(s - u)\ \mathrm{d}u \Big]^{\frac{1}{2}}\ \mathrm{d}s$$

$$\leq (1 + \|X(\xi)\|_{\mathcal{H}^2}) \Big(\int_0^T u^{-2\alpha} K^2(u)\ \mathrm{d}u \Big)^{\frac{1}{2}} \int_0^T s^{\alpha - 1}\ \mathrm{d}s < \infty,$$

we have by Theorem 6.3.2 (factorization formula) that P-a.s.

$$\int_0^t S(t-s)B(X(\xi)(s))\ dW(s)$$

$$= \frac{\sin \alpha\pi}{\pi} \int_0^t (t-s)^{\alpha-1} S(t-s) \int_0^s (s-u)^{-\alpha} S(s-u) B(X(\xi)(u))\ dW(u)\ ds$$

$$= \frac{\sin \alpha\pi}{\pi} R_\alpha \Big(\underbrace{\int_0^{\cdot} (\cdot-u)^{-\alpha} S(\cdot-u) B(X(\xi)(u))\ dW(u)}_{=:\ Y_\alpha} \Big)(t).$$

Since the mapping $\varphi : \Omega_T \times [0,T] \to L_2(U,H)$ given by

$$\varphi(u,\omega,s) := 1_{[0,s[}(u)(s-u)^{-\alpha} S(s-u) B(X(\xi)(u,\omega))$$

fulfills the conditions of Theorem 6.1.4, the process Y_α can be understood as a $\mathcal{B}([0,T]) \otimes \mathcal{F}_T$-measurable version of $\int_0^t (t-u)^{-\alpha} S(t-u) B(X(\xi)(u))\ dW(u)$, $t \in [0,T]$ (see the proof of [26, Theorem 4.18, p. 109]).

To get the P-a.s. continuity of the stochastic integral we have to show that P-a.s.

$$Y_\alpha \in L^p([0,T];H).$$

By Theorem 6.1.2 we can estimate $E\big(\|Y_\alpha(t)\|^p\big)$ independently of $t \in [0,T]$ in the following way

$$E\big(\|Y_\alpha(t)\|^p\big)$$

$$\leqslant \Big(\frac{p}{2}(p-1)\Big)^{\frac{p}{2}} \Big(\int_0^t (t-s)^{-2\alpha} \big(E(\|S(t-s)B(X(\xi)(s))\|_{L_2}^p)\big)^{\frac{2}{p}}\ ds \Big)^{\frac{p}{2}}$$

$$\leqslant \Big(\frac{p}{2}(p-1)\Big)^{\frac{p}{2}} \Big(\int_0^t (t-s)^{-2\alpha} K^2(t-s) \big(E((1+\|X(\xi)(s)\|^p))\big)^{\frac{2}{p}}\ ds \Big)^{\frac{p}{2}}$$

$$\leqslant \Big(\frac{p}{2}(p-1)\Big)^{\frac{p}{2}} \Big(\int_0^t (t-s)^{-2\alpha} K^2(t-s)(1+\|X(\xi)\|_{\mathcal{H}^p})^2\ ds \Big)^{\frac{p}{2}}$$

$$\leqslant \Big(\frac{p}{2}(p-1)\Big)^{\frac{p}{2}} (1+\|X(\xi)\|_{\mathcal{H}^p})^p \Big(\int_0^T s^{-2\alpha} K^2(s)\ ds \Big)^{\frac{p}{2}} < \infty.$$

Finally, we obtain by the real Fubini theorem that

$$E\Big(\int_0^T \|Y_\alpha(t)\|^p\ dt \Big) = \int_0^T E\big(\|Y_\alpha(t)\|^p\big)\ dt < \infty.$$

$\square$

Appendix A
The Bochner Integral

Let $(X, \| \, \|)$ be a Banach space, $\mathcal{B}(X)$ the Borel σ-field of X and $(\Omega, \mathcal{F}, \mu)$ a measure space with finite measure μ.

A.1 Definition of the Bochner Integral

Step 1: As a first step we want to define the integral for simple functions which are defined as follows. Set

$$\mathcal{E} := \left\{ f : \Omega \to X \,\middle|\, f = \sum_{k=1}^{n} x_k 1_{A_k}, \, x_k \in X, \, A_k \in \mathcal{F}, \, 1 \leqslant k \leqslant n, \, n \in \mathbb{N} \right\}$$

and define a semi-norm $\| \, \|_{\mathcal{E}}$ on the vector space $\mathcal{E}$ by

$$\|f\|_{\mathcal{E}} := \int \|f\| \, d\mu, \quad f \in \mathcal{E}.$$

To get that $(\mathcal{E}, \| \, \|_{\mathcal{E}})$ is a normed vector space we consider equivalence classes with respect to $\| \, \|_{\mathcal{E}}$. For simplicity we will not change the notation.

For $f \in \mathcal{E}, f = \sum_{k=1}^{n} x_k 1_{A_k}$, A_k's *pairwise disjoint* (such a representation is called *normal* and always exists, because $f = \sum_{k=1}^{n} x_k 1_{A_k}$, where $f(\Omega) = \{x_1, \ldots, x_k\}$, $x_i \neq x_j$, and $A_k := \{f = x_k\}$) and we now define the Bochner integral to be

$$\int f \, d\mu := \sum_{k=1}^{n} x_k \mu(A_k).$$

© Springer International Publishing Switzerland 2015
W. Liu, M. Röckner, *Stochastic Partial Differential Equations: An Introduction*,
Universitext, DOI 10.1007/978-3-319-22354-4

(Exercise: This definition is independent of representations, and *hence* linear.) In this way we get a mapping

$$\text{int} : \left(\mathcal{E}, \|\ \|_\mathcal{E}\right) \to \left(X, \|\ \|\right)$$
$$f \mapsto \int f \, d\mu$$

which is linear and uniformly continuous since $\left\| \int f \, d\mu \right\| \leq \int \|f\| \, d\mu$ for all $f \in \mathcal{E}$.

Therefore we can extend the mapping int to the abstract completion of $\mathcal{E}$ with respect to $\|\ \|_\mathcal{E}$ which we denote by $\overline{\mathcal{E}}$.

Step 2:   We give an explicit representation of $\overline{\mathcal{E}}$.

Definition A.1.1 A function $f : \Omega \to X$ is called strongly measurable if it is $\mathcal{F}/\mathcal{B}(X)$-measurable and $f(\Omega) \subset X$ is separable.

Definition A.1.2 Let $1 \leq p < \infty$. Then we define

$$\mathcal{L}^p(\Omega, \mathcal{F}, \mu; X) := \mathcal{L}^p(\mu; X)$$
$$:= \left\{ f : \Omega \to X \ \middle|\ f \text{ is strongly measurable with}\right.$$
$$\left. \text{respect to } \mathcal{F}, \text{ and } \int \|f\|^p \, d\mu < \infty \right\}$$

and the semi-norm

$$\|f\|_{L^p} := \left(\int \|f\|^p \, d\mu \right)^{\frac{1}{p}}, \quad f \in \mathcal{L}^p(\Omega, \mathcal{F}, \mu; X).$$

The space of all equivalence classes in $\mathcal{L}^p(\Omega, \mathcal{F}, \mu; X)$ with respect to $\|\ \|_{L^p}$ is denoted by $L^p(\Omega, \mathcal{F}, \mu; X) := L^p(\mu; X)$.

Claim $L^1(\Omega, \mathcal{F}, \mu; X) = \overline{\mathcal{E}}$.

Step 2.a: $\left(L^1(\Omega, \mathcal{F}, \mu; X), \|\ \|_{L^1}\right)$ is complete.

The proof is just a modification of the proof of the Fischer–Riesz theorem by the following proposition.

Proposition A.1.3 *Let $(\Omega, \mathcal{F})$ be a measurable space and let X be a Banach space. Then:*

(i) *the set of $\mathcal{F}/\mathcal{B}(X)$-measurable functions from Ω to X is closed under the formation of pointwise limits, and*

(ii) *the set of strongly measurable functions from Ω to X is closed under the formation of pointwise limits.*

Proof Simple exercise or see [18, Proposition E.1, p. 350]. □

Step 2.b $\mathcal{E}$ is a dense subset of $L^1(\Omega, \mathcal{F}, \mu; X)$ with respect to $\| \ \|_{L^1}$.
This will follow from the following lemma.

Lemma A.1.4 *Let E be a metric space with metric d and let $f : \Omega \to E$ be strongly measurable. Then there exists a sequence f_n, $n \in \mathbb{N}$, of simple E-valued functions (i.e. f_n is $\mathcal{F}/\mathcal{B}(E)$-measurable and takes only a finite number of values) such that for arbitrary $\omega \in \Omega$ the sequence $d(f_n(\omega), f(\omega))$, $n \in \mathbb{N}$, is monotonely decreasing to zero.*

Proof ([26, Lemma 1.1, p. 16]) Let $\{e_k \mid k \in \mathbb{N}\}$ be a countable dense subset of $f(\Omega)$. For $m \in \mathbb{N}$ and $\omega \in \Omega$ define

$$d_m(\omega) := \min\{d(f(\omega), e_k) \mid k \leq m\} \quad (= \text{dist}(f(\omega), \{e_k, k \leq m\})),$$
$$k_m(\omega) := \min\{k \leq m \mid d_m(\omega) = d(f(\omega), e_k)\},$$
$$f_m(\omega) := e_{k_m(\omega)}.$$

Obviously f_m, $m \in \mathbb{N}$, are simple functions since they are $\mathcal{F}/\mathcal{B}(E)$-measurable (exercise) and

$$f_m(\Omega) \subset \{e_1, e_2, \ldots, e_m\}.$$

Moreover, by the density of $\{e_k \mid k \in \mathbb{N}\}$, the sequence $d_m(\omega)$, $m \in \mathbb{N}$, is monotonically decreasing to zero for arbitrary $\omega \in \Omega$. Since $d(f_m(\omega), f(\omega)) = d_m(\omega)$ the assertion follows. □

Let now $f \in L^1(\Omega, \mathcal{F}, \mu; X)$. By Lemma A.1.4 above we get the existence of a sequence of simple functions f_n, $n \in \mathbb{N}$, such that

$$\|f_n(\omega) - f(\omega)\| \downarrow 0 \quad \text{for all } \omega \in \Omega \text{ as } n \to \infty.$$

Hence $f_n \xrightarrow{n \to \infty} f$ in $\| \ \|_{L^1}$ by Lebesgue's dominated convergence theorem.

A.2 Properties of the Bochner Integral

Proposition A.2.1 (Bochner Inequality) *Let $f \in L^1(\Omega, \mathcal{F}, \mu; X)$. Then*

$$\left\| \int f \, d\mu \right\| \leq \int \|f\| \, d\mu.$$

Proof We know the assertion is true for $f \in \mathcal{E}$, i.e. int $: \mathcal{E} \to X$ is linear, continuous with $\|\mathrm{int} f\| \leq \|f\|_{\mathcal{E}}$ for all $f \in \mathcal{E}$, so the same is true for its unique continuous extension $\overline{\mathrm{int}} : \overline{\mathcal{E}} = L^1(\mu; X) \to X$, i.e. for all $f \in L^1(X, \mu)$

$$\left\| \int f \, \mathrm{d}\mu \right\| = \left\| \overline{\mathrm{int}} f \right\| \leq \|f\|_{\overline{\mathcal{E}}} = \int \|f\| \, \mathrm{d}\mu.$$

$\square$

Proposition A.2.2 *Let* $f \in L^1(\Omega, \mathcal{F}, \mu; X)$. *Then*

$$\int L \circ f \, \mathrm{d}\mu = L\left(\int f \, \mathrm{d}\mu \right)$$

holds for all $L \in L(X, Y)$, *where* Y *is another Banach space.*

Proof Simple exercise or see [18, Proposition E.11, p. 356]. $\square$

Proposition A.2.3 (Fundamental Theorem of Calculus) *Let* $-\infty < a < b < \infty$ *and* $f \in C^1([a, b]; X)$. *Then*

$$f(t) - f(s) = \int_s^t f'(u) \, \mathrm{d}u := \begin{cases} \int 1_{[s,t]}(u) f'(u) \, \mathrm{d}u & \text{if } s \leq t \\ -\int 1_{[t,s]}(u) f'(u) \, \mathrm{d}u & \text{otherwise} \end{cases}$$

for all $s, t \in [a, b]$ *where* $\mathrm{d}u$ *denotes the Lebesgue measure on* $\mathcal{B}(\mathbb{R})$.

Proof

Claim 1: If we set $F(t) := \int_s^t f'(u) \, \mathrm{d}u$, $t \in [a, b]$, we get that $F'(t) = f'(t)$ for all $t \in [a, b]$.

For that we have to prove that

$$\left\| \frac{1}{h} \big(F(t + h) - F(t) \big) - f'(t) \right\|_X \xrightarrow{h \to 0} 0.$$

To this end we fix $t \in [a, b]$ and take an arbitrary $\varepsilon > 0$. Since f' is continuous on $[a, b]$ there exists a $\delta > 0$ such that $\left\| f'(u) - f'(t) \right\|_X < \varepsilon$ for all $u \in [a, b]$ with $|u - t| < \delta$. Then we obtain that

$$\left\| \frac{1}{h} \big(F(t + h) - F(t) \big) - f'(t) \right\|_X = \left\| \frac{1}{h} \int_t^{t+h} \big(f'(u) - f'(t) \big) \, \mathrm{d}u \right\|_X$$

$$\leq \frac{1}{|h|} \int_{t \wedge (t+h)}^{t \vee (t+h)} \left\| f'(u) - f'(t) \right\|_X \, \mathrm{d}u < \varepsilon$$

if $t + h \in [a, b]$ and $|h| < \delta$.

Claim 2: If $\tilde{F} \in C^1([a, b]; X)$ is a further function with $\tilde{F}' = F' = f'$ then there exists a constant $c \in X$ such that $F - \tilde{F} = c$.

Let us first assume that $F(a) = 0 = \tilde{F}(a)$. Then for all $L \in X^* = L(X, \mathbb{R})$ we define $g_L := L(F - \tilde{F})$. Then $g_L' = 0$ and therefore g_L is constant c_L. But $g_L(a) = L(F(a) - \tilde{F}(a)) = 0$, so $c_L = 0$. Since X^* separates the points of X, by the Hahn–Banach theorem (see [3, Satz 4.2, p. 114]) this implies that $F - \tilde{F} = 0$. In the general case we apply the above to $F - F(a)$ and $\tilde{F} - \tilde{F}(a)$ to obtain the assertion. $\qquad \square$

Appendix B
Nuclear and Hilbert–Schmidt Operators

Let $(U, \langle\,,\,\rangle_U)$ and $(H, \langle\,,\,\rangle)$ be two separable Hilbert spaces. The space of all bounded linear operators from U to H is denoted by $L(U, H)$; for simplicity we write $L(U)$ instead of $L(U, U)$. If we speak of the adjoint operator of $L \in L(U, H)$ we write $L^* \in L(H, U)$. An element $L \in L(U)$ is called symmetric if $\langle Lu, v \rangle_U = \langle u, Lv \rangle_U$ for all $u, v \in U$. In addition, $L \in L(U)$ is called nonnegative if $\langle Lu, u \rangle \geqslant 0$ for all $u \in U$.

Definition B.0.1 (Nuclear Operator)   An element $T \in L(U, H)$ is said to be a nuclear operator if there exists a sequence $(a_j)_{j \in \mathbb{N}}$ in H and a sequence $(b_j)_{j \in \mathbb{N}}$ in U such that

$$Tx = \sum_{j=1}^{\infty} a_j \langle b_j, x \rangle_U \quad \text{for all } x \in U$$

and

$$\sum_{j \in \mathbb{N}} \|a_j\| \cdot \|b_j\|_U < \infty.$$

The space of all nuclear operators from U to H is denoted by $L_1(U, H)$.

If $U = H$ and $T \in L_1(U, H)$ is nonnegative and symmetric, then T is called *trace class*.

Proposition B.0.2   *The space $L_1(U, H)$ endowed with the norm*

$$\|T\|_{L_1(U,H)} := \inf\Big\{ \sum_{j \in \mathbb{N}} \|a_j\| \cdot \|b_j\|_U \; \Big| \; Tx = \sum_{j=1}^{\infty} a_j \langle b_j, x \rangle_U, \; x \in U \Big\}$$

is a Banach space.

© Springer International Publishing Switzerland 2015
W. Liu, M. Röckner, *Stochastic Partial Differential Equations: An Introduction*,
Universitext, DOI 10.1007/978-3-319-22354-4

Proof [60, Corollary 16.25, p. 154].   □

Definition B.0.3 Let $T \in L(U)$ and let e_k, $k \in \mathbb{N}$, be an orthonormal basis of U. Then we define

$$\operatorname{tr} T := \sum_{k \in \mathbb{N}} \langle Te_k, e_k \rangle_U$$

if the series is convergent.

One has to notice that this definition could depend on the choice of the orthonormal basis. But there is the following result concerning nuclear operators.

Remark B.0.4 If $T \in L_1(U)$ then $\operatorname{tr} T$ is well-defined independently of the choice of the orthonormal basis e_k, $k \in \mathbb{N}$. Moreover we have that

$$|\operatorname{tr} T| \leq \|T\|_{L_1(U)}.$$

Proof Let $(a_j)_{j \in \mathbb{N}}$ and $(b_j)_{j \in \mathbb{N}}$ be sequences in U such that

$$Tx = \sum_{j \in \mathbb{N}} a_j \langle b_j, x \rangle_U$$

for all $x \in U$ and $\sum_{j \in \mathbb{N}} \|a_j\|_U \cdot \|b_j\|_U < \infty$.

Then we get for any orthonormal basis e_k, $k \in \mathbb{N}$, of U that

$$\langle Te_k, e_k \rangle_U = \sum_{j \in \mathbb{N}} \langle e_k, a_j \rangle_U \cdot \langle e_k, b_j \rangle_U$$

and therefore

$$\sum_{k \in \mathbb{N}} |\langle Te_k, e_k \rangle_U| \leq \sum_{j \in \mathbb{N}} \sum_{k \in \mathbb{N}} |\langle e_k, a_j \rangle_U \cdot \langle e_k, b_j \rangle_U|$$

$$\leq \sum_{j \in \mathbb{N}} \left(\sum_{k \in \mathbb{N}} |\langle e_k, a_j \rangle_U|^2 \right)^{\frac{1}{2}} \cdot \left(\sum_{k \in \mathbb{N}} |\langle e_k, b_j \rangle_U|^2 \right)^{\frac{1}{2}}$$

$$= \sum_{j \in \mathbb{N}} \|a_j\|_U \cdot \|b_j\|_U < \infty.$$

This implies that we can exchange the summation to get that

$$\sum_{k \in \mathbb{N}} \langle Te_k, e_k \rangle_U = \sum_{j \in \mathbb{N}} \sum_{k \in \mathbb{N}} \langle e_k, a_j \rangle_U \cdot \langle e_k, b_j \rangle_U = \sum_{j \in \mathbb{N}} \langle a_j, b_j \rangle_U,$$

and the assertion follows.   □

Definition B.0.5 (Hilbert–Schmidt Operator)   A bounded linear operator T : $U \to H$ is called Hilbert–Schmidt if

$$\sum_{k\in\mathbb{N}} \|Te_k\|^2 < \infty$$

where e_k, $k \in \mathbb{N}$, is an orthonormal basis of U.

The space of all Hilbert–Schmidt operators from U to H is denoted by $L_2(U, H)$.

Remark B.0.6

(i) The definition of Hilbert–Schmidt operator and the number

$$\|T\|^2_{L_2(U,H)} := \sum_{k\in\mathbb{N}} \|Te_k\|^2$$

does not depend on the choice of the orthonormal basis e_k, $k \in \mathbb{N}$, and we have that $\|T\|_{L_2(U,H)} = \|T^*\|_{L_2(H,U)}$. For simplicity we also write $\|T\|_{L_2}$ instead of $\|T\|_{L_2(U,H)}$.

(ii) $\|T\|_{L(U,H)} \leqslant \|T\|_{L_2(U,H)}$.

(iii) Let G be another Hilbert space and $S_1 \in L(H, G)$, $S_2 \in L(G, U)$, $T \in L_2(U, H)$. Then $S_1 T \in L_2(U, G)$ and $TS_2 \in L_2(G, H)$ and

$$\|S_1 T\|_{L_2(U,G)} \leqslant \|S_1\|_{L(H,G)} \|T\|_{L_2(U,H)},$$

$$\|TS_2\|_{L_2(G,H)} \leqslant \|T\|_{L_2(U,H)} \|S_2\|_{L(G,U)}.$$

Proof

(i) If e_k, $k \in \mathbb{N}$, is an orthonormal basis of U and f_k, $k \in \mathbb{N}$, is an orthonormal basis of H we obtain by the Parseval identity that

$$\sum_{k\in\mathbb{N}} \|Te_k\|^2 = \sum_{k\in\mathbb{N}}\sum_{j\in\mathbb{N}} |\langle Te_k, f_j\rangle|^2 = \sum_{j\in\mathbb{N}} \|T^*f_j\|^2_U$$

and therefore the assertion follows.

(ii) Let $x \in U$ and f_k, $k \in \mathbb{N}$, be an orthonormal basis of H. Then we get that

$$\|Tx\|^2 = \sum_{k\in\mathbb{N}} \langle Tx, f_k\rangle^2 \leqslant \|x\|^2_U \sum_{k\in\mathbb{N}} \|T^*f_k\|^2_U = \|T\|^2_{L_2(U,H)} \cdot \|x\|^2_U.$$

(iii) Let e_k, $k \in \mathbb{N}$ be an orthonormal basis of U. Then

$$\sum_{k\in\mathbb{N}} \|S_1 Te_k\|^2_G \leqslant \|S_1\|^2_{L(H,G)} \|T\|^2_{L_2(U,H)}.$$

Furthermore, since $(TS_2)^* = S_2^* T^*$, it follows by the above and (i) that $TS_2 \in L_2(G, H)$ and

$$
\begin{aligned}
\|TS_2\|_{L_2(G,H)} &= \|(TS_2)^*\|_{L_2(H,G)} \\
&= \|S_2^* T^*\|_{L_2(H,G)} \\
&\leqslant \|S_2\|_{L(G,U)} \cdot \|T\|_{L_2(U,H)},
\end{aligned}
$$

since a bounded operator has the same norm as its adjoint.   □

Proposition B.0.7 *Let $S, T \in L_2(U, H)$ and let e_k, $k \in \mathbb{N}$, be an orthonormal basis of U. If we define*

$$
\langle T, S \rangle_{L_2} := \sum_{k \in \mathbb{N}} \langle Se_k, Te_k \rangle
$$

we obtain that $\big(L_2(U, H), \langle \, , \, \rangle_{L_2}\big)$ is a separable Hilbert space.

If f_k, $k \in \mathbb{N}$, is an orthonormal basis of H we get that $f_j \otimes e_k := f_j \langle e_k, \cdot \rangle_U$, $j, k \in \mathbb{N}$, is an orthonormal basis of $L_2(U, H)$.

Proof We have to prove the completeness and the separability.

1. $L_2(U, H)$ is complete:

   Let T_n, $n \in \mathbb{N}$, be a Cauchy sequence in $L_2(U, H)$. Then it is clear that it is also a Cauchy sequence in $L(U, H)$. Because of the completeness of $L(U, H)$ there exists an element $T \in L(U, H)$ such that $\|T_n - T\|_{L(U,H)} \longrightarrow 0$ as $n \to \infty$. But by Fatou's lemma we also have for any orthonormal basis e_k, $k \in \mathbb{N}$, of U that

$$
\begin{aligned}
\|T_n - T\|_{L_2}^2 &= \sum_{k \in \mathbb{N}} \langle (T_n - T)e_k, (T_n - T)e_k \rangle \\
&= \sum_{k \in \mathbb{N}} \liminf_{m \to \infty} \big\| (T_n - T_m)e_k \big\|^2 \\
&\leqslant \liminf_{m \to \infty} \sum_{k \in \mathbb{N}} \big\| (T_n - T_m)e_k \big\|^2 = \liminf_{m \to \infty} \|T_n - T_m\|_{L_2}^2 < \varepsilon
\end{aligned}
$$

   for all $n \in \mathbb{N}$ big enough. Therefore the assertion follows.

2. $L_2(U, H)$ is separable:

   If we define $f_j \otimes e_k := f_j \langle e_k, \cdot \rangle_U, j, k \in \mathbb{N}$, then it is clear that $f_j \otimes e_k \in L_2(U, H)$ for all $j, k \in \mathbb{N}$ and for arbitrary $T \in L_2(U, H)$ we get that

$$
\langle f_j \otimes e_k, T \rangle_{L_2} = \sum_{n \in \mathbb{N}} \langle e_k, e_n \rangle_U \cdot \langle f_j, Te_n \rangle = \langle f_j, Te_k \rangle.
$$

Therefore it is obvious that $f_j \otimes e_k, j, k \in \mathbb{N}$, is an orthonormal system. In addition, $T = 0$ if $\langle f_j \otimes e_k, T \rangle_{L_2} = 0$ for all $j, k \in \mathbb{N}$, and therefore $\mathrm{span}(f_j \otimes e_k \mid j, k \in \mathbb{N})$ is a dense subspace of $L_2(U, H)$.   □

Proposition B.0.8 *Let $\big(G, \langle \ , \ \rangle_G\big)$ be a further separable Hilbert space. If $T \in L_2(U, H)$ and $S \in L_2(H, G)$ then $ST \in L_1(U, G)$ and*

$$\|ST\|_{L_1(U,G)} \leqslant \|S\|_{L_2} \cdot \|T\|_{L_2}.$$

Proof Let $f_k, k \in \mathbb{N}$, be an orthonormal basis of H. Then we have that

$$STx = \sum_{k \in \mathbb{N}} \langle Tx, f_k \rangle Sf_k, \quad x \in U$$

and therefore

$$\|ST\|_{L_1(U,G)} \leqslant \sum_{k \in \mathbb{N}} \|T^* f_k\|_U \cdot \|Sf_k\|_G$$

$$\leqslant \Big(\sum_{k \in \mathbb{N}} \|T^* f_k\|_U^2\Big)^{\frac{1}{2}} \cdot \Big(\sum_{k \in \mathbb{N}} \|Sf_k\|_G^2\Big)^{\frac{1}{2}} = \|S\|_{L_2} \cdot \|T\|_{L_2}.$$

□

Remark B.0.9 Let $e_k, k \in \mathbb{N}$, be an orthonormal basis of U. If $T \in L(U)$ is symmetric and nonnegative with $\sum_{k \in \mathbb{N}} \langle Te_k, e_k \rangle_U < \infty$ then $T \in L_1(U)$.

Proof The result is obvious by the previous proposition and the fact that there exists a nonnegative and symmetric $T^{\frac{1}{2}} \in L(U)$ such that $T = T^{\frac{1}{2}} T^{\frac{1}{2}}$ (see Proposition 2.3.4). Then $T^{\frac{1}{2}} \in L_2(U)$.   □

Proposition B.0.10 *Let $L \in L(H)$ and $B \in L_2(U, H)$. Then $LBB^* \in L_1(H)$, $B^* LB \in L_1(U)$ and we have that*

$$\mathrm{tr}\, LBB^* = \mathrm{tr}\, B^* LB.$$

Proof We know by Remark B.0.6(iii) and Proposition B.0.8 that $LBB^* \in L_1(H)$ and $B^* LB \in L_1(U)$. Let $e_k, k \in \mathbb{N}$, be an orthonormal basis of U and let $f_k, k \in \mathbb{N}$, be an orthonormal basis of H. Then the Parseval identity implies that

$$\sum_{k \in \mathbb{N}} \sum_{n \in \mathbb{N}} |\langle f_k, Be_n \rangle \cdot \langle f_k, LBe_n \rangle|$$

$$\leqslant \sum_{n \in \mathbb{N}} \Big(\sum_{k \in \mathbb{N}} |\langle f_k, Be_n \rangle|^2\Big)^{\frac{1}{2}} \cdot \Big(\sum_{k \in \mathbb{N}} |\langle f_k, LBe_n \rangle|^2\Big)^{\frac{1}{2}}$$

$$= \sum_{n \in \mathbb{N}} \|Be_n\| \cdot \|LBe_n\| \leqslant \|L\|_{L(H)} \cdot \|B\|_{L_2}^2.$$

Therefore, below it is allowed to interchange the sums to obtain that

$$\operatorname{tr} LBB^* = \sum_{k \in \mathbb{N}} \langle LBB^* f_k, f_k \rangle = \sum_{k \in \mathbb{N}} \langle B^* f_k, B^* L^* f_k \rangle_U$$

$$= \sum_{k \in \mathbb{N}} \sum_{n \in \mathbb{N}} \langle B^* f_k, e_n \rangle_U \cdot \langle B^* L^* f_k, e_n \rangle_U = \sum_{n \in \mathbb{N}} \sum_{k \in \mathbb{N}} \langle f_k, B e_n \rangle \cdot \langle f_k, LB e_n \rangle$$

$$= \sum_{n \in \mathbb{N}} \langle B e_n, LB e_n \rangle = \sum_{n \in \mathbb{N}} \langle e_n, B^* LB e_n \rangle_U = \operatorname{tr} B^* LB.$$

$\square$

Appendix C
The Pseudo Inverse of Linear Operators

Let $(U, \langle\ ,\ \rangle_U)$ and $(H, \langle\ ,\ \rangle)$ be two Hilbert spaces.

Definition C.0.1 (Pseudo Inverse) Let $T \in L(U, H)$ and $\mathrm{Ker}(T) := \{x \in U \mid Tx = 0\}$. The pseudo inverse of T is defined as

$$T^{-1} := \left(T\,|_{\mathrm{Ker}(T)^\perp}\right)^{-1} : T\left(\mathrm{Ker}(T)^\perp\right) = T(U) \to \mathrm{Ker}(T)^\perp.$$

(Note that T is one-to-one on $\mathrm{Ker}(T)^\perp$.)

Remark C.0.2

(i) There is an equivalent way of defining the pseudo inverse of a linear operator $T \in L(U, H)$. For $x \in T(U)$ one sets $T^{-1}x \in U$ to be the solution of minimal norm of the equation $Ty = x$, $y \in U$.
(ii) If $T \in L(U, H)$ then $T^{-1} : T(U) \to \mathrm{Ker}(T)^\perp$ is linear and bijective.

Proposition C.0.3 *Let $T \in L(U)$ and T^{-1} be the pseudo inverse of T.*

(i) If we define an inner product on $T(U)$ by

$$\langle x, y \rangle_{T(U)} := \langle T^{-1}x, T^{-1}y \rangle_U \quad \text{for all } x, y \in T(U),$$

then $\left(T(U), \langle\ ,\ \rangle_{T(U)}\right)$ is a Hilbert space.
(ii) Let e_k, $k \in \mathbb{N}$, be an orthonormal basis of $(\mathrm{Ker}\,T)^\perp$. Then Te_k, $k \in \mathbb{N}$, is an orthonormal basis of $\left(T(U), \langle\ ,\ \rangle_{T(U)}\right)$.

Proof $T : (\mathrm{Ker}\,T)^\perp \to T(U)$ is bijective and an isometry if $(\mathrm{Ker}\,T)^\perp$ is equipped with $\langle\ ,\ \rangle_U$ and $T(U)$ with $\langle\ ,\ \rangle_{T(U)}$. $\qquad\square$

Now we want to present a result about the images of linear operators. To this end we need the following lemma.

© Springer International Publishing Switzerland 2015
W. Liu, M. Röckner, *Stochastic Partial Differential Equations: An Introduction*,
Universitext, DOI 10.1007/978-3-319-22354-4

Lemma C.0.4 *Let* $T \in L(U, H)$. *Then the set* $T\overline{B_c(0)}$ $(= \{Tu \mid u \in U, \|u\|_U \leqslant c\})$, $c \geqslant 0$, *is convex and closed.*

Proof Since T is linear it is obvious that the set is convex.

Since a convex subset of a Hilbert space is closed (with respect to the norm) if and only if it is weakly closed, it suffices to show that $T\overline{B_c(0)}$ is weakly closed. Since $T : U \to H$ is linear and continuous (with respect to the norms on U, H respectively) it is also obviously continuous with respect to the weak topologies on U, H respectively. But by the Banach–Alaoglu theorem (see e.g. [68, Theorem IV.21, p. 115]) closed balls in a Hilbert space are weakly compact. Hence $\overline{B_c(0)}$ is weakly compact, and so is its continuous image, i.e. $T\overline{B_c(0)}$ is weakly compact, therefore weakly closed. $\square$

Proposition C.0.5 *Let* $(U_1, \langle \, , \, \rangle_1)$ *and* $(U_2, \langle \, , \, \rangle_2)$ *be two Hilbert spaces. In addition, let* $T_1 \in L(U_1, H)$ *and* $T_2 \in L(U_2, H)$. *Then the following statements hold.*

(i) *If there exists a constant* $c \geqslant 0$ *such that* $\|T_1^* x\|_1 \leqslant c\|T_2^* x\|_2$ *for all* $x \in H$ *then* $\{T_1 u \mid u \in U_1, \|u\|_1 \leqslant 1\} \subset \{T_2 v \mid v \in U_2, \|v\|_2 \leqslant c\}$. *In particular, this implies that* $\operatorname{Im} T_1 \subset \operatorname{Im} T_2$.

(ii) *If* $\|T_1^* x\|_1 = \|T_2^* x\|_2$ *for all* $x \in H$ *then* $\operatorname{Im} T_1 = \operatorname{Im} T_2$ *and* $\|T_1^{-1} x\|_1 = \|T_2^{-1} x\|_2$ *for all* $x \in \operatorname{Im} T_1$, *where* T_i^{-1} *is the pseudo inverse of* T_i, $i = 1, 2$.

Proof [26, Proposition B.1, p. 407]

(i) Assume that there exists a $u_0 \in U_1$ such that

$$\|u_0\|_1 \leqslant 1 \quad \text{and} \quad T_1 u_0 \notin \{T_2 v \mid v \in U_2, \|v\|_2 \leqslant c\}.$$

By Lemma C.0.4 we know that the set $\{T_2 v \mid v \in U_2, \|v\|_2 \leqslant c\}$ is closed and convex. Therefore, we get by the separation theorem (see [3, 5.11 Trennungssatz, p. 166]) that there exists an $x \in H$, $x \neq 0$, such that

$$1 < \langle x, T_1 u_0 \rangle \quad \text{and} \quad \langle x, T_2 v \rangle \leqslant 1 \quad \text{for all } v \in U_2 \text{ with } \|v\|_2 \leqslant c.$$

Thus $\|T_1^* x\|_1 > 1$ and $c\|T_2^* x\|_2 = \sup_{\|v\|_2 \leqslant c} |\langle T_2^* x, v \rangle_2| \leqslant 1$, a contradiction.

(ii) By (i) we know that $\operatorname{Im} T_1 = \operatorname{Im} T_2$. It remains to verify that

$$\|T_1^{-1} x\|_1 = \|T_2^{-1} x\|_2 \quad \text{for all } x \in \operatorname{Im} T_1.$$

If $x = 0$ then $\|T_1^{-1} 0\|_1 = 0 = \|T_2^{-1} 0\|_2$.

If $x \in \operatorname{Im} T_1 \setminus \{0\}$ then there exist $u_1 \in (\operatorname{Ker} T_1)^\perp$ and $u_2 \in (\operatorname{Ker} T_2)^\perp$ such that $x = T_1 u_1 = T_2 u_2$. We have to show that $\|u_1\|_1 = \|u_2\|_2$.

Assume that $\|u_1\|_1 > \|u_2\|_2 (> 0)$. Then (i) implies that

$$\frac{x}{\|u_2\|_2} = T_2\left(\frac{u_2}{\|u_2\|_2}\right)$$

$$\in \{T_2 v \mid v \in U_2, \|v\|_2 \leq 1\} = \{T_1 u \mid u \in U_1, \|u\|_1 \leq 1\}.$$

Furthermore,

$$\frac{x}{\|u_2\|_2} = T_1\left(\frac{u_1}{\|u_2\|_2}\right) \quad \text{and} \quad \left\|\frac{u_1}{\|u_2\|_2}\right\|_1 > 1.$$

Both together imply that there exists a $\tilde{u}_1 \in U_1$, $\|\tilde{u}_1\|_1 \leq 1$, so that for $\tilde{u}_2 :=$
$\frac{u_1}{\|u_2\|_2} \in (\mathrm{Ker}\, T_1)^\perp$ we have

$$T_1 \tilde{u}_1 = \frac{x}{\|u_2\|_2} = T_1 \tilde{u}_2, \quad \text{i.e. } \tilde{u}_1 - \tilde{u}_2 \in \mathrm{Ker}\, T_1.$$

Therefore,

$$0 = \langle \tilde{u}_1 - \tilde{u}_2, \tilde{u}_2 \rangle_1 = \langle \tilde{u}_1, \tilde{u}_2 \rangle_1 - \|\tilde{u}_2\|_1^2$$

$$\leq \|\tilde{u}_1\|_1 \|\tilde{u}_2\|_1 - \|\tilde{u}_2\|_1^2 \leq \left(1 - \|\tilde{u}_2\|_1\right) \|\tilde{u}_2\|_1 < 0.$$

This is a contradiction. $\qquad\qquad\qquad\qquad\qquad\qquad\qquad\qquad\qquad\qquad\square$

Corollary C.0.6 *Let $T \in L(U, H)$ and set $Q := TT^* \in L(H)$. Then we have*

$$\mathrm{Im}\, Q^{\frac{1}{2}} = \mathrm{Im}\, T \quad \text{and} \quad \left\|Q^{-\frac{1}{2}}x\right\| = \|T^{-1}x\|_U \text{ for all } x \in \mathrm{Im}\, T,$$

where $Q^{-\frac{1}{2}}$ is the pseudo inverse of $Q^{\frac{1}{2}}$.

Proof Since by Lemma 2.3.4 $Q^{\frac{1}{2}}$ is symmetric we have for all $x \in H$ that

$$\left\|\left(Q^{\frac{1}{2}}\right)^* x\right\|^2 = \left\|Q^{\frac{1}{2}} x\right\|^2 = \langle Qx, x \rangle = \langle TT^* x, x \rangle = \|T^* x\|_U^2.$$

Therefore the assertion follows by Proposition C.0.5. $\qquad\qquad\qquad\qquad\qquad\square$

Appendix D
Some Tools from Real Martingale Theory

We need the following Burkholder–Davis inequality for real-valued continuous local martingales.

Proposition D.0.1 *Let* $(N_t)_{t\in[0,T]}$ *be a real-valued continuous local martingale on a probability space* $(\Omega, \mathcal{F}, P)$ *with respect to a normal filtration* $(\mathcal{F}_t)_{t\in[0,T]}$ *with* $N_0 = 0$.

(i) Then for all stopping times $\tau (\leqslant T)$

$$E(\sup_{t\in[0,\tau]} |N_t|) \leqslant 3E(\langle N\rangle_\tau^{1/2}).$$

(ii) If $E(\langle N\rangle_T^{1/2}) < \infty$, *then* $(N_t)_{t\in[0,T]}$ *is a martingale.*

Proof See e.g. [56, p. 75, line 1] for (*i*). Now we prove (*ii*).

Let $\tau_N : \Omega \to [0, T]$, $N \in \mathbb{N}$, be stopping times such that $(N_{t\wedge\tau_N})_{t\in[0,T]}$ is a martingale and $\lim_{N\to\infty} \tau_N = T$. Then for each $t \in [0, T]$

$$\lim_{N\to\infty} N_{t\wedge\tau_N} = N_t \quad P\text{-a.s.}$$

and by (i)

$$\sup_{N\in\mathbb{N}} |N_{t\wedge\tau_N}| \leq \sup_{s\in[0,T]} |N_s| \in L^1(\Omega; \mathbb{R}).$$

Hence by Lebesgue's dominated convergence theorem for each $t \in [0, T]$

$$\lim_{N\to\infty} N_{t\wedge\tau_N} = N_t \text{ in } L^1(\Omega; \mathbb{R})$$

and assertion (ii) follows. $\qquad\square$

© Springer International Publishing Switzerland 2015
W. Liu, M. Röckner, *Stochastic Partial Differential Equations: An Introduction*,
Universitext, DOI 10.1007/978-3-319-22354-4

Corollary D.0.2 *Let $\varepsilon, \delta \in]0, \infty[$. Then for N as in Proposition D.0.1*

$$P(\sup_{t\in[0,T]} |N_t| \geq \varepsilon) \leq \frac{3}{\varepsilon} E(\langle N \rangle_T^{1/2} \wedge \delta) + P(\langle N \rangle_T^{1/2} > \delta).$$

Proof Let

$$\tau := \inf\{t \geq 0 | \langle N \rangle_t^{1/2} > \delta\} \wedge T.$$

Then $\tau (\leq T)$ is an $\mathcal{F}_t$-stopping time. Hence by Proposition D.0.1

$$P\left(\sup_{t\in[0,T]} |N_t| \geq \varepsilon \right)$$

$$=P\left(\sup_{t\in[0,T]} |N_t| \geq \varepsilon, \tau = T \right) + P\left(\sup_{t\in[0,T]} |N_t| \geq \varepsilon, \tau < T \right)$$

$$\leq \frac{3}{\varepsilon} E(\langle N \rangle_\tau^{1/2}) + P\left(\sup_{t\in[0,T]} |N_t| \geq \varepsilon, \langle N \rangle_T^{1/2} > \delta \right)$$

$$\leq \frac{3}{\varepsilon} E(\langle N \rangle_T^{1/2} \wedge \delta) + P(\langle N \rangle_T^{1/2} > \delta).$$

$\square$

Appendix E
Weak and Strong Solutions:
The Yamada–Watanabe Theorem

The main reference for this chapter is [71].

Let H be a separable Hilbert space, with inner product $\langle \cdot, \cdot \rangle_H$ and norm $\| \ \|_H$. Let V, E be separable Banach spaces with norms $\| \ \|_V$ and $\| \ \|_E$, such that

$$V \subset H \subset E$$

continuously and densely. For a topological space X let $\mathcal{B}(X)$ denote its Borel σ-algebra. By Kuratowski's theorem we have that $V \in \mathcal{B}(H)$, $H \in \mathcal{B}(E)$ and $\mathcal{B}(V) = \mathcal{B}(H) \cap V$, $\mathcal{B}(H) = \mathcal{B}(E) \cap H$.

Setting $\|x\|_V := \infty$ if $x \in H \setminus V$, we extend $\| \ \|_V$ to a function on H. We recall that by Exercise 4.2.3 this extension is $\mathcal{B}(H)$-measurable and lower semicontinuous. Hence the following path space is well-defined:

$$\mathbb{B} := \left\{ w \in C(\mathbb{R}_+; H) \ \middle| \ \int_0^T \|w(t)\|_V dt < \infty \text{ for all } T \in [0, \infty) \right\},$$

equipped with the metric

$$\rho(w_1, w_2) := \sum_{k=1}^{\infty} 2^{-k} \left[\left(\int_0^k \|w_1(t) - w_2(t)\|_V dt + \sup_{t \in [0,k]} \|w_1(t) - w_2(t)\|_H \right) \wedge 1 \right].$$

Obviously, $(\mathbb{B}, \rho)$ is a complete separable metric space. Let $\mathcal{B}_t(\mathbb{B})$ denote the σ-algebra generated by all maps $\pi_s : \mathbb{B} \to H$, $s \in [0, t]$, where $\pi_s(w) := w(s)$, $w \in \mathbb{B}$.

Let $(U, \langle \ , \ \rangle_U)$ be another separable Hilbert space and let $L_2(U, H)$ denote the space of all Hilbert–Schmidt operators from U to H equipped with the usual Hilbert–Schmidt norm $\| \ \|_{L_2}$.

Let $b : \mathbb{R}_+ \times \mathbb{B} \to E$ and $\sigma : \mathbb{R}_+ \times \mathbb{B} \to L_2(U, H)$ be $\mathcal{B}(\mathbb{R}_+) \otimes \mathcal{B}(\mathbb{B})/\mathcal{B}(E)$ and $\mathcal{B}(\mathbb{R}_+) \otimes \mathcal{B}(\mathbb{B})/\mathcal{B}(L_2(U, H))$-measurable respectively such that for each $t \in \mathbb{R}_+$

$$b(t, \cdot) \text{ is } \mathcal{B}_t(\mathbb{B})/\mathcal{B}(E)\text{-measurable}$$

© Springer International Publishing Switzerland 2015
W. Liu, M. Röckner, *Stochastic Partial Differential Equations: An Introduction*,
Universitext, DOI 10.1007/978-3-319-22354-4

and

$$\sigma(t, \cdot) \text{ is } \mathcal{B}_t(\mathbb{B})/\mathcal{B}(L_2(U, H))\text{-measurable.}$$

As usual we call $(\Omega, \mathcal{F}, P, (\mathcal{F}_t))$ a stochastic basis if $(\Omega, \mathcal{F}, P)$ is a complete probability space and $(\mathcal{F}_t)$ is a right continuous filtration on Ω augmented by the P-zero sets. Let $\beta_k, k \in \mathbb{N}$, be independent $(\mathcal{F}_t)$-Brownian motions on a stochastic basis $(\Omega, \mathcal{F}, P, (\mathcal{F}_t))$ and define the sequence

$$W(t) := (\beta_k(t))_{k \in \mathbb{N}}, \quad t \in [0, \infty).$$

Below we refer to such a process W on $\mathbb{R}^\infty$ as a *standard $\mathbb{R}^\infty$-Wiener process*. We fix an orthonormal basis $\{e_k, k \in \mathbb{N}\}$ of U and consider W as a cylindrical Wiener process on U, that is, we informally have

$$W(t) = \sum_{k=1}^\infty \beta_k(t) e_k, \quad t \in [0, \infty).$$

We consider the following stochastic evolution equation:

$$dX(t) = b(t, X)dt + \sigma(t, X)dW(t), \quad t \in [0, \infty). \tag{E.1}$$

Definition E.0.1 A pair (X, W), where $X = (X(t))_{t \in [0,\infty)}$ is an $(\mathcal{F}_t)$-adapted process with paths in $\mathbb{B}$ and W is a standard $\mathbb{R}^\infty$-Wiener process on a stochastic basis $(\Omega, \mathcal{F}, P, (\mathcal{F}_t))$, is called a *weak solution* of (E.1) if

(i) For any $T \in [0, \infty)$

$$\int_0^T \|b(s, X)\|_E ds + \int_0^T \|\sigma(s, X)\|_{L_2(U,H)}^2 ds < \infty \quad P\text{-a.e.}$$

(ii) As a stochastic equation on E we have

$$X(t) = X(0) + \int_0^t b(s, X)ds + \int_0^t \sigma(s, X)dW(s), \quad t \in [0, \infty) \quad P\text{-a.e.}$$

Remark E.0.2

(i) By the measurability assumptions on b and σ, it follows that if X is as in Definition E.0.1 then both processes $b(\cdot, X)$ and $\sigma(\cdot, X)$ are $(\mathcal{F}_t)$-adapted.
(ii) We recall that by definition of the H-valued stochastic integral in (ii) we have

$$\int_0^t \sigma(s, X)dW(s) := \int_0^t \sigma(s, X) \circ J^{-1} d\bar{W}(s), \quad t \in [0, \infty),$$

where J is any one-to-one Hilbert–Schmidt operator from U into another Hilbert space $(\bar{U}\langle\,,\,\rangle_{\bar{U}})$ and

$$\bar{W}(t) := \sum_{k=1}^{\infty} \beta_k(t) J e_k, \quad t \in [0, \infty). \tag{E.2}$$

By Sect. 2.5 this definition of the stochastic integral is independent of the choice of J and $(\bar{U},\langle\,,\,\rangle_{\bar{U}})$. We recall that for $s \in [0, \infty)$, $w \in \mathbb{B}$

$$\sigma(s, w) \circ J^{-1} \in L_2(Q^{1/2}(\bar{U}), H)$$

with $\|\sigma(s, w) \circ J^{-1}\|_{L_2(Q^{1/2}(\bar{U}), H)} = \|\sigma(s, w)\|_{L_2(U, H)}$,

where $Q := JJ^*$, and that $\bar{W}$ is a Q-Wiener process on $\bar{U}$.

Below we shall fix one such J and $(\bar{U},\langle\,,\,\rangle_{\bar{U}})$ as in Remark E.0.2(ii) and set

$$\bar{\sigma}(s, w) := \sigma(s, w) \circ J^{-1}, \quad s \in [0, \infty), w \in \mathbb{B},$$

and for any standard $\mathbb{R}^{\infty}$-Wiener process W we define $\bar{W}$ as in (E.2) for the fixed J. Furthermore we define

$$\mathbb{W}_0 := \{w \in C(\mathbb{R}_+, \bar{U}) | w(0) = 0\}$$

equipped with the supremum norm and Borel σ-algebra $\mathcal{B}(\mathbb{W}_0)$. For $t \in \mathbb{R}_+$ let $\mathcal{B}_t(\mathbb{W}_0)$ be the σ-algebra generated by $\pi_s : \mathbb{W}_0 \to \bar{U}, 0 \leq s \leq t, \pi_s(w) := w(s)$.

Definition E.0.3 We say that *weak uniqueness* holds for (E.1) if whenever (X, W) and (X', W') are two weak solutions with stochastic bases $(\Omega, \mathcal{F}, P, (\mathcal{F}_t))$ and $(\Omega', \mathcal{F}', P', (\mathcal{F}'_t))$ such that

$$P \circ X(0)^{-1} = P' \circ X'(0)^{-1},$$

(as measures on $(H, \mathcal{B}(H)))$, then

$$P \circ X^{-1} = P' \circ (X')^{-1}$$

(as measures on $(\mathbb{B}, \mathcal{B}(\mathbb{B})))$.

Definition E.0.4 We say that *pathwise uniqueness* holds for (E.1), if whenever (X, W), (X', W) are two weak solutions on the same stochastic basis $(\Omega, \mathcal{F}, P, (\mathcal{F}_t))$ and with the same standard-$\mathbb{R}^{\infty}$-Wiener process W on $(\Omega, \mathcal{F}, P)$ such that $X(0) = X'(0)$ P-a.e., then P-a.e.

$$X(t) = X'(t), \ t \in [0, \infty).$$

To define strong solutions we need to introduce the following class $\hat{\mathcal{E}}$ of maps:
Let $\hat{\mathcal{E}}$ denote the set of all maps $F : H \times \mathbb{W}_0 \to \mathbb{B}$ such that for every probability
measure μ on $(H, \mathcal{B}(H))$ there exists a $\overline{\mathcal{B}(H) \otimes \mathcal{B}(\mathbb{W}_0)}^{\mu \otimes P^Q} / \mathcal{B}(\mathbb{B})$-measurable map
$F_\mu : H \times \mathbb{W}_0 \to \mathbb{B}$ such that for μ-a.e. $x \in H$

$$F(x, w) = F_\mu(x, w) \text{ for } P^Q\text{-a.e. } w \in \mathbb{W}_0.$$

Here $\overline{\mathcal{B}(H) \otimes \mathcal{B}(\mathbb{W}_0)}^{\mu \otimes P^Q}$ denotes the completion of $\mathcal{B}(H) \otimes \mathcal{B}(\mathbb{W}_0)$ with respect
to $\mu \otimes P^Q$, and P^Q denotes the distribution of the Q-Wiener process on $\bar{U}$ on
$(\mathbb{W}_0, \mathcal{B}(\mathbb{W}_0))$. Of course, F_μ is uniquely determined $\mu \otimes P^Q$-a.e.

Definition E.0.5 A weak solution (X, W) to (E.1) on $(\Omega, \mathcal{F}, P, (\mathcal{F}_t))$ is called a
strong solution if there exists an $F \in \hat{\mathcal{E}}$ such that for $x \in H$, $w \mapsto F(x, w)$ is
$\overline{\mathcal{B}_t(\mathbb{W}_0)}^{P^Q} / \mathcal{B}_t(\mathbb{B})$-measurable for every $t \in [0, \infty)$ and

$$X = F_{P \circ X(0)^{-1}}(X(0), \bar{W}) \quad P\text{-a.e.,}$$

where $\overline{\mathcal{B}_t(\mathbb{W}_0)}^{P^Q}$ denotes the completion with respect to P^Q in $\mathcal{B}(\mathbb{W}_0)$.

Definition E.0.6 Equation (E.1) is said to have a *unique strong solution* if there
exists an $F \in \hat{\mathcal{E}}$ satisfying the adaptiveness condition in Definition E.0.5 and such
that:

1. For every standard $\mathbb{R}^\infty$-Wiener process on a stochastic basis $(\Omega, \mathcal{F}, P, (\mathcal{F}_t))$ and
   any $\mathcal{F}_0 / \mathcal{B}(H)$-measurable $\xi : \Omega \to H$ the $\mathbb{B}$-valued process

$$X := F_{P \circ \xi^{-1}}(\xi, \bar{W})$$

   is $(\mathcal{F}_t)$-adapted and satisfies (i), (ii) in Definition E.0.1, i.e. $(F(\xi, \bar{W}), W)$ is a
   weak solution to (E.1), and in addition $X(0) = \xi$ P-a.e.
2. For any weak solution (X, W) to (E.1) we have

$$X = F_{P \circ X(0)^{-1}}(X(0), \bar{W}) \ P\text{-a.e.}$$

Remark E.0.7 Since $X(0)$ of a weak solution is P-independent of $\bar{W}$, thus

$$P \circ (X(0), \bar{W})^{-1} = \mu \otimes P^Q,$$

we have that the existence of a unique strong solution for (E.1) implies that weak
uniqueness also holds.

Now we can formulate the main result of this section.

Theorem E.0.8 *Let σ and b be as above. Then Eq. (E.1) has a unique strong solution if and only if both of the following properties hold:*

(i) For every probability measure μ on $(H, \mathcal{B}(H))$ there exists a weak solution (X, W) of (E.1) such that μ is the distribution of $X(0)$.

(ii) Pathwise uniqueness holds for (E.1).

Proof Suppose (E.1) has a unique strong solution. Then (ii) obviously holds. To show (i) one only has to take the probability space $(\mathbb{W}_0, \mathcal{B}(\mathbb{W}_0), P^Q)$ and consider $(H \times \mathbb{W}_0, \overline{\mathcal{B}(H) \otimes \mathcal{B}(\mathbb{W}_0)}^{\mu \otimes P^Q}, \mu \otimes P^Q)$ with filtration

$$\bigcap_{\varepsilon > 0} \sigma(\mathcal{B}(H) \otimes \mathcal{B}_{t+\varepsilon}(\mathbb{W}_0), \mathcal{N}), \quad t \geq 0,$$

where $\mathcal{N}$ denotes all $\mu \otimes P^Q$-zero sets in $\overline{\mathcal{B}(H) \otimes \mathcal{B}(\mathbb{W}_0)}^{\mu \otimes P^Q}$. Let $\xi : H \times \mathbb{W}_0 \to H$ and $W : H \times \mathbb{W}_0 \to \mathbb{W}_0$ be the canonical projections. Then $X := F_{P \circ \xi^{-1}}(\xi, W)$ is the desired weak solution in (i). $\square$

Now let us suppose that (i) and (ii) hold. The proof that then there exists a unique strong solution for (E.1) is quite technical. We structure it through a series of lemmas.

Lemma E.0.9 *Let $(\Omega, \mathcal{F})$ be a measurable space such that $\{\omega\} \in \mathcal{F}$ for all $\omega \in \Omega$ and such that*

$$D := \{(\omega, \omega) | \omega \in \Omega\} \in \mathcal{F} \otimes \mathcal{F}$$

(which is the case, e.g. if Ω is a Polish space and $\mathcal{F}$ its Borel σ-algebra). Let P_1, P_2 be probability measures on $(\Omega, \mathcal{F})$ such that $P_1 \otimes P_2(D) = 1$. Then $P_1 = P_2 = \delta_{\omega_0}$ for some $\omega_0 \in \Omega$.

Proof Let $f : \Omega \to [0, \infty)$ be $\mathcal{F}$-measurable. Then

$$\int f(\omega_1) P_1(d\omega_1) = \iint f(\omega_1) P_1(d\omega_1) P_2(d\omega_2)$$

$$= \iint 1_D(\omega_1, \omega_2) f(\omega_1) P_1(d\omega_1) P_2(d\omega_2)$$

$$= \iint 1_D(\omega_1, \omega_2) f(\omega_2) P_1(d\omega_1) P_2(d\omega_2) = \int f(\omega_2) P_2(d\omega_2),$$

so $P_1 = P_2$. Furthermore,

$$1 = \iint 1_D(\omega_1, \omega_2)P_1(d\omega_1)P_2(d\omega_2) = \int P_1(\{\omega_2\})P_2(d\omega_2),$$

hence $1 = P_1(\{\omega_2\})$ for P_2-a.e. $\omega_2 \in \Omega$. Therefore, $P_1 = \delta_{\omega_0}$ for some $\omega_0 \in \Omega$. □

Fix a probability measure μ on $(H, \mathcal{B}(H))$ and let (X, W) with stochastic basis $(\Omega, \mathcal{F}, P, (\mathcal{F}_t))$ be a weak solution to (E.1) with initial distribution μ. Define a probability measure P_μ on $(H \times \mathbb{B} \times \mathbb{W}_0, \mathcal{B}(H) \otimes \mathcal{B}(\mathbb{B}) \otimes \mathcal{B}(\mathbb{W}_0))$ by

$$P_\mu := P \circ (X(0), X, \bar{W})^{-1}.$$

Lemma E.0.10 *There exists a family $K_\mu((x, w), dw_1), x \in H, w \in \mathbb{W}_0$, of probability measures on $(\mathbb{B}, \mathcal{B}(\mathbb{B}))$ having the following properties:*

(i) For every $A \in \mathcal{B}(\mathbb{B})$ the map

$$H \times \mathbb{W}_0 \ni (x, w) \mapsto K_\mu((x, w), A)$$

is $\mathcal{B}(H) \otimes \mathcal{B}(\mathbb{W}_0)$-measurable.

(ii) For every $\mathcal{B}(H) \otimes \mathcal{B}(\mathbb{B}) \otimes \mathcal{B}(\mathbb{W}_0)$-measurable map $f : H \times \mathbb{B} \times \mathbb{W}_0 \to [0, \infty)$ we have

$$\int_{H \times \mathbb{B} \times \mathbb{W}_0} f(x, w_1, w)P_\mu(dx, dw_1, dw)$$

$$= \int_H \int_{\mathbb{W}_0} \int_{\mathbb{B}} f(x, w_1, w)K_\mu((x, w), dw_1)P^Q(dw)\mu(dx).$$

(iii) If $t \in [0, \infty)$ and $f : \mathbb{B} \to [0, \infty)$ is $\mathcal{B}_t(\mathbb{B})$-measurable, then

$$H \times \mathbb{W}_0 \ni (x, w) \mapsto \int_{\mathbb{B}} f(w_1)K_\mu((x, w), dw_1)$$

is $\overline{\mathcal{B}(H) \otimes \mathcal{B}_t(\mathbb{W}_0)}^{\mu \otimes P^Q}$-measurable, where $\overline{\mathcal{B}(H) \otimes \mathcal{B}_t(\mathbb{W}_0)}^{\mu \otimes P^Q}$ denotes the completion with respect to $\mu \otimes P^Q$ in $\mathcal{B}(H) \otimes \mathcal{B}(\mathbb{W}_0)$.

Proof Let $\Pi : H \times \mathbb{B} \times \mathbb{W}_0 \to H \times \mathbb{W}_0$ be the canonical projection. Since $X(0)$ is $\mathcal{F}_0$-measurable, hence P-independent of $\bar{W}$, it follows that

$$P_\mu \circ \Pi^{-1} = P \circ (X(0), \bar{W})^{-1} = \mu \otimes P^Q.$$

Hence by the existence result on regular conditional distributions (cf. e.g. [47, Corollary to Theorem 3.3 on p. 15]), the existence of the family $K_\mu((x, w), dw_1)$, $x \in H$, $w \in \mathbb{W}_0$, satisfying (i) and (ii) follows.

To prove (iii) it suffices to show that for $t \in [0, \infty)$ and for all $A_0 \in \mathcal{B}(H)$, $A_1 \in \mathcal{B}_t(\mathbb{B})$, $A \in \mathcal{B}_t(\mathbb{W}_0)$ and

$$A' := \{\pi_{r_1} - \pi_t \in B_1, \pi_{r_2} - \pi_{r_1} \in B_2, \ldots, \pi_{r_k} - \pi_{r_{k-1}} \in B_k\} (\subset \mathbb{W}_0),$$

$$t \leq r_1 < \ldots < r_k, B_1, \ldots, B_k \in \mathcal{B}(\bar{U}),$$

$$\int_{A_0} \int_{\mathbb{W}_0} 1_{A \cap A'}(w) K_\mu((x, w), A_1) P^Q(dw) \mu(dx)$$

$$= \int_{A_0} \int_{\mathbb{W}_0} 1_{A \cap A'}(w) \mathbb{E}_{\mu \otimes P^Q}(K_\mu(\cdot, A_1)|\mathcal{B}(H) \otimes \mathcal{B}_t(\mathbb{W}_0)) P^Q(dw) \mu(dx), \qquad (E.3)$$

since the system of all $A \cap A'$, $A \in \mathcal{B}_t(\mathbb{W}_0)$, A' as above generates $\mathcal{B}(\mathbb{W}_0)$. But by part (ii) above, the left-hand side of (E.3) is equal to

$$\int_{H \times \mathbb{B} \times \mathbb{W}_0} 1_{A_0}(x) 1_{A \cap A'}(w) 1_{A_1}(w_1) P_\mu(dx, dw_1, dw)$$

$$= \int_\Omega 1_{A_0}(X(0)) 1_{A_1}(X) 1_A(\bar{W}) 1_{A'}(\bar{W}) dP. \qquad (E.4)$$

But $1_{A'}(\bar{W})$ is P-independent of $\mathcal{F}_t$, since W is a standard $\mathbb{R}^\infty$-Wiener process on $(\Omega, \mathcal{F}, P, \mathcal{F}_t)$, so the right-hand side of (E.4) is equal to

$$\int_\Omega 1_{A'}(\bar{W}) dP \cdot \int_\Omega 1_{A_0}(X(0)) 1_{A_1}(X) 1_A(\bar{W}) dP$$

$$= P^Q(A') \int_{H \times \mathbb{B} \times \mathbb{W}_0} 1_{A_0}(x) 1_A(w) 1_{A_1}(w_1) P_\mu(dx, dw_1, dw)$$

$$= P^Q(A') \int_{A_0} \int_A K_\mu((x, w), A_1) P^Q(dw) \mu(dx)$$

$$= P^Q(A') \int_{A_0} \int_A \mathbb{E}_{\mu \otimes P^Q}(K_\mu(\cdot, A_1)|\mathcal{B}(H) \otimes \mathcal{B}_t(\mathbb{W}_0))((x, w)) P^Q(dw) \mu(dx)$$

$$= \int_{A_0} \int_{\mathbb{W}_0} 1_{A \cap A'}(w) \mathbb{E}_{\mu \otimes P^Q}(K_\mu(\cdot, A_1)|\mathcal{B}(H) \otimes \mathcal{B}_t(\mathbb{W}_0))((x, w)) P^Q(dw) \mu(dx),$$

since A' is P^Q-independent of $\mathcal{B}_t(\mathbb{W}_0)$. $\square$

For convenient labelling subsequently, up to and including the proof of Lemma E.0.14, we rename our weak solution as $(X^{(1)}, W^{(1)}) := (X, W)$ and define $(\Omega^{(1)}, \mathcal{F}^{(1)}, P^{(1)}, (\mathcal{F}_t^{(1)})), \bar{W}^{(1)}, P_\mu^{(1)}, K_\mu^{(1)}$ correspondingly. Now take another weak solution $(X^{(2)}, W^{(2)})$ of (E.1) with the same initial distribution μ on a stochastic basis $(\Omega^{(2)}, \mathcal{F}^{(2)}, P^{(2)}, (\mathcal{F}_t^{(2)}))$. Define $\bar{W}^{(2)}, P_\mu^{(2)}$ correspondingly and let $K_\mu^{(2)}$ be constructed correspondingly as in Lemma E.10. For $x \in H$ define a measure Q_x on

$$(H \times \mathbb{B} \times \mathbb{B} \times \mathbb{W}_0, \mathcal{B}(H) \otimes \mathcal{B}(\mathbb{B}) \otimes \mathcal{B}(\mathbb{B}) \otimes \mathcal{B}(\mathbb{W}_0))$$

by

$$Q_x(A) := \int_H \int_{\mathbb{B}} \int_{\mathbb{B}} \int_{\mathbb{W}_0} 1_A(z, w_1, w_2, w)$$
$$K_\mu^{(1)}((z, w), dw_1) K_\mu^{(2)}((z, w), dw_2) P^Q(dw) \delta_x(dz).$$

Define the stochastic basis

$$\tilde{\Omega} := H \times \mathbb{B} \times \mathbb{B} \times \mathbb{W}_0$$
$$\tilde{\mathcal{F}}^x := \overline{\mathcal{B}(H) \otimes \mathcal{B}(\mathbb{B}) \otimes \mathcal{B}(\mathbb{B}) \otimes \mathcal{B}(\mathbb{W}_0)}^{Q_x}$$
$$\tilde{\mathcal{F}}_t^x := \bigcap_{\varepsilon > 0} \sigma(\mathcal{B}(H) \otimes \mathcal{B}_{t+\varepsilon}(\mathbb{B}) \otimes \mathcal{B}_{t+\varepsilon}(\mathbb{B}) \otimes \mathcal{B}_{t+\varepsilon}(\mathbb{W}_0), \mathcal{N}_x),$$

where

$$\mathcal{N}_x := \{N \in \tilde{\mathcal{F}}^x | Q_x(N) = 0\},$$

and define maps

$$\Pi_0 : \tilde{\Omega} \to H, \ (x, w_1, w_2, w) \mapsto x,$$
$$\Pi_i : \tilde{\Omega} \to \mathbb{B}, \ (x, w_1, w_2, w) \mapsto w_i \in \mathbb{B}, \quad i = 1, 2,$$
$$\Pi_3 : \tilde{\Omega} \to \mathbb{W}_0, \ (x, w_1, w_2, w) \mapsto w \in \mathbb{W}_0.$$

Then, obviously,

$$Q_x \circ \Pi_0^{-1} = \delta_x \tag{E.5}$$

and for $i = 1, 2$

$$Q_x \circ \Pi_3^{-1} = P^Q(= P \circ (\bar{W}^{(i)})^{-1}). \tag{E.6}$$

Lemma E.0.11 *There exists an $N_0 \in \mathcal{B}(H)$ with $\mu(N_0) = 0$ such that for all $x \in N_0^c$ we have that Π_3 is an $(\tilde{\mathcal{F}}_t^x)$-Wiener process on $(\tilde{\Omega}, \tilde{\mathcal{F}}^x, Q_x)$ taking values in $\bar{U}$.*

Proof By definition Π_3 is $(\tilde{\mathcal{F}}_t^x)$-adapted for every $x \in H$. Furthermore, for $0 \leqslant s < t$, $y \in H$, and $A_0, \tilde{A}_0 \in \mathcal{B}(H)$, $A_i \in \mathcal{B}_s(\mathbb{B})$, $i = 1, 2$, $A_3 \in \mathcal{B}_s(\mathbb{W}_0)$,

$$\int_{\tilde{A}_0} \mathbb{E}_{Q_x}(\exp(i\langle y, \Pi_3(t) - \Pi_3(s)\rangle_{\tilde{U}}) 1_{A_0 \times A_1 \times A_2 \times A_3}) \mu(\mathrm{d}x)$$

$$= \int_{\tilde{A}_0} \int_{\mathbb{W}_0} \exp(i\langle y, w(t) - w(s)\rangle_{\tilde{U}}) 1_{A_0}(x) 1_{A_3}(w)$$

$$K_\mu^{(1)}((x, w), A_1) K_\mu^{(2)}((x, w), A_2) P^Q(\mathrm{d}w) \mu(\mathrm{d}x)$$

$$= \int_{\tilde{A}_0} \int_{\mathbb{W}_0} \exp(i\langle y, w(t) - w(s)\rangle_{\tilde{U}}) P^Q(\mathrm{d}w) Q_x(A_0 \times A_1 \times A_2 \times A_3) \mu(\mathrm{d}x),$$

where we used Lemma E.0.10(iii) in the last step. Now the assertion follows by (E.6), a monotone class argument and the same reasoning as in the proof of Proposition 2.1.13.   □

Lemma E.0.12 *There exists an* $N_1 \in \mathcal{B}(H)$, $N_0 \subset N_1$, *with* $\mu(N_1) = 0$ *such that for all* $x \in N_1^c$, (Π_1, Π_3) *and* (Π_2, Π_3) *with stochastic basis* $(\tilde{\Omega}, \tilde{\mathcal{F}}^x, Q_x, (\tilde{\mathcal{F}}_t^x))$ *are weak solutions of* (E.1) *such that*

$$\Pi_1(0) = \Pi_2(0) = x \quad Q_x\text{-}a.e.,$$

therefore, $\Pi_1 = \Pi_2 \ Q_x\text{-}a.e.$

Proof ([26, Theorem 4.18, p. 109]) For $i = 1, 2$ consider the set $A_i \in \tilde{\mathcal{F}}^x$ defined by

$$A_i := \left\{ \Pi_i(t) - \Pi_0 = \int_0^t b(s, \Pi_i) \mathrm{d}s + \int_0^t \bar{\sigma}(s, \Pi_i) \mathrm{d}\Pi_3(s) \quad \forall\, t \in [0, \infty) \right\}$$

$$\cap \left\{ \int_0^T \|b(t, \Pi_i)\|_E \mathrm{d}t + \int_0^T \|\sigma(t, \Pi_i)\|_{L_2(U,H)}^2 \mathrm{d}t < \infty \quad \forall\, T \in [0, \infty) \right\}.$$

Define $A \in \mathcal{B}(H) \otimes \mathcal{B}(\mathbb{B}) \otimes \mathcal{B}(\mathbb{W}_0)$ analogously with Π_i replaced by the canonical projection from $H \times \mathbb{B} \times \mathbb{W}_0$ onto the second and Π_0, Π_3 by the canonical projection onto the first and third coordinate respectively. Then by Lemma E.0.10(ii) for $i = 1, 2$

$$\int_H Q_x(A_i)\, \mu(\mathrm{d}x)$$

$$= \int_H \int_{\mathbb{W}_0} \int_{\mathbb{B}} \int_{\mathbb{B}} 1_{A_i}(x, w_1, w_2, w) K_\mu^{(1)}((x, w), \mathrm{d}w_1) K_\mu^{(2)}((x, w), \mathrm{d}w_2) P^Q(\mathrm{d}w) \mu(\mathrm{d}x)$$

$$= P_\mu^{(i)}(A) = P^{(i)}(\{(X^{(i)}(0), X^{(i)}, \bar{W}^{(i)}) \in A\}) = 1. \tag{E.7}$$

Since all measures in the left-hand side of (E.7) are probability measures, it follows that for μ-a.e. $x \in H$

$$1 = Q_x(A_i) = Q_x(A_{i,x}),$$

where for $i = 1, 2$

$$A_{i,x} := \left\{ \Pi_i(t) - x = \int_0^t b(s, \Pi_i)ds + \int_0^t \bar{\sigma}(s, \Pi_i)d\Pi_3(s) \; \forall t \in [0, \infty) \right\}$$

$$\cap \left\{ \int_0^T \|b(t, \Pi_i)\|_E + \|\sigma(t, \Pi_i)\|^2_{L_2(U,H)} dt < \infty \; \forall T \in [0, \infty) \right\}.$$

Hence the first assertion follows. The second then follows by the pathwise uniqueness assumption in condition (ii) of the theorem.   □

Lemma E.0.13 *There exists a $\overline{\mathcal{B}(H) \otimes \mathcal{B}(\mathbb{W}_0)}^{\mu \otimes P^Q} / \mathcal{B}(\mathbb{B})$-measurable map*

$$F_\mu : H \times \mathbb{W}_0 \to \mathbb{B}$$

such that

$$K_\mu^{(1)}((x, w), \cdot) = K_\mu^{(2)}((x, w), \cdot) = \delta_{F_\mu(x,w)}$$

$$(= \text{Dirac measure on } \mathcal{B}(\mathbb{B}) \text{ with mass in } F_\mu(x, w))$$

for $\mu \otimes P^Q$-a.e. $(x, w) \in H \times \mathbb{W}_0$. Furthermore, F_μ is $\overline{\mathcal{B}(H) \otimes \mathcal{B}_t(\mathbb{W}_0)}^{\mu \otimes P^Q} / \mathcal{B}_t(\mathbb{B})$-measurable for all $t \in [0, \infty)$, where $\overline{\mathcal{B}(H) \otimes \mathcal{B}_t(\mathbb{W}_0)}^{\mu \otimes P^Q}$ denotes the completion with respect to $\mu \otimes P^Q$ in $\mathcal{B}(H) \otimes \mathcal{B}(\mathbb{W}_0)$.

Proof By Lemma E.0.12 for all $x \in N_1^c$, we have

$$1 = Q_x(\{\Pi_1 = \Pi_2\})$$

$$= \int_{\mathbb{W}_0} \int_\mathbb{B} \int_\mathbb{B} 1_D(w_1, w_2)K_\mu^{(1)}((x, w), dw_1)K_\mu^{(2)}((x, w), dw_2)P^Q(dw),$$

where $D := \{(w_1, w_1) \in \mathbb{B} \times \mathbb{B} | w_1 \in \mathbb{B}\}$. Hence by Lemma E.0.9 there exists an $N \in \mathcal{B}(H) \otimes \mathcal{B}(\mathbb{W}_0)$ such that $\mu \otimes P^Q(N) = 0$ and for all $(x, w) \in N^c$ there exists an $F_\mu(x, w) \in \mathbb{B}$ such that

$$K_\mu^{(1)}((x, w), dw_1) = K_\mu^{(2)}((x, w), dw_1) = \delta_{F_\mu(x,w)}(dw_1).$$

Set $F_\mu(x, w) := 0$, if $(x, w) \in N$. Let $A \in \mathcal{B}(\mathbb{B})$. Then for $i = 1, 2$

$$\{F_\mu \in A\} = (\{F_\mu \in A\} \cap N) \cup (\{K_\mu^{(i)}(\cdot, A) = 1\} \cap N^c)$$

and the measurability properties of F_μ follow from Lemma E.0.10. $\square$

We note here that, of course, F_μ depends on the two weak solutions $(X^{(1)}, W^{(1)})$ and $(X^{(2)}, W^{(2)})$ chosen(!) above.

Lemma E.0.14 *We have*

$$X^{(i)} = F_\mu(X^{(i)}(0), \bar{W}^{(i)}) \quad P^{(i)}\text{-a.e. for both } i = 1 \text{ and } i = 2.$$

In particular, any two weak solutions with initial law μ have the same distribution on $\mathbb{B}$.

Proof By Lemmas E.0.10 and E.0.13 for both $i = 1$ and $i = 2$ we have

$$P^{(i)}(\{X^{(i)} = F_\mu(X^{(i)}(0), \bar{W}^{(i)})\})$$

$$= \int_H \int_{\mathbb{W}_0} \int_{\mathbb{B}} 1_{\{w_1 = F_\mu(x,w)\}}(x, w_1, w)\delta_{F_\mu(x,w)}(dw_1)P^Q(dw)\mu(dx)$$

$$= 1.$$

$\square$

Let W' be another standard $\mathbb{R}^\infty$-Wiener process on a stochastic basis $(\Omega', \mathcal{F}', P', (\mathcal{F}'_t))$ and $\xi : \Omega' \to H$ an $\mathcal{F}'_0/\mathcal{B}(H)$-measurable map such that $\mu = P' \circ \xi^{-1}$. Set

$$X' := F_\mu(\xi, \bar{W}').$$

Lemma E.0.15 *(X', W') is a weak solution to (E.1) with $X'(0) = \xi$ P'-a.s. In particular, if $(\tilde{X}, W')$ is a weak solution to (E.1) on $(\Omega', \mathcal{F}', P', (\mathcal{F}'_t))$ with $\tilde{X}(0) = \xi$ P'-a.e., then*

$$\tilde{X} = F_\mu(\xi, \bar{W}') \quad P'\text{-a.e.}$$

Proof By the measurability properties of F_μ (cf. Lemma E.0.13) it follows that X' is adapted. We have

$$P'(\{\xi = X'(0)\}) = P'(\{\xi = F_\mu(\xi, \bar{W}')(0)\})$$

$$= \mu \otimes P^Q(\{(x, w) \in H \times \mathbb{W}_0 | x = F_\mu(x, w)(0)\})$$

$$= P(\{X(0) = F_\mu(X(0), \bar{W})(0)\}) = 1,$$

where we used Lemma E.0.14 in the last step.

To see that (X', W') is a weak solution we consider the set $A \in \mathcal{B}(H) \otimes \mathcal{B}(\mathbb{B}) \otimes \mathcal{B}(\mathbb{W}_0)$ defined in the proof of Lemma E.0.12. We have to show that

$$P'(\{(X'(0), X', \bar{W}') \in A\}) = 1.$$

But since $X'(0) = \xi$ is P'-independent of $\bar{W}'$, we have

$$\int 1_A(X'(0), F_\mu(X'(0), \bar{W}'), \bar{W}') dP'$$

$$= \int_H \int_{\mathbb{W}_0} 1_A(x, F_\mu(x, w), w) P^Q(dw) \mu(dx)$$

$$= \int_H \int_{\mathbb{W}_0} \int_{\mathbb{B}} 1_A(x, w_1, w) \delta_{F_\mu(x,w)}(dw_1) P^Q(dw) \mu(dx)$$

$$= \int 1_A(x, w_1, w) P_\mu(dx, dw_1, dw)$$

$$= P(\{(X(0), X, \bar{W}) \in A\}) = 1,$$

where we used Lemmas E.0.10 and E.0.13 in the second to last step. The last part of the assertion now follows from condition (ii) in Theorem E.0.8. $\square$

Remark E.0.16 We stress that so far we have only used that we have (at least) one weak solution to (E.1) with the fixed initial distribution μ, and that pathwise uniqueness holds for all solutions with initial distribution μ or with a deterministic starting point in a set of full μ-measure.

To complete the proof we still have to construct $F \in \hat{\mathcal{E}}$ (for which we shall use our assumption (i) in Theorem E.0.8 in full strength, i.e. that we have a weak solution for every initial distribution) and to check the adaptiveness conditions on it. Below we shall also apply what we have just obtained above to δ_x replacing μ. So, for each $x \in H$ we have a function $F_{\delta_x} : H \times \mathbb{W}_0 \to \mathbb{B}$. Now define

$$F(x, w) := F_{\delta_x}(x, w), \quad x \in H, \ w \in \mathbb{W}_0. \tag{E.8}$$

The proof of Theorem E.0.8 is then completed by the following lemma.

Lemma E.0.17 *Let μ be a probability measure on $(H, \mathcal{B}(H))$ and $F_\mu : H \times \mathbb{W}_0 \to \mathbb{B}$ as constructed in Lemma E.0.13. Then for μ-a.e. $x \in H$*

$$F(x, \cdot) = F_\mu(x, \cdot) \quad P^Q - a.e.$$

Furthermore, $F(x, \cdot)$ is $\overline{\mathcal{B}_t(\mathbb{W}_0)}^{P^Q}/\mathcal{B}_t(\mathbb{B})$-measurable for all $x \in H$, $t \in [0, \infty)$, where $\overline{\mathcal{B}_t(\mathbb{W}_0)}^{P^Q}$ denotes the completion of $\mathcal{B}_t(\mathbb{W}_0)$ with respect to P^Q in $\mathcal{B}(\mathbb{W}_0)$.

In particular, (by Lemmas E.0.13 and E.0.15) Condition 1 and (by the last part of Lemma E.0.15) Condition 2 in Definition E.0.6 hold.

Proof Let

$$\bar{\Omega} := H \times \mathbb{B} \times \mathbb{W}_0$$

$$\bar{\mathcal{F}} := \mathcal{B}(H) \otimes \mathcal{B}(\mathbb{B}) \otimes \mathcal{B}(\mathbb{W}_0)$$

and let $x \in H$. Define a measure $\bar{Q}_x$ on $(\bar{\Omega}, \bar{\mathcal{F}})$ by

$$\bar{Q}_x(A) := \int_H \int_{\mathbb{W}_0} \int_{\mathbb{B}} 1_A(z, w_1, w) K_\mu((z, w), dw_1) P^Q(dw) \delta_x(dz)$$

with K_μ as in Lemma E.0.10. Consider the stochastic basis $(\bar{\Omega}, \bar{\mathcal{F}}^x, \bar{Q}_x, (\bar{\mathcal{F}}_t^x))$ where

$$\bar{\mathcal{F}}^x := \overline{\mathcal{B}(H) \otimes \mathcal{B}(\mathbb{B}) \otimes \mathcal{B}(\mathbb{W}_0)}^{\bar{Q}_x},$$

$$\bar{\mathcal{F}}_t^x := \bigcap_{\varepsilon > 0} \sigma(\mathcal{B}(H) \otimes \mathcal{B}_{t+\varepsilon}(\mathbb{B}) \otimes \mathcal{B}_{t+\varepsilon}(\mathbb{W}_0), \bar{\mathcal{N}}_x),$$

where $\bar{\mathcal{N}}_x := \{N \in \bar{\mathcal{F}}^x | \bar{Q}_x(N) = 0\}$. As in the proof of Lemma E.0.12 one shows that for x outside a μ-zero set $N_1 \in \mathcal{B}(H)$, (Π, Π_3) on $(\bar{\Omega}, \bar{\mathcal{F}}^x, \bar{Q}_x, (\bar{\mathcal{F}}_t^x))$ is a weak solution to (E.1) with $\Pi(0) = x$ $\bar{Q}_x$-a.e. Here

$$\Pi_0 : H \times \mathbb{B} \times \mathbb{W}_0 \to H, \ (x, w_1, w) \mapsto x,$$

$$\Pi : H \times \mathbb{B} \times \mathbb{W}_0 \to \mathbb{B}, \ (x, w_1, w) \mapsto w_1,$$

$$\Pi_3 : H \times \mathbb{B} \times \mathbb{W}_0 \to \mathbb{W}_0, \ (x, w_1, w) \mapsto w.$$

By Lemma E.0.15 $(F_{\delta_x}(x, \Pi_3), \Pi_3)$ on the stochastic basis $(\bar{\Omega}, \bar{\mathcal{F}}^x, \bar{Q}_x, (\bar{\mathcal{F}}_t^x))$ is a weak solution to (E.1) with

$$F_{\delta_x}(x, \Pi_3)(0) = x \ \bar{Q}_x - a.s.$$

for every $x \in H$. Hence by our pathwise uniqueness assumption (ii), it follows that for all $x \in N_1^c$

$$F_{\delta_x}(x, \Pi_3) = \Pi \quad \bar{Q}_x - a.s. \tag{E.9}$$

For all $A \in \mathcal{B}(H) \otimes \mathcal{B}(\mathbb{B}) \otimes \mathcal{B}(\mathbb{W}_0)$ by Lemma E.0.13

$$\int_H \int_{\mathbb{W}_0} \int_{\mathbb{B}} 1_A(x, w_1, w) \delta_{F_\mu(x,w)}(dw_1) P^Q(dw) \mu(dx) = \int_H \bar{Q}_x(A) \mu(dx).$$

But for each $x \in N_1^c$ by (E.9)

$$\bar{Q}_x(A) = \int_{\bar{\Omega}} 1_A(\Pi_0, F_{\delta_x}(x, \Pi_3), \Pi_3) d\bar{Q}_x$$

$$= \int_{\mathbb{W}_0} 1_A(x, F_{\delta_x}(x, w), w) P^Q(dw)$$

$$= \int_{\mathbb{W}_0} \int_{\mathbb{B}} 1_A(x, w_1, w) \delta_{F_{\delta_x}(x,w)}(dw_1) P^Q(dw). \tag{E.10}$$

Since $x \mapsto \bar{Q}_x(A) = \int_{\mathbb{W}_0} \int_{\mathbb{B}} 1_A(x, w_1, w) K_\mu((x, w), dw_1) P^Q(dw)$ is $\overline{\mathcal{B}(H)}^\mu$-measurable, so is the right-hand side of (E.10). Therefore, we can integrate with respect to μ and obtain

$$\int_H \int_{\mathbb{W}_0} \int_{\mathbb{B}} 1_A(x, w_1, w) \delta_{F_\mu(x,w)}(dw_1) P^Q(dw) \mu(dx)$$

$$= \int_H \int_{\mathbb{W}_0} \int_{\mathbb{B}} 1_A(x, w_1, w) \delta_{F_{\delta_x}(x,w)}(dw_1) P^Q(dw) \mu(dx),$$

which implies the first assertion.

Let $x \in H$, $t \in [0, \infty)$, $A \in \mathcal{B}_t(\mathbb{B})$, and define

$$\bar{F}_{\delta_x} := 1_{\{x\} \times \mathbb{W}_0} F_{\delta_x}.$$

Then

$$\bar{F}_{\delta_x} = F_{\delta_x} \quad \delta_x \otimes P^Q - a.e.,$$

hence by the last part of Lemma E.0.13.

$$\{\bar{F}_{\delta_x} \in A\} \in \overline{\mathcal{B}(H) \otimes \mathcal{B}_t(\mathbb{W}_0)}^{\delta_x \otimes P^Q}. \tag{E.11}$$

But

$$\{\bar{F}_{\delta_x} \in A\} = \{x\} \times \{F_{\delta_x}(x, \cdot) \in A\} \cup (H \backslash \{x\}) \times \{0 \in A\},$$

so by (E.11) it follows that

$$\{F_{\delta_x}(x, \cdot) \in A\} \in \overline{\mathcal{B}_t(\mathbb{W}_0)}^{P^Q}.$$

$\square$

Remark E.0.18 For a detailed proof of the Yamada–Watanabe Theorem in the framework of the "semigroup (or mild solution) approach" to SPDEs, we refer to [63].

Appendix F
Continuous Dependence of Implicit Functions on a Parameter

In this section we fix two Banach spaces $(E, \| \ \|_E)$ and $(\Lambda, \| \ \|_\Lambda)$. For the whole section we consider a mapping $G : \Lambda \times E \to E$ with the following property: There exists an $\alpha \in [0, 1[$ such that

$$\|G(\lambda, x) - G(\lambda, y)\|_E \leqslant \alpha \|x - y\|_E \qquad \text{for all } \lambda \in \Lambda \text{ and all}$$
$$x, y \in E.$$

Then we get by the contraction theorem that there exists exactly one mapping $\varphi : \Lambda \to E$ such that $\varphi(\lambda) = G(\lambda, \varphi(\lambda))$ for all $\lambda \in \Lambda$.

Theorem F.0.1 (Continuity of the Implicit Function)

(i) If we assume in addition that the mapping $\lambda \mapsto G(\lambda, x)$ is continuous from Λ to E for all $x \in E$ we get that $\varphi : \Lambda \to E$ is continuous.

(ii) If there exists an $L \geqslant 0$ such that
$\|G(\lambda, x) - G(\tilde{\lambda}, x)\|_E \leqslant L \|\lambda - \tilde{\lambda}\|_\Lambda$ *for all $x \in E$*
then the mapping $\varphi : \Lambda \to E$ is Lipschitz continuous.

Proof

(i) We fix $\lambda_0 \in \Lambda$. Then for any other $\lambda \in \Lambda$

$$\varphi(\lambda) - \varphi(\lambda_0) = G(\lambda, \varphi(\lambda)) - G(\lambda_0, \varphi(\lambda_0))$$
$$= [G(\lambda, \varphi(\lambda)) - G(\lambda, \varphi(\lambda_0))] + [G(\lambda, \varphi(\lambda_0)) - G(\lambda_0, \varphi(\lambda_0))].$$

Because of the contraction property we obtain that

$$\|\varphi(\lambda) - \varphi(\lambda_0)\|_E \leqslant \alpha \|\varphi(\lambda) - \varphi(\lambda_0)\|_E + \|G(\lambda, \varphi(\lambda_0)) - G(\lambda_0, \varphi(\lambda_0))\|_E$$

© Springer International Publishing Switzerland 2015
W. Liu, M. Röckner, *Stochastic Partial Differential Equations: An Introduction*,
Universitext, DOI 10.1007/978-3-319-22354-4

and hence

$$\|\varphi(\lambda) - \varphi(\lambda_0)\|_E \leq \frac{1}{1-\alpha} \|G(\lambda, \varphi(\lambda_0)) - G(\lambda_0, \varphi(\lambda_0))\|_E.$$

Therefore we get the result (i).

(ii) In the same way as in (i) we obtain that for arbitrary λ and $\tilde{\lambda} \in \Lambda$

$$\|\varphi(\lambda) - \varphi(\tilde{\lambda})\|_E \leq \frac{1}{1-\alpha} \|G(\lambda, \varphi(\tilde{\lambda})) - G(\tilde{\lambda}, \varphi(\tilde{\lambda}))\|_E \leq \frac{L}{1-\alpha} \|\lambda - \tilde{\lambda}\|_\Lambda,$$

where we used the additional Lipschitz property of the mapping G in the last step.   □

Appendix G
Strong, Mild and Weak Solutions

In this chapter we only state the results and refer to [26, 34] for the proofs.

As in previous chapters let $(U, \| \ \|_U)$ and $(H, \| \ \|)$ be separable Hilbert spaces. We take $Q = I$ and fix a cylindrical Q-Wiener process $W(t)$, $t \geqslant 0$, in U on a probability space $(\Omega, \mathcal{F}, P)$ with a normal filtration $\mathcal{F}_t$, $t \geqslant 0$. Moreover, we fix $T > 0$ and consider the following type of stochastic differential equations in H:

$$dX(t) = [CX(t) + F(X(t))] \, dt + B(X(t)) \, dW(t), \quad t \in [0, T],$$

$$X(0) = \xi, \tag{G.1}$$

where:

- $C : D(C) \to H$ is the infinitesimal generator of a C_0-semigroup $S(t)$, $t \geqslant 0$, of linear operators on H,
- $F : H \to H$ is $\mathcal{B}(H)/\mathcal{B}(H)$-measurable,
- $B : H \to L(U, H)$,
- ξ is a H-valued, $\mathcal{F}_0$-measurable random variable.

Definition G.0.1 (Mild Solution) An H-valued predictable process $X(t)$, $t \in [0, T]$, is called a *mild solution* of (G.1) if

$$X(t) = S(t)\xi + \int_0^t S(t - s)F(X(s)) \, ds$$

$$+ \int_0^t S(t - s)B(X(s)) \, dW(s) \quad P\text{-a.s.} \tag{G.2}$$

for each $t \in [0, T]$. In particular, the appearing integrals have to be well-defined.

© Springer International Publishing Switzerland 2015
W. Liu, M. Röckner, *Stochastic Partial Differential Equations: An Introduction*,
Universitext, DOI 10.1007/978-3-319-22354-4

Definition G.0.2 (Analytically Strong Solutions) A $D(C)$-valued predictable process $X(t)$, $t \in [0, T]$, (i.e. $(s, \omega) \mapsto X(s, \omega)$ is $\mathcal{P}_T / \mathcal{B}(H)$- measurable) is called an *analytically strong solution* of (G.1) if

$$X(t) = \xi + \int_0^t CX(s) + F(X(s))\, ds + \int_0^t B(X(s))\, dW(s) \quad P\text{-a.s.} \qquad (G.3)$$

for each $t \in [0, T]$. In particular, the integrals on the right-hand side have to be well-defined, that is, $CX(t)$, $F(X(t))$, $t \in [0, T]$, are P-a.s. Bochner integrable and $B(X) \in \mathcal{N}_W$.

Definition G.0.3 (Analytically Weak Solution) An H-valued predictable process $X(t)$, $t \in [0, T]$, is called an *analytically weak solution* of (G.1) if

$$\langle X(t), \zeta \rangle = \langle \xi, \zeta \rangle + \int_0^t \langle X(s), C^* \zeta \rangle + \langle F(X(s)), \zeta \rangle\, ds$$

$$+ \int_0^t \langle \zeta, B(X(s)) dW(s) \rangle \quad P\text{-a.s.} \qquad (G.4)$$

for each $t \in [0, T]$ and $\zeta \in D(C^*)$. Here $(C^*, D(C^*))$ is the adjoint of $(C, D(C))$ on H. In particular, as in Definitions G.0.2 and G.0.1, the appearing integrals have to be well-defined.

Proposition G.0.4 (Analytically Weak Versus Analytically Strong Solutions)

(i) *Every analytically strong solution of* (G.1) *is also an analytically weak solution.*
(ii) *Let $X(t)$, $t \in [0, T]$, be an analytically weak solution of* (G.1) *with values in $D(C)$ such that $B(X(t))$ takes values in $L_2(U, H)$ for all $t \in [0, T]$. We further assume that*

$$P\left(\int_0^T \|CX(t)\|\, dt < \infty \right) = 1$$

$$P\left(\int_0^T \|F(X(t))\|\, dt < \infty \right) = 1$$

$$P\left(\int_0^T \|B(X(t))\|_{L_2}^2\, dt < \infty \right) = 1.$$

Then the process is also an analytically strong solution.

Proposition G.0.5 (Analytically Weak Versus Mild Solutions)

(i) *Let* $X(t)$, $t \in [0, T]$, *be an analytically weak solution of* (G.1) *such that* $B(X(t))$ *takes values in* $L_2(U, H)$ *for all* $t \in [0, T]$. *Besides we assume that*

$$P\left(\int_0^T \|X(t)\| \, dt < \infty \right) = 1$$

$$P\left(\int_0^T \|F(X(t))\| \, dt < \infty \right) = 1$$

$$P\left(\int_0^T \|B(X(t))\|_{L_2}^2 \, dt < \infty \right) = 1.$$

Then the process is also a mild solution.

(ii) *Let* $X(t)$, $t \in [0, T]$, *be a mild solution of* (G.1) *such that the mappings*

$$(t, \omega) \mapsto \int_0^t S(t - s) F(X(s, \omega)) \, ds$$

$$(t, \omega) \mapsto \int_0^t S(t - s) B(X(s)) \, dW(s)(\omega)$$

have predictable versions. In addition, we require that

$$P(\int_0^T \|F(X(t))\| \, dt < \infty) = 1$$

$$\int_0^T E(\int_0^t \|\langle S(t - s) B(X(s)), C^* \zeta \rangle\|_{L_2(U, \mathbb{R})}^2 \, ds) \, dt < \infty$$

for all $\zeta \in D(C^*)$.

Then the process is also an analytically weak solution.

Remark G.0.6 The precise relation of mild and analytically weak solutions with the variational solutions from Definition 4.2.1 is obviously more difficult to describe in general. We shall concentrate just on the following quite typical special case:

Consider the situation of Sect. 4.2, but with A and B independent of t and ω. Assume that there exist a self-adjoint operator $(C, D(C))$ on H such that $-C \geq \text{const.} > 0$ and $F : H \to H \, \mathcal{B}(H)/\mathcal{B}(H)$-measurable such that

$$A(x) = Cx + F(x), \quad x \in V,$$

and

$$V := D((-C)^{\frac{1}{2}}),$$

equipped with the graph norm of $(-C)^{\frac{1}{2}}$. Then it is easy to see that C extends to a continuous linear operator form V to V^*, again denoted by C such that for $x \in V$, $y \in D(C)$

$$_{V^*}\langle Cx, y \rangle_V = \langle x, Cy \rangle. \tag{G.5}$$

Now let X be a (variational) solution in the sense of Definition 4.2.1, then it follows immediately from (G.5) that X is an analytically weak solution in the sense of Definition G.0.3.

Appendix H
Some Interpolation Inequalities

For the following see [61, Lemma 2.1 and the subsequent remark].

Lemma H.0.1 *Let $\Lambda \subset \mathbb{R}^d$ be open, $d \in \mathbb{N}$ and $L^p := L^p(\Lambda)$ for $p \in [1, \infty)$.*

(i) If $d = 2$, then $\|\varphi\|_{L^4}^4 \leqslant 4\|\varphi\|_{L^2}^2 \|\nabla\varphi\|_{L^2}^2$ for all $\varphi \in H_0^{1,2}(\Lambda)$.

(ii) If $d = 3$, then $\|\varphi\|_{L^4}^4 \leqslant 8\|\varphi\|_{L^2} \|\nabla\varphi\|_{L^2}^3$ for all $\varphi \in H_0^{1,2}(\Lambda)$.

(iii) If $d = 2$, then $\|\varphi\psi\|_{L^1} \leqslant \|D_1\varphi\|_{L^1} \|D_2\psi\|_{L^1}$ for all $\varphi, \psi \in L^2(\Lambda) \cap H_0^{1,1}(\Lambda)$, where $D_i := \frac{\partial}{\partial x_i}$.

Proof Obviously, by a density argument it suffices to show assertions (i)–(iii) for $\varphi \in C_0^1(\Lambda)$, since by Sobolev embedding $H_0^{1,2} \subset L^p(\Lambda)$ for all $p \in [1, \infty[$ if $d = 2$, and $H_0^{1,2}(\Lambda) \subset L^{\frac{2d}{d-2}}(\Lambda)$ if $d \geqslant 3$. Therefore, fix $\varphi \in C_0^1(\Lambda)$.

(i) Using the chain rule and the fundamental theorem of calculus we obtain

$$\varphi^2(x, y) = \int_{-\infty}^x D_1\left(\varphi^2(s, y)\right) \, \mathrm{d}s = \int_{-\infty}^x 2\varphi(s, y)D_1\varphi(s, y)\mathrm{d}s$$

and hence

$$\int\int \varphi^4(x, y) \, \mathrm{d}x \, \mathrm{d}y = \int\int \varphi^2(x, y)\, \varphi^2(x, y) \, \mathrm{d}x \, \mathrm{d}y$$

$$\leqslant 4 \int\int \left[\int |\varphi(s, y)|\, |D_1\varphi(s, y)| \, \mathrm{d}s\right]$$

$$\cdot \left[\int |\varphi(x, t)|\, |D_2\varphi(x, t)| \, \mathrm{d}t\right] \mathrm{d}x \, \mathrm{d}y$$

© Springer International Publishing Switzerland 2015
W. Liu, M. Röckner, *Stochastic Partial Differential Equations: An Introduction*,
Universitext, DOI 10.1007/978-3-319-22354-4

$$= 4 \int \int |\varphi\,(s,y)|\,|D_1\varphi\,(s,y)|\,\mathrm{d}s\,\mathrm{d}y$$

$$\cdot \int \int |\varphi\,(x,t)|\,|D_2\varphi\,(x,t)|\,\mathrm{d}t\,\mathrm{d}x$$

$$\leqslant 4\|\varphi\|_{L^2}\,\underbrace{\|D_1\varphi\|_{L^2}}_{\leqslant\|\nabla\varphi\|_{L^2}}\,\|\varphi\|_{L^2}\,\underbrace{\|D_2\varphi\|_{L^2}}_{\leqslant\|\nabla\varphi\|_{L^2}}$$

$$\leqslant 4\|\varphi\|_{L^2}^2\|\nabla\varphi\|_{L^2}^2,$$

where we used Fubini's theorem and the Cauchy–Schwarz inequality.

(ii) Fix $z \in \mathbb{R}$. Again by the chain rule and the fundamental theorem of calculus we have

$$\varphi^2\,(x,y,z) = \int_{-\infty}^{z} D_3\varphi^2\,(x,y,s)\,\mathrm{d}s \leqslant 2\int_{\mathbb{R}} |\varphi\,(x,y,s)|\,|D_3\varphi\,(x,y,s)|\,\mathrm{d}s.$$

Hence by (i)

$$\int \int \varphi^4\,(x,y,z)\,\mathrm{d}x\,\mathrm{d}y \leqslant 4\int \int \varphi^2\,(x,y,z)\,\mathrm{d}x\,\mathrm{d}y \int \int |\nabla_{x,y}\varphi\,(x,y,z)|^2\,\mathrm{d}x\,\mathrm{d}y$$

$$\leqslant 4\int \int \int 2\,|\varphi\,(x,y,s)|\,|D_3\varphi\,(x,y,s)|\,\mathrm{d}s\mathrm{d}x\mathrm{d}y$$

$$\cdot \int \int |\nabla_{x,y}\varphi\,(x,y,z)|^2\,\mathrm{d}x\,\mathrm{d}y.$$

Integrating over $z \in \mathbb{R}$ implies

$$\|\varphi\|_{L^4}^4 \leqslant 8\underbrace{\int \int \int |\varphi\,(x,y,z)|\,|D_3\varphi\,(x,y,z)|\,\mathrm{d}x\,\mathrm{d}y\,\mathrm{d}z}_{\leqslant\|\varphi\|_{L^2}\|D_3\varphi\|_{L^2}\leqslant\|\varphi\|_{L^2}\|\nabla\varphi\|_{L^2}}$$

$$\cdot \underbrace{\int \int \int |\nabla_{x,y}\varphi\,(x,y,z)|^2\,\mathrm{d}x\,\mathrm{d}y\,\mathrm{d}z}_{\leqslant\|\nabla\varphi\|_{L^2}^2}$$

$$\leqslant 8\|\varphi\|_{L^2}\|\nabla\varphi\|_{L^2}^3.$$

(iii) By the fundamental theorem of calculus we have

$$\varphi\,(x,y) = \int_{-\infty}^{x} D_1\varphi\,(s,y)\,\mathrm{d}s = \int_{-\infty}^{y} D_2\varphi\,(x,t)\,\mathrm{d}t$$

and by Fubini's theorem we get for all $\psi \in C_0^1(\Lambda)$

$$
\int \int \varphi(x, y) \, \psi(x, y) \, dx \, dy
$$

$$
\leq \int \int \left[\int |D_1\varphi(s, y)| \, ds \right] \left[\int |D_2\psi(x, t)| \, dt \right] dx \, dy
$$

$$
= \int \int |D_1\varphi(s, y)| \, ds \, dy \int \int |D_2\psi(x, t)| \, dt \, dx
$$

$$
= \|D_1\varphi\|_{L^1} \|D_2\psi\|_{L^1} .
$$

$\square$

Appendix I
Girsanov's Theorem in Infinite Dimensions with Respect to a Cylindrical Wiener Process

In this section, which is an extended version of [23, Appendix A.1], we consider the Girsanov theorem for stochastic differential equations on Hilbert spaces of type (I.2) below with cylindrical Wiener noise. We shall give a complete and reasonably self-contained proof of this well-known folklore result (see, for instance, [26, 32, 36]). The proof is reduced to the Girsanov theorem of general real-valued continuous local martingales (see [70, (1.7) Theorem, page 329]).

We consider the following situation: We are given a negative definite self-adjoint operator $A : D(A) \subset H \to H$ on a separable Hilbert space $(H, \langle \cdot, \cdot \rangle)$ with $(-A)^{-1+\delta}$ being of trace class, for some $\delta \in]0, 1[$, a measurable map $F : H \to H$ of at most linear growth and W a cylindrical Wiener process over H defined on a filtered probability space $(\Omega, \mathcal{F}, (\mathcal{F}_t), P)$ represented in terms of the eigenbasis $(e_k)_{k\in\mathbb{N}}$ of $(A, D(A))$ through a sequence

$$W(t) = (W_k(t)e_k)_{k\in\mathbb{N}}, \quad t \geq 0, \tag{I.1}$$

where $W_k, k \in \mathbb{N}$, are independent real-valued Brownian motions starting at zero on $(\Omega, \mathcal{F}, (\mathcal{F}_t), P)$. Consider the stochastic equations

$$dX(t) = (AX(t) + F(X(t)))dt + dW(t), \quad t \in [0, T], \quad X(0) = x, \tag{I.2}$$

and

$$dZ(t) = AZ(t)dt + dW(t), \quad t \in [0, T], \quad Z(0) = x, \tag{I.3}$$

for some $T > 0$.

Theorem I.0.2 *Let $x \in H$. Then (I.2) has a unique weak mild solution and its law P_x on $C([0, T]; H)$ is equivalent to the law Q_x of the solution to (I.3) (which is just the classical Ornstein–Uhlenbeck process). If F is bounded x may be replaced by an $\mathcal{F}_0$-measurable H-valued random variable.*

© Springer International Publishing Switzerland 2015
W. Liu, M. Röckner, *Stochastic Partial Differential Equations: An Introduction*,
Universitext, DOI 10.1007/978-3-319-22354-4

The rest of this section is devoted to the proof of this theorem. We first need some preparation and start by recalling that because $\mathrm{Tr}[(-A)^{-1+\delta}] < \infty$, $\delta \in]0, 1[$, the stochastic convolution

$$W_A(t) := \int_0^t e^{(t-s)A} \, dW(s), \quad t \geq 0, \tag{I.4}$$

is a well defined $\mathcal{F}_t$-adapted stochastic process ("OU process") with continuous paths in H and

$$Z(t, x) := e^{tA}x + W_A(t), \quad t \in [0, T], \tag{I.5}$$

is the unique mild solution of (I.2). Let

$$Q_x := P \circ Z(\cdot, x)^{-1}, \tag{I.6}$$

and $b(t), t \geq 0$, be a progressively measurable H-valued process on $(\Omega, \mathcal{F}, (\mathcal{F}_t), P)$ such that

$$E\left[\int_0^T |b(s)|^2 ds\right] < \infty. \tag{I.7}$$

Define

$$Y(t) := \int_0^t \langle b(s), dW(s)\rangle := \sum_{k \geq 1} \int_0^t \langle b(s), e_k\rangle \, dW_k(s), \quad t \in [0, T]. \tag{I.8}$$

Lemma I.0.3 *The series on the r.h.s. of (I.8) converges in $L^2(\Omega, P; C([0, T]; \mathbb{R}))$. Hence the stochastic integral $Y(t)$ is well defined and a continuous real-valued martingale, which is square integrable.*

Proof We have for all $n, m \in \mathbb{N}$, $n > m$, by Doob's inequality

$$E\left[\sup_{t \in [0,T]} \left|\sum_{k=m}^n \int_0^t \langle e_k, b(s)\rangle \, dW_k(s)\right|^2\right] \leq 2E\left[\left|\sum_{k=m}^n \int_0^T \langle e_k, b(s)\rangle \, dW_k(s)\right|^2\right]$$

$$= 2 \sum_{k,l=m}^n E\left[\int_0^T \langle e_k, b(s)\rangle \, dW_k(s) \int_0^T \langle e_l, b(s)\rangle \, dW_l(s)\right]$$

$$= 2 \sum_{k=m}^n E\left[\int_0^T \langle e_k, b(s)\rangle^2 \, ds\right] \rightarrow 0,$$

as $m, n \rightarrow \infty$ because of (I.7). Hence the series on the right-hand side of (I.8) converges in $L^2(\Omega, P; C([0, T]; \mathbb{R}))$ and the assertion follows. $\qquad \square$

Remark I.0.4 It can be shown that if $\int_0^t \langle b(s), dW(s) \rangle$, $t \in [0, T]$, is defined as usual, using approximations by elementary functions (see [67, Lemma 2.4.2]) the resulting process is the same.

It is now easy to calculate the corresponding variation process $\langle \int_0^{\cdot} \langle b(s), dW(s) \rangle \rangle_t$, $t \in [0, T]$.

Lemma I.0.5 *We have*

$$\langle Y \rangle_t = \left\langle \int_0^{\cdot} \langle b(s), dW(s) \rangle \right\rangle_t = \int_0^t |b(s)|^2 \, ds, \quad t \in [0, T].$$

Proof We have to show that

$$Y^2(t) - \int_0^t |b(s)|^2 ds, \quad t \in [0, T],$$

is a martingale, i.e. for all bounded $\mathcal{F}_t$-stopping times τ we have

$$E[Y^2(\tau)] = E\left[\int_0^{\tau} |b(s)|^2 \, ds \right],$$

which follows immediately as in the proof of Lemma I.0.3.   □

Define the measure

$$\widetilde{P} := e^{Y(T) - \frac{1}{2} \langle Y \rangle_T} \cdot P, \tag{I.9}$$

on $(\Omega, \mathcal{F})$, which is equivalent to P. Since $\mathcal{E}(t) := e^{Y(t) - \frac{1}{2} \langle Y \rangle_t}$, $t \in [0, T]$, is a nonnegative local martingale, it follows by Fatou's Lemma that $\mathcal{E}$ is a supermartingale, and since $\mathcal{E}(0) = 1$, we have

$$E[\mathcal{E}(t)] \leq E[\mathcal{E}(0)] = 1.$$

Hence $\widetilde{P}$ is a sub-probability measure.

Proposition I.0.6 *Suppose that $\widetilde{P}$ is a probability measure i.e.*

$$E[\mathcal{E}(T)] = 1. \tag{I.10}$$

Then

$$\widetilde{W}_k(t) := W_k(t) - \int_0^t \langle e_k, b(s) \rangle \, ds, \quad t \in [0, T], \ k \in \mathbb{N},$$

are independent real-valued Brownian motions starting at 0 on $(\Omega, \mathcal{F}, (\mathcal{F}_t), \widetilde{P})$ i.e.

$$\widetilde{W}(t) := (\widetilde{W}_k(t)e_k)_{k\in\mathbb{N}}, \quad t \in [0, T],$$

is a cylindrical Wiener process over H on $(\Omega, \mathcal{F}, (\mathcal{F}_t), \widetilde{P})$.

Proof By the classical Girsanov Theorem (for general real-valued martingales, see [70, (1.7) Theorem, page 329]) it follows that for every $k \in \mathbb{N}$

$$W_k(t) - \langle W_k, Y \rangle_t, \quad t \in [0, T],$$

is a local martingale under $\widetilde{P}$. Set

$$Y_n(t) := \sum_{k=1}^{n} \int_0^t \langle e_k, b(s) \rangle \, dW_k(s), \quad t \in [0, T], \ n \in \mathbb{N}.$$

Then, because by Cauchy–Schwartz's inequality

$$|\langle W_k, Y - Y_n \rangle_t| \leq \langle W_k \rangle_t^{1/2} \langle Y - Y_n \rangle_t^{1/2}, \quad t \in [0, T], \ n \in \mathbb{N},$$

and since

$$E[\langle Y - Y_n \rangle_t] = E[(Y(t) - Y_n(t))^2] \to 0 \quad \text{as } n \to \infty,$$

by Lemma I.0.3, we conclude that (selecting a subsequence if necessary) P-a.s. for all $t \in [0, T]$

$$\langle W_k, Y \rangle_t = \lim_{n\to\infty} \langle W_k, Y_n \rangle_t = \int_0^t \langle e_k, b(s) \rangle \, ds,$$

since $\langle W_k, W_l \rangle_t = 0$ if $k \neq l$, by independence. Hence each $\widetilde{W}_k$ is a local martingale under $\widetilde{P}$.

It remains to show that for every $n \in \mathbb{N}$, $(\widetilde{W}_1, ..., \widetilde{W}_n)$ is, under $\widetilde{P}$, an n-dimensional Brownian motion. But P-a.s. for $l \neq k$

$$\langle \widetilde{W}_l, \widetilde{W}_k \rangle_t = \langle W_l, W_k \rangle_t = \delta_{l,k} \, t, \quad t \in [0, T].$$

Since P is equivalent to $\widetilde{P}$, this also holds $\widetilde{P}$-a.s. Hence by Lèvy's Characterization Theorem [70, (3.6) Theorem, page 150] it follows that $(\widetilde{W}_1, ..., \widetilde{W}_n)$ is an n-dimensional Brownian motion in $\mathbb{R}^n$ for all n, under $\widetilde{P}$.   □

Proposition I.0.7 *Let $W_A(t)$, $t \in [0, T]$, be defined as in (I.4). Then there exists an $\epsilon > 0$ such that*

$$E\left[\exp\left\{\epsilon \sup_{t \in [0,T]} |W_A(t)|\right\}^2\right] < \infty.$$

Proof Consider the distribution $Q_0 := P \circ W_A^{-1}$ of W_A on $E := C([0, T]; H)$. If Q_0 is a Gaussian measure on E, the assertion follows by Fernique's Theorem (see [31]). To show that Q_0 is a Gaussian measure on E we have to show that for each l in the dual space E' of E we have that $Q_0 \circ l^{-1}$ is Gaussian on $\mathbb{R}$. We prove this in two steps.

Step 1. Let $t_0 \in [0, T]$, $h \in H$ and $\ell(\omega) := \langle h, \omega(t_0) \rangle$ for $\omega \in E$. To see that $Q_0 \circ \ell^{-1}$ is Gaussian on $\mathbb{R}$, consider a sequence $\delta_k \in C([0, T]; \mathbb{R}), k \in \mathbb{N}$, such that $\delta_k(t)dt$ converges weakly to the Dirac measure ϵ_{t_0}. Then for all $\omega \in E$

$$\ell(\omega) = \lim_{k \to \infty} \int_0^T \langle h, \omega(s) \rangle \, \delta_k(s) \, ds = \lim_{k \to \infty} \int_0^T \langle h\delta_k(s), \omega(s) \rangle \, ds$$

$$= \lim_{k \to \infty} \langle h\delta_k, \omega \rangle_{L^2([0,T];H)} .$$

Since (e.g. by Da Prato [19, Proposition 2.15]) the law of W_A in $L^2([0, T]; H)$ is Gaussian, it follows that the distribution of ℓ is Gaussian.

Step 2. Let $\ell \in E'$ be arbitrary. The following argument is taken from [24, Proposition A.2]. Let $\omega \in E$, then we can consider its Bernstein approximation

$$\beta_n(\omega)(t) := \sum_{k=1}^n \binom{n}{k} \varphi_{k,n}(t)\omega(Tk/n), \quad n \in \mathbb{N}, \ t \in [0, T],$$

where $\varphi_{k,n}(t) := (tT)^k(1 - tT)^{n-k}$. But the linear map

$$H \ni x \to \ell(x\varphi_{k,n}) \in \mathbb{R}$$

is continuous on H, hence there exists an $h_{k,n} \in H$ such that

$$\ell(x\varphi_{k,n}) = \langle h_{k,n}, x \rangle, \quad x \in H.$$

Since $\beta_n(\omega) \to \omega$ uniformly for all $\omega \in E$ as $n \to \infty$, it follows that for all $\omega \in E$

$$\ell(\omega) = \lim_{n \to \infty} \ell(\beta_n(\omega)) = \lim_{n \to \infty} \sum_{k=1}^n \binom{n}{k} \langle h_{k,n}, \omega(Tk/n) \rangle.$$

Hence it follows by Step 1 that $Q_0 \circ \ell^{-1}$ is Gaussian. $\qquad \square$

Now we turn to SDE (I.2). We are going to apply the above with

$$b(t) := \frac{1}{2} F(e^{tA} x + W_A(t)), \quad t \in [0, T],$$

and define

$$M := e^{\int_0^T \langle F(e^{tA}x + W_A(t)), dW(t) \rangle - \frac{1}{2} \int_0^T |F(e^{tA}x + W_A(t))|^2 dt},$$

$$\widetilde{P}_x := MP.$$

(I.11)

Proposition I.0.8 $\widetilde{P}_x$ *is a probability measure on* $(\Omega, \mathcal{F})$, *i.e.* $E(M) = 1$.

Proof As before we set $Z(t, x) := e^{tA} x + W_A(t)$, $t \in [0, T]$. By Proposition I.0.7 the arguments below are standard, (see e.g. [49, Corollaries 5.14 and 5.16, pages 199/200]). Since F is of at most linear growth, by Proposition I.0.7 we can find $N \in \mathbb{N}$ large enough such that for all $0 \leq i \leq N$ and $t_i := \frac{iT}{N}$

$$E \left[e^{\frac{1}{2} \int_{t_{i-1}}^{t_i} |F(e^{tA}x + W_A(t))|^2 dt} \right] < \infty.$$

Defining $F_i(e^{tA} x + W_A(t)) := 1_{(t_{i-1}, t_i]}(t) F(e^{tA} x + W_A(t))$ it follows from Novikov's criterion [70, (1.16) Corollary, page 333] that for all $1 \leq i \leq N$

$$\mathcal{E}_i(t) := e^{\int_0^t \langle F_i(e^{sA}x + W_A(s)), dW(s) \rangle - \frac{1}{2} \int_0^t |F_i(e^{sA}x + W_A(s))|^2 ds}, \quad t \in [0, T],$$

is an $\mathcal{F}_t$-martingale under P. But then since $\mathcal{E}_i(t_{i-1}) = 1$, by the martingale property of each $\mathcal{E}_i$ we can conclude that

$$E \left[e^{\int_0^t \langle F(e^{sA}x + W_A(s)), dW(s) \rangle - \frac{1}{2} \int_0^t |F(e^{sA}x + W_A(s))|^2 ds} \right]$$

$$= E \left[\mathcal{E}_N(t_N) \mathcal{E}_{N-1}(t_{N-1}) \cdots \mathcal{E}_1(t_1) \right]$$

$$= E \left[\mathcal{E}_N(t_{N-1}) \mathcal{E}_{N-1}(t_{N-1}) \cdots \mathcal{E}_1(t_1) \right]$$

$$= E \left[\mathcal{E}_{N-1}(t_{N-1}) \cdots \mathcal{E}_1(t_1) \right]$$

$$\cdots$$

$$= E[\mathcal{E}_1(t_1)] = E[\mathcal{E}_1(t_0)] = 1.$$

□

Remark I.0.9 It is obvious from the previous proof that x may always be replaced by an $\mathcal{F}_0$-measurable H-valued map which is exponentially integrable, and by any

$\mathcal{F}_0$-measurable H-valued map if F is bounded. The same holds for the rest of the proof of Theorem I.0.2, i.e., the following two propositions.

Proposition I.0.10 *We have $\widetilde{P}_x$-a.s.*

$$Z(t,x) = e^{tA}x + \int_0^t e^{(t-s)A}F(Z(s,x))\,ds + \int_0^t e^{(t-s)A}\,d\widetilde{W}(s), \quad t \in [0,T], \quad (\text{I}.12)$$

where $\widetilde{W}$ is the cylindrical Wiener process under $\widetilde{P}_x$ introduced in Proposition I.0.6 with $b(s) := F(Z(s,x))$, which applies because of Proposition I.0.8, i.e. under $\widetilde{P}_x$, $Z(\cdot,x)$ is a mild solution of

$$dZ(t) = (AZ(t) + F(Z(t)))dt + d\widetilde{W}(t), \quad t \in [0,T], \ Z(0) = x.$$

Proof Since F is of at most linear growth and because of Proposition I.0.7, to prove (I.12) it is enough to show that for all $k \in \mathbb{N}$ and $F_k := \langle e_k, F \rangle$, $Z_k := \langle e_k, Z \rangle$, $x_k := \langle e_k, x \rangle$ for $x \in H$ we have, since $Ae_k = -\lambda_k e_k$, that

$$dZ_k(t,x) = (-\lambda_k Z_k(t,x) + F_k(Z(t,x)))dt + d\widetilde{W}_k(t), \quad t \in [0,T], \ Z_k(0) = x_k.$$

But this is obvious by the definition of $\widetilde{W}_k$. $\qquad\qquad\square$

Proposition I.0.10 settles the existence part of Theorem I.0.2. Now let us turn to the uniqueness part and complete the proof of Theorem I.0.2.

Proposition I.0.11 *The weak solution to (I.2) constructed above is unique and its law is equivalent to Q_x.*

Proof Let $X(t,x)$, $t \in [0,T]$, be a weak mild solution to (I.2) on a filtered probability space $(\Omega, \mathcal{F}, (\mathcal{F}_t), P)$ for a cylindrical process of type (I.1). Hence

$$X(t,x) = e^{tA}x + W_A(t) + \int_0^t e^{(t-s)A}F(X(s,x))\,ds,$$

for some cylindrical Wiener process $W(t) = (W_k(t)\,e_k)_{k\in\mathbb{N}}$, $t \in [0,T]$, on some filtered probability space $(\Omega, \mathcal{F}, (\mathcal{F}_t), P)$. Since F is at most of linear growth, it follows from Gronwall's inequality that for some constant $C \geq 0$

$$\sup_{t\in[0,T]} |X(t,x)| \leq C_1 \left(1 + \sup_{t\in[0,T]} |e^{tA}x + W_A(t)|\right).$$

Hence by Proposition I.0.7

$$E\left[\exp\left\{\epsilon \sup_{t\in[0,T]} |X(t,x)|\right\}^2\right] < \infty. \qquad (\text{I}.13)$$

Define

$$M := e^{-\int_0^T \langle F(X(s,x)), dW(s) \rangle - \frac{1}{2} \int_0^T |F(X(s,x))|^2 ds}$$

and

$$\widetilde{P} := M \cdot P.$$

Then by exactly the same arguments as above

$$E[M] = 1.$$

Hence by Proposition I.0.6 defining

$$\widetilde{W}_k(t) := W_k(t) + \int_0^t \langle e_k, F(X(s,x)) \rangle \, ds, \quad t \in [0,T], \ k \in \mathbb{N},$$

we obtain that $\widetilde{W}(t) := (\widetilde{W}_k(t)e_k)_{k \in \mathbb{N}}$ is a cylindrical Wiener process under $\widetilde{P}$ and thus

$$\widetilde{W}_A(t) := \int_0^t e^{(t-s)A} \, d\widetilde{W}(s) = W_A(t) + \int_0^t e^{(t-s)A} F(X(s,x)) \, ds, \quad t \in [0,T],$$

and therefore,

$$X(t,x) = e^{tA}x + \widetilde{W}_A(t), \quad t \in [0,T],$$

is an Ornstein–Uhlenbeck process under $\widetilde{P}$ starting at x. Consequently,

$$\widetilde{P} \circ X(\cdot, x)^{-1} = Q_x. \tag{I.14}$$

But since it is easy to see that,

$$\int_0^T \langle F(X(s,x)), dW(s) \rangle = \int_0^T \langle F(X(s,x)), d\widetilde{W}(s) \rangle - \int_0^T |F(X(s,x))|^2 \, ds,$$

it follows that

$$P = e^{\int_0^T \langle F(X(s,x)), d\widetilde{W}(s) \rangle - \frac{1}{2} \int_0^T |F(X(s,x))|^2 ds} \cdot \widetilde{P}.$$

For $k \in \mathbb{N}$ define $\widehat{W}_k : C([0,T];H) \to \mathbb{R}$ by

$$\widehat{W}_k(w)(t) := \langle e_k, \ e^{tA}x - w(t) \rangle + \lambda_k \int_0^t \langle e_k, e^{sA}x - w(s) \rangle \, ds, \quad w \in C([0,T];H), \ t \in [0,T].$$

Then, since for $k \in \mathbb{N}$, $t \in [0, T]$

$$
\widetilde{W}_k(t) = \left\langle e_k, \widetilde{W}_A(t) \right\rangle + \lambda_k \int_0^t \left\langle e_k, \widetilde{W}_A(s) \right\rangle ds
$$

and

$$
X(t, x) = e^{tA} x + \widetilde{W}_A(t),
$$

it follows that

$$
\widehat{W}_k(X(\cdot, x)(\omega))(t) = \widetilde{W}_k(\omega)(t), \quad k \in \mathbb{N}, \ t \in [0, T], \ \omega \in \Omega. \tag{I.15}
$$

Hence $(\widehat{W}_k(t))_{t \in [0,T]}$, $k \in \mathbb{N}$, are independent Brownian motions on $(C([0, T]; H),$ $\mathcal{F}^X, (\mathcal{F}_t^X), Q_x)$, where $\mathcal{F}^X, \mathcal{F}_t^X$ are the image σ-algebras under $X(\cdot, x)$ of $\mathcal{F}$ and $\mathcal{F}_t$ respectively. It is easy to check that $\mathcal{F}_t^X$, $t \in [0, T]$, is a normal filtration and that

$$
\mathcal{F}_t^\pi \subset \mathcal{F}_t^X \ \text{ for all } t \in [0, T], \tag{I.16}
$$

where $\mathcal{F}_t^\pi$ is the normal filtration associated to the process $\pi(t) : C([0, T]; H) \to H$ defined by

$$
\pi(t)(w) := w(t), \quad t \in [0, T], \ w \in C([0, T]; H),
$$

with respect to the measure Q_x on $C([0, T]; H)$ equipped with its Borel σ-algebra $\mathcal{F}_T^\pi (= \sigma(\{\pi(t) | t \in [0, T]\}))$, i.e. for $t \in [0, T]$

$$
\mathcal{F}_t^\pi := \bigcap_{s > t} \sigma(\sigma(\pi(r) | r \leqslant s) \cup \mathcal{N}^{Q_x}),
$$

where $\mathcal{N}^{Q_x} := \{N \in \mathcal{F}_T^\pi | Q_x(N) = 0\}$.

Since each $\widehat{W}_k(t)$, $t \in [0, T]$, is $(\sigma(\pi(r) | r \leqslant t))_{t \in [0,T]}$ adapted, it is an $(\mathcal{F}_t^\pi)$-Brownian motion due to (I.16). Hence the stochastic integral in the following definition is well-defined:

$$
\rho_x(w) := e^{\int_0^T \left\langle F(\pi(s)), d\widehat{W}(s) \right\rangle (w) - \frac{1}{2} \int_0^T |F(w(s))|^2 ds}, \quad w \in C([0, T]; H), \tag{I.17}
$$

where

$$
\widehat{W}(t) := (\widehat{W}_k(t) e_k)_{k \in \mathbb{N}}, \ t \in [0, T],
$$

is the cylindrical Wiener process on H corresponding to $\widehat{W}_k(t)$, $t \in [0, T]$, $k \in \mathbb{N}$. We emphasize that thus $\rho_x : C([0, T]; H) \to]0, \infty[$ is an $\mathcal{F}_T^\pi (\subset \mathcal{F}_T^X)$-measurable function which is defined independently of $X(\cdot, x)$, $(\Omega, \mathcal{F}, (\mathcal{F}_t), P)$ and W. The definition of the stochastic integral in (I.17) is analogous to (I.8). Hence, due to the general construction of $\mathbb{R}$-valued stochastic integrals, it follows that $\widetilde{P}$-a.s.

$$\rho_x(X(\cdot, x)) = e^{\int_0^T \langle F(X(s,x)), \, dW(s)\rangle - \frac{1}{2}\int_0^T |F(X(s,x))|^2 ds} \, .$$

Since, therefore, $P = \rho_x(X(\cdot, x)) \, \widetilde{P}$, we deduce from (I.14) that

$$P \circ X(\cdot, x)^{-1} = \rho_x Q_x \, .$$

$\square$

References

1. Alberverio, S., Ma, Z.M., Röckner, M.: Quasi regular Dirichlet forms and the stochastic quantization problem. In: A Festschrift in Honor of Masatoshi Fukushima. Interdisciplinary Mathematical Sciences, vol. 17, pp. 27–58. World Scientific, Singapore (2015)
2. Albeverio, S., Röckner, M.: Stochastic differential equations in infinite dimensions: solutions via Dirichlet forms. Probab. Theor. Relat. Fields **89**, 347–386 (1991)
3. Alt, H.W.: Lineare Funktionalanalysis. Springer, Berlin (1992)
4. Aronson, D.G.: The porous medium equation. In: Nonlinear Diffusion Problems (Montecatini Terme, 1985). Lecture Notes in Mathematics, vol. 1224, pp. 1–46. Springer, Berlin (1986)
5. Bauer, H.: Measure and Integration Theory. de Gruyter Studies in Mathematics, vol. 26. Walter de Gruyter & Co., Berlin (2001)
6. Blömker, D.: Amplitude Equations for Stochastic Partial Differential Equations. Interdisciplinary Mathematical Sciences, vol. 3. World Scientific, Hackensack (2007)
7. Blömker, D., Flandoli, F., Romito, M.: Markovianity and ergodicity for a surface growth PDE. Ann. Probab. **37**, 275–313 (2009)
8. Blömker, D., Romito, M.: Regularity and blow up in a surface growth model. Dyn. Partial Differ. Equ. **6**, 227–252 (2009)
9. Blömker, D., Romito, M.: Local existence and uniqueness in the largest critical space for a surface growth model. Nonlinear Differ. Equ. Appl. **19**, 365–381 (2012)
10. Bogachev, V.I., Krylov, N.V., Röckner, M., Shaposhnikov, S.: Fokker-Planck Kolmogorov equations. Regular and Chaotic Dynamics, Moscow-Izhevsk (2013, in Russian)
11. Brezis, H.: Équations et inéquations non linéaires dans les espaces vectoriels en dualité. Ann. Inst. Fourier (Grenoble) **18**, 115–175 (1968)
12. Browder, F.E.: Pseudo-monotone operators and nonlinear elliptic boundary value problems on unbounded domains. Proc. Nat. Acad. Sci. USA **74**, 2659–2661 (1977)
13. Carmona, R., Rozovskiĭ, B.: Stochastic Partial Differential Equations: Six Perspectives. Mathematical Surveys and Monographs, vol. 64. American Mathematical Society, Providence (1999)
14. Cerrai, S., Freidlin, M.: On the Smoluchowski-Kramers approximation for a system with an infinite number of degrees of freedom. Probab. Theor. Relat. Fields **135**, 363–394 (2006)
15. Cerrai, S., Freidlin, M.: Smoluchowski-Kramers approximation for a general class of SPDEs. Probab. Theor. Relat. Fields **135**, 363–394 (2006)
16. Chow, P.-L.: Stochastic Partial Differential Equations. Chapman & Hall/CRC Applied Mathematics and Nonlinear Science Series. Chapman & Hall/CRC, Boca Raton (2007)
17. Chueshov, I., Millet, A.: Stochastic 2D hydrodynamical type systems: well posedness and large deviations. Appl. Math. Optim. **61**, 379–420 (2010)

© Springer International Publishing Switzerland 2015
W. Liu, M. Röckner, *Stochastic Partial Differential Equations: An Introduction*,
Universitext, DOI 10.1007/978-3-319-22354-4

18. Cohn, D.L.: Measure Theory. Birkhäuser, Boston (1980)
19. Da Prato, G.: Kolmogorov Equations for Stochastic PDEs. Advanced Courses in Mathematics
    – CRM Barcelona. Birkhaeuser, Basel (2004)
20. Da Prato, G.: Introduction to Stochastic Analysis and Malliavin Calculus, 2nd edn. Appunti.
    Scuola Normale Superiore di Pisa (Nuova Serie), vol. 7. Edizioni della Normale, Pisa (2008)
21. Da Prato, G., Debussche, A.: Stochastic Cahn-Hilliard equation. Nonlinear Anal. **26**, 241–263
    (1996)
22. Da Prato, G., Flandoli, F., Priola, E., Röckner, M.: Strong uniqueness for stochastic evolution
    equations in Hilbert spaces with bounded measurable drift. Ann. Prob. **41**, 3306–3344 (2013)
23. Da Prato, G., Flandoli, F., Priola, E., Röckner, M.: Strong uniqueness for stochastic evolution
    equations with unbounded measurable drift term. J. Theor. Probab. (to appear, 2014)
24. Da Prato, G., Lunardi, A.: Maximal L^2-regularity for Dirichlet problems in Hilbert spaces. J.
    Math. Pures Appl. **99**(9), 741–765 (2013)
25. Da Prato, G., Röckner, M., Rozovskiĭ, B.L., Wang, F.-Y.: Strong solutions of stochastic
    generalized porous media equations: existence, uniqueness, and ergodicity. Commun. Partial
    Differ. Equ. **31**, 277–291 (2006)
26. Da Prato, G., Zabczyk, J.: Stochastic Equations in Infinite Dimensions. Encyclopedia of
    Mathematics and Applications. Cambridge University Press, Cambridge (1992) [see also
    Second Edition 2014]
27. Da Prato, G., Zabczyk, J.: Ergodicity for Infinite-Dimensional Systems. London Mathematical
    Society Lecture Note Series, vol. 229. Cambridge University Press, Cambridge (1996)
28. Dalang, R., Khoshnevisan, D., Müller, C., Nualart, D., Xiao, Y.: A Minicourse on Stochastic
    Partial Differential Equations. Lecture Notes in Mathematics, 1962. Springer, Berlin (2009)
29. Doob, J.L.: Stochastic Processes. A Wiley-Interscience Publication, New York (1953)
30. Elezović, N., Mikelić, A.: On the stochastic Cahn-Hilliard equation. Nonlinear Anal. **16**, 1169–
    1200 (1991)
31. Fernique, X.: Régularité des Trajectoires des Fonctions Aleátoires Gaussiennes. In: Hennequin,
    P.L. (ed.) Ecoles d'Eté de Probabilités de Saint-Flour IV-1975. Lecture Notes in Mathematics,
    vol. 480, pp. 1–97. Springer, Berlin (1975)
32. Ferrario, B.: Uniqueness and absolute continuity for semilinear SPDE's. In: Dalang, R.C.,
    Dozzi, M., Russo, F. (eds.) Seminar on Stochastic Analysis, Random Fields and Applications
    VII. Progress in Probability, vol. 87, Birkhäuser/Springer Basel AG, Basel (2013)
33. Freidlin, M.: Some remarks on the Smoluchowski-Kramers approximation. J. Stat. Phys. **117**,
    617–634 (2004)
34. Frieler, K., Knoche, C.: Solutions of stochastic differential equations in infinite dimensional
    Hilbert spaces and their dependence on initial data. Diploma Thesis, Bielefeld University,
    BiBoS-Preprint E02-04-083 (2001)
35. Fritz, P., Hairer, M.: A Course on Rough Paths, with an Introduction to Regularity Structures.
    Springer, New York (2014)
36. Gatarek, D., Goldys, B.: On solving stochastic evolution equations by the change of drift with
    application to optimal control. In: Stochastic Partial Differential Equations and Applications
    (Trento, 1990). Pitman Research Notes in Mathematics Series, vol. 268, pp. 180–190.
    Longman Sci. Tech., Harlow (1992)
37. Gawarecki, L., Mandrekar, V.: Stochastic differential equations in infinite dimensions with
    applications to stochastic partial differential equations. In: Probability and Its Applications
    (New York). Springer, Heidelberg (2011)
38. Gess, B., Liu, W., Röckner, M.: Random attractors for a class of stochastic partial differential
    equations driven by general additive noise. J. Differ. Equ. **251**, 1225–1253 (2011)
39. Gilbarg, D., Trudinger, N.S.: Elliptic Partial Differential Equations of Second Order, 2nd edn.
    Grundlehren der Mathematischen Wissenschaften [Fundamental Principles of Mathematical
    Sciences], vol. 224. Springer, Berlin (1983)
40. Gyöngy, I.: On stochastic equations with respect to semimartingales, III. Stochastics **7**, 231–
    254 (1982)

41. Gyöngy, I., Krylov, N.V.: On stochastic equations with respect to semimartingales, I. Stochastics **4**, 1–21 (1980/1981)

42. Gyöngy, I., Krylov, N.V.: On stochastics equations with respect to semimartingales, II. Itô formula in Banach spaces. Stochastics **6**, 153–173 (1981/1982)

43. Hairer, M.: Solving the KPZ equation. Ann. Math. **2**, 559–664 (2013)

44. Hairer, M.: A theory of regularity structures. Invent. Math. **198**, 269–504 (2014)

45. Heywood, J.: On a conjecture concerning the Stokes problem in nonsmooth domains. In: Mathematical Fluid Mechanics. Advances in Mathematical Fluid Mechanics, pp. 195–205. Birkhäuser, Basel (2001)

46. Holden, H., Øksendal, B., Ubøe, J., Zhang, T.: Stochastic partial differential equations. In: A Modeling, White Noise Functional Approach, 2nd edn. Springer, New York (2010)

47. Ikeda, N., Watanabe, S.: Stochastic Differential Equations and Diffusion Processes. North-Holland Mathematical Library, vol. 24. North-Holland, Amsterdam (1981)

48. Kallianpur, G., Xiong, J.: Stochastic Differential Equations in Infinite-Dimensional Spaces. Institute of Mathematical Statistics Lecture Notes-Monograph Series, vol. 26. Institute of Mathematical Statistics, Hayward (1995)

49. Karatzas, I., Shreve, S.E.: Brownian Motion and Stochastic Calculus. Graduate Texts in Mathematics, vol. 113. Springer, New York (1988)

50. Kotelenez, P.: Stochastic ordinary and stochastic partial differential equations. Transition from microscopic to macroscopic equations. In: Stochastic Modelling and Applied Probability, vol. 58. Springer, New York (2008)

51. Kramers, H.: Brownian motion in a field of force and the diffusion model of chemical reactions. Physica **7**, 284–304 (1940)

52. Krylov, N.V.: On Kolmogorov's equations for finite-dimensional diffusions. In: Stochastic PDE's and Kolmogorov Equations in Infinite Dimensions (Cetraro, 1998). Lecture Notes in Mathematics, vol. 1715, pp. 1–63. Springer, Berlin (1999)

53. Krylov, N.V., Röckner, M., Zabczyk, J.: Stochastic PDE's and Kolmogorov Equations in Infinite Dimensions. Lecture Notes in Mathematics, vol. 1715. Springer, Berlin (1999)

54. Krylov, N.V., Rozovskiĭ, B.L.: Stochastic evolution equations. In: Current Problems in Mathematics. Akad. Nauk SSSR, vol. 14 (Russian), pp. 71–147, 256. Vsesoyuz. Inst. Nauchn. i Tekhn. Informatsii, Moscow (1979)

55. Leray, J.: Sur le mouvement d'un liquide visqueux emplissant l'espace. Acta Math. **63**, 193–248 (1934)

56. Lipster, R., Shiryaev, A.: Theory of Martingales. Kluwer Academic, Dordrecht (1989)

57. Liu, W.: Existence and uniqueness of solutions to nonlinear evolution equations with locally monotone operators. Nonlinear Anal. **74**, 7543–7561 (2011)

58. Liu, W., Röckner, M.: SPDE in Hilbert space with locally monotone coefficients. J. Funct. Anal. **259**, 2902–2922 (2010)

59. Liu, W., Röckner, M.: Local and global well-posedness of SPDE with generalized coercivity conditions. J. Differ. Equ. **254**, 725–755 (2013)

60. Meise, R., Vogt, D.: Einführung in die Funktionalanalysis. Vieweg, New York (1992)

61. Menaldi, J.-L., Sritharan, S.: Stochastic 2D Navier-Stokes equation. Appl. Math. Optim. **46**, 31–53 (2002)

62. Novick-Cohen, A.: The Cahn-Hilliard equation: mathematical and modeling perspectives. Adv. Math. Sci. Appl. **8**, 965–985 (1998)

63. Ondreját, M.: Uniqueness for stochastic evolution equations in Banach spaces. Dissertationes Math. (Rozprawy Mat.) **426**, 1–63 (2004)

64. Pardoux, É.: Sur des équations aux dérivées partielles stochastiques monotones. C. R. Acad. Sci. Paris Sér. A-B **275**, A101–A103 (1972)

65. Pardoux, É.: Équations aux dérivées partielles stochastiques de type monotone. In: Séminaire sur les Équations aux Dérivées Partielles (1974–1975), III, Exp. No. 2, p. 10. Collège de France, Paris (1975)

66. Peszat, S., Zabczyk, J.: Stochastic Partial Differential Equations with Lévy Noise. An Evolution Equation.. Encyclopedia of Mathematics and Its Applications, vol. 113. Cambridge University Press, Cambridge (2007)
67. Prévôt, C., Röckner, M.: A Concise Course on Stochastic Partial Differential Equations. Lecture Notes in Mathematics, vol. 1905. Springer, Berlin (2007)
68. Reed, M., Simon, B.: Methods of Modern Mathematical Physics. Academic, London (1972)
69. Ren, J., Röckner, M., Wang, F.: Stochastic generalized porous media and fast diffusion equations. J. Differ. Equ. **238**, 118–152 (2007)
70. Revuz, D., Yor, M.: Continuous Martingales and Brownian Motion. Grundlehren der Mathematischen Wissenschaften [Fundamental Principles of Mathematical Sciences], vol. 293. Springer, Berlin (1999)
71. Röckner, M., Schmuland, B., Zhang, X.: Yamada-Watanabe Theorem for stochastic evolution equations in infinite dimensions. Cond. Matt. Phys. **11**, 247–259 (2008)
72. Röckner, M., Zhang, T.-S.: Stochastic 3D tamed Navier-Stokes equation: existence, uniqueness and small time large deviation principles. J. Differ. Equ. **252**, 716–744 (2012)
73. Röckner, M., Zhang, X.: Stochastic tamed 3D Navier-Stokes equations: existence, uniqueness and ergodicity. Probab. Theor. Relat. Fields **145**, 211–267 (2009)
74. Röckner, M., Zhang, X.: Tamed 3D Navier-Stokes equation: existence, uniqueness and regularity. Infin. Dimens. Anal. Quantum Probab. Relat. Top. **12**, 525–549 (2009)
75. Rozovskiĭ, B.: Stochastic Evolution Systems. Mathematics and Its Applications, vol. 35. Kluwer Academic, Dordrecht (1990)
76. Smoluchowski, M.: Drei Vorträge über Diffusion Brownsche Bewegung und Koagulation von Kolloidteilchen. Physik Zeit. **17**, 557–585 (1916)
77. Sohr, H.: The Navier-Stokes Equations. Birkhäuser Advanced Texts: Basler Lehrbücher [Birkhäuser Advanced Texts: Basel Textbooks], An Elementary Functional Analytic Approach. Birkhäuser, Basel (2001)
78. Taira, K.: Analytic Semigroups and Semilinear Initial Boundary Value Problems. Cambridge University Press, Cambridge (1995)
79. Temam, R.: Navier-Stokes Equations. Theory and Numerical Analysis AMS Chelsea Publishing, Providence (2001) [Reprint of the 1984 edition]
80. Walsh, J.B.: An introduction to stochastic partial differential equations. In: École d'été de Probabilités de Saint-Flour, XIV—1984. Lecture Notes in Mathematics, vol. 1180, pp. 265–439. Springer, Berlin (1986)
81. Weizsäcker, H., Winkler, G.: Stochastic Integrals: An Introduction. Vieweg, New York (1990)
82. Zeidler, E.: Nonlinear Functional Analysis and Its Applications, II/B. Nonlinear Monotone Operators. Springer, New York (1990) [Translated from the German by the author and Leo F. Boron]
83. Zhang, X.: A tamed 3D Navier-Stokes equation in uniform C^2-domains. Nonlinear Anal. **71**, 3093–3112 (2009)

Index

© Springer International Publishing Switzerland 2015
W. Liu, M. Röckner, *Stochastic Partial Differential Equations: An Introduction*,
Universitext, DOI 10.1007/978-3-319-22354-4

Printed in the United States
By Bookmasters